Integrated Design and Cost Management for Civil Engineers

Integrated Design and Cost Management for Civil Engineers

Andrew Whyte

CRC Press
Taylor & Francis Group
Boca Raton   London   New York

CRC Press is an imprint of the
Taylor & Francis Group, an **informa** business

A SPON PRESS BOOK

CRC Press
Taylor & Francis Group
6000 Broken Sound Parkway NW, Suite 300
Boca Raton, FL 33487-2742

Printed on acid-free paper
Version Date: 20140508

International Standard Book Number-13: 978-0-415-80921-4 (Paperback)

Library of Congress Cataloging-in-Publication Data

Whyte, Andrew, 1967-
  Integrated design and cost management for civil engineers / author, Andrew Whyte.
    pages cm
  Includes bibliographical references and index.
  ISBN 978-0-415-80921-4 (paperback)
  1. Engineering economy. 2. Public works--Costs. 3. Construction industry--Management. 4. Building--Cost control. I. Title.

TA177.4.W49 2014
624.068'1--dc23   2014016989

Visit the Taylor & Francis Web site at
http://www.taylorandfrancis.com

and the CRC Press Web site at
http://www.crcpress.com

Contents

List of figures

List of tables

Preface

A successful civil engineer needs to be able to provide his or her client with practical solutions to specific problems. Not only does the solution need to be technically fit-for-use but it also needs to be cost-effective and time-efficient. Using case studies to illustrate the principles and processes it describes, this book is a guide to designing, costing and scheduling a civil engineering project to suit a client's brief. The procedures presented emphasise correct quantification and planning of works to give reliable cost and time predictions towards minimising the risk of losing business through cost blowouts or losing profits through underestimation. Methods are framed within the necessary local ethical and legal requirements. The main territories applicable are Australasia and Southeast Asia and the wider sphere of Commonwealth Nations, although the principles are internationally relevant. Guiding you through the complete process of project design, costing and tendering, this book is the ideal bridge between studying civil engineering and practising it in a commercial context.

This text guides civil engineers in their quest for practicable solutions that:

- Are cost effective   (Chapter 2)
- Can be completed within a reasonable timeline   (Chapter 3)
- Conform to relevant quality controls   (Chapter 4)
- Are framed within appropriate contract documents   (Chapter 5)
- Satisfy ethical professional procedures, and   (Chapter 6)
- Address the client's brief through a structured approach   (Chapter 7)
  to integrated design and cost management.

Acknowledgements

This work acknowledges with many thanks:
- Expert advice received from David Scott, Cong Bui, Ian Chandler, Lau Hieng Ho, Phil Evans, Joan Squelch, Vanissorn Vimonsatit, Ranjan Sarukkalige, Navid Nikraz, Ommid Nikraz, Martin Edge, Seaton Baxter, Robert Pollock, Assem Al-Hajj, Richard Laing, Graeme Castle, Doug Gordon, Ravi Dhir, Tom Dyer and Mick Mawdesley.
- Input suggestions from Anna Pham, Liam Gayner, Philip Gajda, Colleen Smythe, Carlo Cammarano, Andrew Crew, Aziz Albishri, Maryam Alavitoussi, Paul Brandis, Faisal Alazzaz, Graeme Bikaun, Abdullah Almusharraf, Erin Macpherson, Tachella Atmodjo, Ting Sim Nee, Shariful Malik, Nicholas Marshall and Mark Luca.
- Applied knowledge interpretations by Nicholas Loke, Chen Shok Yin, Sun Yini, Jarrad Coffey, Nicholas Teraci, Kai Teraci, Tim Bird, Mat B. De Gersigny and Glenn Hood.
- Valuable industry reflections received from Bob Hunt, Paul Vogel, Gerry Hofmann, Richard Choy, Tom Engelke, Allan Williams, Fran Evans and on-going support from Jenny Lojitin, Maia L.W. and the many colleagues and friends who offered help and encouragement.

Author

Dr. Andrew Whyte, head of the Civil Engineering Department at Curtin University in Perth, Australia, having worked in industrial and academic environments in Europe and the Asia-Pacific region, has gained a wide-ranging knowledge of construction management, design team integration, whole-life assessment of structures and asset development using local technologies and low-cost and sustainable methods.

Chapter 1

Introduction

1.1 CIVIL ENGINEERING ATTRIBUTES

A civil engineer must develop and be able to apply a multitude of skill bases. Professional and technical attributes might be argued to encompass designing civil engineering solutions, construction procurement, site-work arrangement, as well as the allocation, time-scheduling and cost-control of all relevant resources. The civil engineer is charged with the integration of design and construction, based on solid engineering foundations couched within a code of ethics that seeks to protect and enhance society's well-being.

The information presented in this text shall seek to assist today's civil engineer to

- Investigate and integrate technical and professional skills and responsibilities to practise confidently and competently
- Evaluate design options with respect to constructability and sustainability
- Communicate and justify built-asset solutions effectively, using a variety of media, and describing how a preferred design/construction option maintains consistency with (inter)national standards' provisions
- Develop and apply skills and attributes that allow the professional to function effectively within, and lead, building design teams
- Demonstrate an understanding of the responsibilities of ethical professional engineers in addressing quality, cost and time variables in the planning, procurement and contractual administration of projects

The learned bodies of civil engineering globally (under an international umbrella commonly known as the Washington Accord[1],*) agree that (graduate) practitioners must address technical compliance and also

---

* Further reading/references annotated and listed at the end of the book.

communicate their rationale for practicability of solutions (see Engineer's attributes in chart below).

Quality reference material is required to assist both new civil engineering graduates and experienced design practitioners to maintain relevancy in appropriate design justifications, guide work tasks, control costs and structure project timelines.

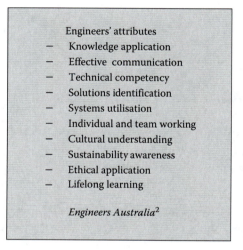

Engineers' attributes
— Knowledge application
— Effective communication
— Technical competency
— Solutions identification
— Systems utilisation
— Individual and team working
— Cultural understanding
— Sustainability awareness
— Ethical application
— Lifelong learning

Engineers Australia[2]

This text seeks to add to the existing body of knowledge to assist civil engineers in designing fit-for-use solutions within established time frames and reliable budgets.

1.2 DESIGN, CONSTRUCTION AND MANAGEMENT OF CIVIL ENGINEERING PROJECTS

Civil engineers work in multidisciplinary building design teams; their role depends largely on alignment with either design activities, or alternatively, on on-site construction work. Relationships between those who design and those who construct are traditionally described in a contract, which formalises obligations and responsibilities.

A process of tendering is undertaken to allocate a project to a builder (see Traditional tendering process). Traditionally, a client, with a need for some building work, shall seek out a team of construction and civil engineering designers and engage them to prepare a design solution.

The design solution is presented in the form of three specific types of documents: *drawings* (which describe *what* is to be built), *specifications* (which describe *how* the building materials interact) and *bills of quantities* (which describe *how much* work is involved).

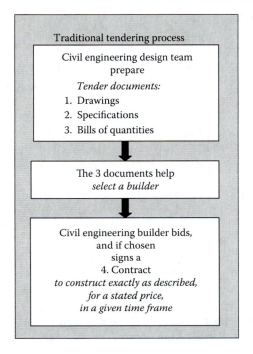

These tender documents are then used to invite offers from construction firms that might be interested to build the design. The client's team reviews all offers received, then choose a builder whom they consider to have submitted the best bid. An agreement is then cemented in a *contract*, which sets out the obligations and responsibilities of both the client and the builder; the builder, having signed the contract, becomes the *contractor*, and the project commences.

The civil engineer has an active involvement at all stages of a building project, and may be expected to be able to apply appropriate skillsets to project designing, constructing and managing.

Roles that might be undertaken by the civil engineer include

Design team and/or project administration role
- Project administrator/superintendent
- Client's representative/project manager
- The design team leader
- Specialist design consultant attached to the design team

Construction team role
- Main contractor (builder who signs a contract to construct the works)
- Specialist nominated subcontractor (nominated directly by the client's team)

- Specialist domestic subcontractor (selected by the main contractor)
- Specialist supply of components or services (or both)

Project parties are required to interact and integrate their activities towards the successful design and construction of the client's required built-asset.

A typical/basic representation of party relationships involved in the realisation of a traditionally procured project[2] is shown in Figure 1.1.

The interrelationships of the parties engaged to build a solution to address a client's needs must integrate respective design, construction and management inputs.

The civil engineer and the other practitioners engaged in a project must collectively seek practicable solutions that

- Are cost-effective	(Chapter 2)
- Can be completed within a reasonable timeline	(Chapter 3)
- Conform to relevant quality controls	(Chapter 4)
- Are framed within appropriate contract documents	(Chapter 5)
- Satisfy ethical professional procedures	(Chapter 6)
- Address the client's brief through a structured approach to integrated design and cost management	(Chapter 7)

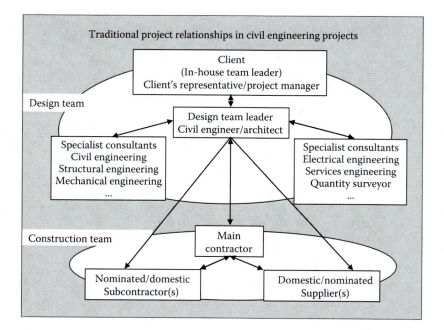

Figure 1.1 Traditional relationships.

To illustrate an integration of the essential elements in civil engineering design solutions, the following chapter breakdown describes a structured presentation of the components of cost, time, quality, contract and ethics.

1.3 CHAPTER BREAKDOWN

Chapter 2 discusses cost planning and control and highlights the main methods of civil engineering cost estimation and monitoring as well as applicable time–value of money and life cycle costs analyses techniques and applications:

- Approximate estimating (Section 2.2.1)
- Preliminary estimating (Section 2.2.2)
- Detailed estimating (Section 2.2.3)
- Cash flow and income monitoring (Section 2.3)
- Time–value of money and engineering economics (Section 2.4)
- Life cycle cost analyses with applicable case studies (Section 2.5)

Chapter 3 reviews time schedules and the extent to which project period predictions and engineering method statements are influenced by risk. This chapter also extends discussions of time towards project management approaches such as value management, critical chain, agile methods as well as looking at project delay variables:

- Scheduling: charts, networks, critical path and evaluation (Section 3.1)
- Rescheduling: programme crash-costing and resources aggregation (Section 3.2)
- Scheduling risk (Section 3.3)
- Linear time-and-distance scheduling (Section 3.4)
- Method statements (Section 3.5)
- Value management (Section 3.6)
- Critical chain project management (Section 3.7)
- Agile management (Section 3.8)
- Delays and fluctuations in projects (Section 3.9)

Chapter 4 presents quality control approaches and particularly the need to prepare and conform to specific quality management systems and implement safe working practices. The chapter then recognises that continuous improvement is a fundamental extension of quality management and reviews prefabrication and defects prediction as a means to improve productivity:

- Quality systems and quality standards (Section 4.1)
- Quality and contractual requirements (Section 4.2)
- Quality and continuous improvement (Section 4.3)
- Occupational health and safety in construction (Section 4.4)
- Prefabrication and modularisation productivity (Section 4.5)
- Prefabrication, design specification decisions (Section 4.6)
- Predicting defects in civil engineering activities (Section 4.7)

Chapter 5 explains the contract documentation required to bring projects to fruition with particular regard to the clarification of the full range of responsibilities and obligations of the parties contracted to provide sustainable solutions. Detailed discussions are made for specifications, bills of quantities and design drawing standards:

- Contractual and legal practices and procedures (Section 5.1)
- Specifications (Section 5.2)
- Rates schedules and bills of quantities (Section 5.3)
- Design drawings (Section 5.4)

Chapter 6 considers engineering ethics and professional practice, and details a range of subject fields that allow the satisfaction of an ethical imperative for continuing professional development, which includes the acknowledgement of professional tradition, leadership and appropriate communication approaches, awareness of change management structures and recognition of the need for participation within a multidisciplinary team:

- Engineering traditions and (natural) philosophy (Section 6.1)
- Professional engineering ethics and institution membership (Section 6.2)
- Leadership theory (Section 6.3)
- Communication (Section 6.3.4)
- Change management (Section 6.3.5)
- Professional integration in a multidisciplinary team (Section 6.4)

Chapter 7 details an integrated approach towards the provision of cost-effective, time-efficient, fully complaint design solutions able to satisfy a client's brief and suggests a step-by step design presentation approach illustrated by a range of practice examples. This chapter concludes with several illustrative cost and output efficiency tables:

- Integrated design solution step-by-step approach   (Section 7.1)
- Practice examples   (Sections 7.1.1–7.1.3)
- Representative cost and efficiency data   (Section 7.2)

Cost planning and control are central to design realisation. Chapter 2 details cost analysis techniques and applications.

Chapter 2

Cost planning and control

The application of management techniques to control civil engineering work and the establishment of cost principles for projects are essential in today's construction industry. Engineers must organise, plan and assess the cost of a facility at all stages of a project's life cycle; from concept to feasibility through design, construction, commissioning and handover, and then on into operation, maintenance, future refurbishment(s) and ultimately asset decommissioning at the end of the facility's lifetime.

This chapter discusses and highlights the main methods of civil engineering cost planning, estimating and cost control, as well as the time–value of money and life cycle analyses:

- Approximate estimating (Section 2.2.1)
- Preliminary estimating (Section 2.2.2)
- Detailed estimating (Section 2.2.3)
- Cash flow and income monitoring (Section 2.3)
- Time–value of money and engineering economics (Section 2.4)
- Life cycle cost analyses (Section 2.5)

The development of predictive work-breakdown structures and construction-critical paths for the construction phase allow design engineers to estimate the cost of various components of a project reliably, and allow specific-task risk appreciation. The first step is the determination of what constitutes a civil engineering project and how it might be brought to fruition.

2.1 COST PREDICTION AND ESTIMATING IN CIVIL ENGINEERING PROJECTS

Civil engineering projects globally, nationally and regionally[1] are unique one-of activities to achieve a built asset end product. Originally, any and all nonmilitary engineering was deemed to be 'civil' engineering.[2] Current civil engineering applications address structural stability concerns, appropriate building material usage, soils appreciation, water supply systems, transportation infrastructure development, hydrodynamics, environmental control systems, land reclamation, drainage, watercourse manipulation, canal/harbour/dock/marine construction, power generation facilities, sewage disposal and sewerage systems, municipal works and town planning concerns.

All civil engineering projects require the organisation, control and manipulation of a workforce, building materials, as well as plant and machinery to accurately position all construction components. The key resources required to realise a construction project might be easily defined as the three M's, namely, women and men, machines and materials.

Traditionally, civil engineering was classified as one of eight sectors of a national construction industry; today's technical applications, however, blur the historical divisions of public sector, private sector, housing, industrial, commercial, civil engineering, repair and maintenance and demolition. Civil engineering can and does contribute to each and every one of these eight sector divisions.

Civil engineering projects go through several main phases of development, each of which must be planned, organised and costed. Phases include

- *Concept*: The client's recognition and communication of a specific need
- *Feasibility*: A client consultation process to determine the scale of requirement and an assessment of whether funds and motivation exists to continue
- *Design*: The consultant's explicit representation of a facility that will satisfy the client's brief
- *Construction phase*: The building of the designed facility until practical completion and handover
- *Operational phase*: Necessary periodic life cycle maintenance of the existing facility to ensure its continuing operational efficiency
- *End-of-life*: Recognition of the work and residual costs involved in decommissioning/recycling the components of the built asset

Civil engineering and building development processes related to the above phases are often termed as construction stages:

- Briefing stage (related to the concept phase)
- Sketch plan stage (linked with feasibility)
- Working drawings stage (to clarify the design phase)
- Site operations/finalised drawing stages (necessarily linked to construction)
- Life cycle analysis[3] (to encompass costs-in-use/end-of-life residuals)

Opportunities exist at each project phase and development stage to refine both technical design compliance(s), and given an increasingly detailed appreciation of the design of facility to be built, the project's cost estimate is also a dominant concern.

Today's clients and stakeholders demand a clear relationship between the original feasibility estimate and the final cost at practical completion.

Any disparity in the predicted cost compared with the actual expenditure creates much tension (see *West Australia Sunday Times*). Big increases in final cost detract greatly from the (public) perception of the overall usefulness of the built asset.

Cost management processes are increasingly important to civil engineers and the projects that they bring to fruition.

West Australia Sunday Times
7th Aug 2011

WA'S BLOWOUT BILL

$7,000,000,000

Spiralling cost of the state's
yet-to-be-finished projects

Perth Waterfront:
Original cost: $300M
Updated cost: $440M

Woodside Pluto LNP
Original cost: $11.2B
Updated cost: $14.9B

Perth City Link:
Original cost: $500M
Updated cost: $737M

Perth Arena
Original cost: $160M
Updated cost: $520.5M

Oakajee Port and Rail
Original cost: $4.3B
Updated cost: $5.94B

Fiona Stanley Hospital
Original cost: $400M
Updated cost: $1.7B

Section 2.2 presents the principal techniques that allow the civil engineer to accurately predict the cost of their design at various stages of development.

2.2 COST ESTIMATING

Once the client has determined that a need does indeed exist for their proposed built asset, a design team of consultants are given the go-ahead to begin to flesh-out the initial concept. From the feasibility phase onwards,

Table 2.1 Staging

Stage		Cost planning and control of tasks
Briefing	Inception	Confirmation of cost limit
	Feasibility	Profitability
Sketch plans	Outline proposals	Prepare cost plan of possible solution
	Scheme design	Agree on outline cost plan
Working drawings	Detail design	Cost checks
	Production info.	Final cost checks
	Bill of quantities	Detailed cost breakdown
	Tender action	Cost analysis
Site operations	Project planning	Cost scheduling
	Operations on site	Cost monitoring
	Completion	Final account
Life cycle analysis	Feedback	Cost updates/corrections, cost-in-use

more and more detailed elemental designs and related component cost estimates are needed.

Cost planning and control[4] of the various tasks of a civil engineering construction project must progressively (i) confirm the cost limits, (ii) determine the potential profitability, (iii) prepare a cost plan of possible design solutions, (iv) agree on an outline cost plan, (v) conduct cost checks, (vi) prepare a detailed cost breakdown and then (vii) monitor and update all costs and seek to justify variances between predicted variables and actual expenditure. A linkage between the construction stages described in Section 2.1 and the respective cost planning tasks might be tabulated as follows (Table 2.1).

Estimating project cost in a structured and objective fashion at each (progressively detailed) stage becomes an important means to maintain client confidence and ensure that the project remains on-budget.[5]

2.2.1 Approximate estimating

Upon a client's willingness to proceed beyond the project concept phase, the design team is charged with the preparation of an initial estimate that takes into account current costs (of labour, plant and materials), as well as future trends in facility design and specification.

Approximate estimating has a minimum of preparation and is based on the judgments and past experience of a design team with knowledge of similarly appointed constructed assets and structures.

Price books are widely available for a project's approximate estimate. Price books are, in the industry, somewhat of a guilty pleasure; oft consulted but seldom acknowledged by practitioners, who wish to be seen as financial experts in their own right. For the inexperienced, price books are an essential means to confirm cost limits.

Price book texts include[6-11] *Spon's Asia-Pacific Construction Costs Handbook, Rawlinson's Construction Cost Guide, Cordell's Building Cost Guide* and a raft of other international examples such as *Spon's Civil Engineering and Highway Works Price Book, Spon's Railway Construction Price Book* and BCIS Wessex and Laxton's Building Price Books.

Approximate estimates might be regarded as having an accuracy of more or less 20%; in other words, the proposal for a warehouse might attain an initial approximate estimate of $100K, but must be explained (to a client) as being between the $80K and $120K budget range, at the very early stage of development. Traditionally, approximate estimates are given as a monetary rate per (functional) unit.

Depending on the proposed facility, an approximate estimate might be given in terms of $/area, $/user, $/flow rate or $/power output (see Approximate estimate: Typical functional unit rates). Increasingly, stakeholders request that approximate estimates be given as $/m² (or $/ft²) for a facility of a specific size.

APPROXIMATE ESTIMATE: TYPICAL FUNCTIONAL UNIT RATES

Office/residential asset	$/m² (or $/ft²)
School	$/pupil
Power station	$/megawatt
Hospital	$/bed
Bridge	$/unit max flow

Section 2.2.1.1 provides an approximate estimate practice example.

2.2.1.1 Approximate estimating: Practice example

Civil engineers who seek to build confidence in their ability to design within practicable limits might be advised to develop some basic approximate estimating skills across a range of potential developments.

Steps toward (what might be termed *quick-and-dirty*) approximate estimates might be

Step 1: Clarify the basic concept resulting from an initial client briefing.

Step 2: Review current accessible media releases and literature related to similar, local projects (see *West Australia Business News*).

Step 3: Consult locally applicable in-house office final accounts (if available) or price book(s) and cost indices.

Step 4: Reflect, rationalise and seek to identify an approximate estimate based on the initial scope of works within a ≈20% degree of accuracy (and a working acknowledgement of price inflation).

West Australia Business News
11th June 2009

Cost overruns lead to reforms

Project name	Original budget	Updated budget	% increase
New performing arts venue	$42M	$91M	118%

Several development scenarios might be suggested as follows:

(a) Recreation and Arts Building   Theatre ≥500 seats + seating/
                                      stage equipment
(b) Industrial Building   Warehouse, high bay, no heating/
                                      air conditioning
(c) Administrative Building   Offices for letting, high-rise,
                                      air-conditioned
(d) Health Care Building   General hospital, 150 beds

In scenario (a) above, an approximate estimate for a Recreation and Arts Building is approached using a four-step procedure as follows:

Step 1: To prepare the approximate estimate, the design team clarifies that a client requires a theatre of 770 seats, including all necessary performance/stage equipment, located in Perth, Western Australia A theatre of 700 or so seats is deemed to equate to approximately 18,000 m².

Step 2: Following a trawl of the most relevant local media releases related to similar project developments (in this particular case, perhaps the *West Australian Business News* as captioned above), an overall assessment based on the updated figure of $91M is deemed appropriate

770 seats and area of ≈ 18,000 m²
$91M/770 seats ≈ $118K/seat or
$91M/18,000 m² ≈ $5K/m²
Note: 'Original' budget pegged at $42M or ($42M/18,000) ≈ $2.3 k/m²

Step 3:   Identify and apply locally available price book(s) (in this particular case, *Spon's Asia-Pacific Construction Costs Handbook* is deemed appropriate)
500-plus theatre with seats and stage equipment ≈Aus$3.2K/m^2

Step 4:   Reflection and rationalisation of the quick-and-dirty available data set suggests that a range of between $3K–$5K/m^2 (or $58M–$91M) is deemed a suitable range for the approximate estimate of an 18,000 m^2 theatre
Note: In approximate estimating, it is prudent to err on the high side
Thus, $75M ± 20% is a reasonable figure to forward to a client who requests a design team provide a cost for a 770-seat theatre

Approximate estimation Step 3 for industrial, administrative and health Care buildings (scenarios b, c and d) might be summarised in (Table 2.2), alongside civil engineering developments of a station and a dam.

Approximate cost estimates for large-scale civil engineering projects are somewhat reliant on a design teams' recognition of the interplay between the three M's of women and men, machines and materials, as well as the resultant (national/international) location cost weightings of these three interrelated resource variables. Labour costs in particular might be expected to vary greatly across Australasia/Asia. International comparisons and equivalencies for the Asia-Pacific region provide a necessary relevancy, not just for on-site workers but also increasingly for consultancy design/ drafting services. International comparisons and equivalencies may be found in texts provided by the *Australian Bureau of Statistics, the United National Economic and Social Commission for Asia and the Pacific and Spon's Asia-Pacific Cost Handbooks.*[6,12,13]

Table 2.2 Approximate estimates

Facility	Approximate estimates	
	Cost reference A[6]	Cost reference B[9]
Warehouse, high-bay, no heating/air-conditioning	Aus$550/m^2	Aus$425–560/m^2 GB£280–370/m^2
Offices for letting, high-rise, air-conditioned	Aus$2800/m^2	—
General hospital, 150 bed + teaching facilities	Aus$3750/m^2	—
Station railway	—	Aus$2525–4242/m^2 GB£1670–2810/m^2
Dam reinforced concrete	—	Aus$1860–2727/m^2 GB£1230–1800/m^2

2.2.2 Preliminary estimating

Building on the preparation of approximate estimates discussed in Section 2.2.1, civil engineers require preliminary estimates at the 'sketch plan' or 'outline proposal' stage to address the evolving scheme design and ultimately agree on an outline cost plan with the client.

Preliminary estimates are used to help determine the relative costs of alternative designs or alternative scheme proposals to address a client's need(s), and thus guide the team in deciding on a (practicable) detailed design. A preliminary estimate, for example, for an earth dam, might involve a basic assessment of

- Quantities of materials involved, with an identification of cost per unit for rock/clay excavation, transportation and placement of concrete outlets, spillways and the like
- A single-figure percentage additional for feasibility, investigation and design
- Administration and supervision single-figure percentage additions
- A single-figure percentage addition for both contingencies and market variables

Preliminary estimates require some initial design work, sketch plans, general arrangement make ups and elemental division consideration (see Preliminary estimate: Elemental division).

These might be used to allow the calculation of the major quantities of elements (in substructure foundations, and in superstructure frames and finishes).

An overall project cost might then be given a preliminary estimate, calculated from unit rates per approximate volume of hardcore-fill, concrete, structural steelwork and the like.

The accuracy of a preliminary estimate might be pegged at 15% to 20%, although confidence increases with design detail.

PRELIMINARY ESTIMATE: ELEMENTAL DIVISION

1. **Substructure**
   Demolition, site clearance, earthworks, foundations
2. **Superstructure**
   Frame, floors, roof, external walls
3. **Internal finishes**
   Wall, floor, ceiling
4. **Fittings and fixtures**
   Internal architecture, painter work
5. **Services**
   Sanitary, water, heating, air-conditioning, mechanical and electrical services
6. **External works**
   Site organisation, drainage, minor works

Acknowledgment of relative cost weightings of the elements that constitute a project can assist the design team in reviewing specification alternatives. Comparisons of design specifications, and their respective costs, go towards a process of value management (value engineering) in which alternative design and specification options (that are technically compatible) are identified and reviewed in terms of their fit-for-use installation cost, with future cost-in-use an additional essential consideration. The Australian Standard for Value Management (AS 4183-20007), sets out procedures for the choice of a best (cost) alternative from a range of (specification) options.

Section 2.2.2.1 provides a preliminary estimating practice example.

2.2.2.1 Preliminary estimating: Practice example

Preliminary estimates are conducted to further clarify the rough, approximate estimate prepared for a construction project. Six steps involved in a preliminary estimate (to develop the initial approximate estimate) are as follows:

Step 1:   Review the initial approximate estimate (see Section 2.2.1).

Step 2:   Break down the overall concept into its constituent elemental divisions.

Step 3:   Break down the elemental divisions into basic subelements.

Step 4:   Seek to assess a percentage weighting of the value of the contributory elements based either on experience or any number of available textbooks,[5] to extrapolate unit rate approximate estimates for project elements.

Step 5:   Consult locally applicable in-house office unit rates (if available) or price book(s) and cost indices for the respective elements and subelements, then.

Step 6:   Reflect, rationalise and seek to identify a preliminary estimate based on (sub)element-by-(sub)element design specifications within a ≈15% to 20% degree of accuracy (acknowledging market, profitability and price inflation).

A development scenario might involve a request to build a Large Industrial Building Warehouse, of high bay, owner occupation, no heating and no air-conditioning (as Section 2.2.1).

A preliminary estimate for the warehouse may adopt the following approach:

Step 1:   Acknowledge an approximate estimate of Aus$550/m^2 (see Section 2.2.1.1).

Step 2:   Clarify that a large steel portal frame warehouse of 6000 m^2 might be broken down into elemental divisions: 1, substructure; 2, superstructure; 3, internal-finishes; 4, fittings/fixtures; 5, services; and 6, external works.

Step 3: Develop basic design specifications for relevant subelements such that the substructure might be described in terms of earthworks (*in situ* concrete foundations and associated formwork and reinforcement). The superstructure might be specified as having precast concrete floors and a steel frame.

Step 4: Using (newly developed) design criteria, in-house experience and building texts, allocate practicable percentage weightings for the elements such that:

- The substructure might be assumed as 20% of the project's make up (with an extrapolated unit rate of Aus\$550/m^2 × 20% ≈ Aus\$110/m^2).
- The superstructure assessed as making-up 55% (≈Aus\$300/m^2).
- Internal finishes and fittings contributing only a nominal amount.
- Services (5%) and external works (20%; rated at Aus\$30/m^2 and Aus\$110/m^2, respectively).

Step 5: Develop respective subelement unit rates from in-house sources or external price books that give an 'all-in' foundation construction rate of Aus\$100/m^2 or an equivalency for an all-in 450 × 1000-mm-deep foundation rate of GB£110/m.

Step 6: Reflect, rationalise and seek to identify a preliminary estimate based on (sub)element by (sub)element design specifications for a large warehouse of 6000 m^2 within a ≈15% to 20% degree of accuracy of, in this example, Aus\$3.4M (in which the substructure is estimated at Aus\$0.67M).

The preliminary estimating steps (Steps 1–6) are summarised using a tabular approach (Table 2.3).

Having established the elemental cost breakdown for this basic structure, the design team might subsequently conduct a value management (value engineering) exercise towards further practicability assessments of the design at this relatively early stage. From the above preliminary estimate, the dominant cost element for further consideration is the superstructure. Current steel frame/columns may offer opportunities for respecification (within current fit-for-use structural stability considerations) in terms of the somewhat less expensive concrete option; notwithstanding supply chain, build time and life cycle cost considerations, towards a more comprehensive comparison of superstructure design specification alternatives. Subsequent design specifications are then confirmed and a detailed estimate prepared.

2.2.3 Detailed estimating

Confirmation of a project's elemental and subelemental design specifications are finalised following relevant value management reviews conducted

Table 2.3 Preliminary estimates for a large steel portal frame warehouse of 6000 m²

Step 1	Step 2	Step 3	Step 4		Step 5	Step 6
	Element	Subelement	Elemental weighting	Element $ equivalency	Element unit rate[9]	
Approximate estimate, Aus$550/m²	1. Substructure	Demolition, site clearance, earthworks, foundations	20%	Aus$110/m²	450 × 1000 mm deep, in situ concrete foundations all-in excavate formwork, reinforce and dispose	Aus$100/m² and GB£110/m
						Aus$0.7M
	2. Superstructure	Frame, floors, roof, external walls	55%	Aus$300/m²	Steel frame and precast concrete floor	Aus$270–340/m² and GB£180–230/m²
					Flat roof deck and finish	Aus$75–90/m² and GB£50–230/m²
						Aus$1.8M
	3. Internal finishes	Wall, floor, ceiling	Nominal	—	—	—
	4. Fittings/fixtures	Internals, painter-work	Nominal	—	—	—
	5. Services	Water, mechanical and electrical, sanitary	5%	Aus$30/m²	Lighting, general	Aus$55–75/m² and GB£38–52/m²
						Aus$0.2M
	6. External works	Drainage, site organisation, minor works	20%	Aus$110/m²	Drainage	Aus$15–30/m² and GB£10–20/m²
					Water services	Aus$80–105/m² and GB£53–71/m²
						Aus$0.7M
					Total	Aus$3.4M

during the preliminary estimating stage of development. Alongside subsequent detailed final design drawings and final specification choices, a detailed estimate of cost is prepared.

At this stage, the project is definitely going ahead. A detailed estimate is prepared to facilitate the tendering process. On the one hand, the client's design team shall use their detailed estimate to allow future comparison with the builders invited to tender for (and to submit a bid to construct) the construction project. On the other hand, firms keen to be chosen as the contractor to construct the design, will prepare their own builder's detailed estimate of the cost of the work on which a percentage for profit is added.

The detailed estimate, prepared by the client's design team to help choose a builder, is then retained to assist in periodic interim payments for work done on-site and project stages completed. A civil engineering standard method of measurement is adopted to itemise the project into its constituent parts.[14] Computer-aided design/drafting packages used to compile drawings, can help identify (default output) quantities of materials; such outputs are subsequently described in a standard format for civil engineering nomenclature within Bills of Quantities (BQs) or Bills of Approximate Quantities (further detailed discussion of BQ preparation is made in Section 5.3).

BQ item descriptions provide a standard explanation of the nature and quantity of the work to be done, the type and quality of the materials to be used and, alongside the specification document, describe installation with reference to technical (national) standards. Detailed estimating requires a preparation of civil engineering BQs in conjunction with construction method statements (structured descriptions of how to carry out a project's work tasks, which itemises equipment, plant and manpower requirements for scheduled tasks). Although estimating can be done without the compilation of standard work classifications, it is suggested, however, that without sufficient standardised details, any estimating conducted may only be regarded as approximate or preliminary. A typical BQ item, in which Aus$162.50 is the cost to excavate 25 m^3 of general material to a depth of 1.5 m, at a unit rate of Aus$6.50/m^3, can be represented in a BQ document in a standard format with reference to the Civil Engineering Standard Method of Measurement[14] (with CESMM standard identification reference E423), as follows:

Typical BQ item		Unit	Quantities	Unit rate	Total
	Earthworks			Aus$	Aus$
E424	General excavation				
	Material other than topsoil, rock or artificial hard material Maximum depth: 1–2 m	m^3	25	6.50	162.50

- The first column allows a cross-reference to the CESMM standard, where 'E' signifies earthworks, '4' identifies the first division classification of general excavation, '2' describes a second division subcategory of soil type and '4' describes the third division depth category of less than 2 m. The second column describes this in words.
- The 'quantities' column-amount of 25 m³ is measured from a design drawing.
- The 'unit rate' column-amount of Aus$6.50 requires to be built up from an amalgamation of labour, plant and materials cost estimates (often with item-by-item, single-figure percentage additions for overheads and profit, respectively).
  - Labour costs include fitting and fixing materials and goods in position.
  - Hourly rate × output a worker/gang can complete in 1 h.
  - Materials and goods costs include delivery to site, unloading, handling/stacking and storage costs, as well as cutting and any resultant waste.
  - Plant costs are based on equipment to construct, fit and fix the machine hire (or purchase rate) with required.

The efficiency rate for a unit of work done is referred to as gang-hours the time taken to do a unit-amount of a task (Figure 2.1).

The total cost to excavate (Aus$6.50/m³) is broken down into

Plant:
- $40/h × 0.05 h/m³ = $2.00/m³

Labour: Operator + Nominal banksman
- $90/h × 0.05 h/m³ = $4.50/m³

Material installation:
- nil

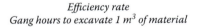

Efficiency rate
Gang hours to excavate 1 m³ of material

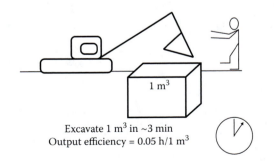

Excavate 1 m³ in ~3 min
Output efficiency = 0.05 h/1 m³

Figure 2.1 Efficiency rates.

Table 2.4 Resource by factor

Resource	Factors to be priced for all-in costs
Labour	Wage/minimum wage, overtime, public holidays, guaranteed weekly earning, bonus payments, tool and clothing allowance, annual holidays, travelling time, fare and lodging, injury and sickness benefits, provident fund contributions, training levy, employer's liability insurance, responsibility (foreman, ganger, craft/skill rating)
Plant	Equipment and machinery continuity/hire/lease/purchase/borrowing costs, time-related costs (cost of fuel, oil), fixed costs (mobilisation cost to bring-to-site and remove-from-site), production rates (output [m³/h] with related environmental conditions [weather, water, slope]), weight restriction measures, support equipment (slings, chains, skips)
Materials	Supplier quotations, delivery requirements, trade discounts, waste allowances, unloading/storage/distribution costs, timing and rate of deliveries, amount to be stored on-site, shrinkage
Other	Percentage additions and add-on costs for specialist subcontractor work, overheads, profit and other on-costs

The resources' (respective labour, plant and materials) all-in costs are themselves an amalgamation of several contributory factors (Table 2.4).

Generally, detailed estimates require work task identification and unit rates from all-in labour, plant and material costs; a practice example is discussed in Section 2.2.3.1.

2.2.3.1 Detailed estimating: Practice example

After confirmation of the project's final design and specifications, and completion of component value management comparisons, a detailed estimation is conducted.

The six steps to addressing detailed estimates (to finalise a project's cost parameters) are as follows:

Step 1: Review the preliminary estimate (see Section 2.2.2) and ensure that the finalised preferred elemental design choices reflect a value management comparison of all potential fit-for-purpose specification alternatives.

Step 2: Ensure that key items have been described in a standard format to reflect civil engineering standard methods of measurement nomenclature (Section 5.3).

Step 3: Develop from in-house sources, external price books and construction texts, a range of work task gang-hour efficiencies and unit rates for respective all-in labour, plant and material costs; wherever possible, validate data by industry supplier quotations and time-and-motion studies of on-site production rates (generally, work efficiencies for method-related tasks might be gleaned from experience or from any available texts).[15]

Step 4: Prepare a tabulated method statement of the task(s); broken down into respective labour, plant and material components where applicable. Overheads and profit are added item-by-item in recognition that each work task has differing associated risks, safety provisions and profitability potential.

To address Operational Health and Safety,[16] the National Standard for Construction Work (Commerce 2011) must be used where all undergo construction induction training in accordance with the Occupational Safety and Health Regulations of 1996 (Worksafe 2009). Stakeholders such as clients (Protection 2007a), designers (Protection 2007b) and main contractors (Commerce 2007) are responsible for exercising the regulations as required.

Step 5: Input costs (based on the data set generated by step 3) into the tabulated work breakdown structure of project tasks (generated by step 4).

Detailed estimates from the client's design team may be used to input profit and overhead percentages collectively at the end.

The contractor's detailed estimate, on the other hand, might input profit and overhead percentages individually; weighting certain items where increased (early) profitability may exist for efficient working practices.

Step 6: Review and update the item descriptions generated in step 2 above.

Input the final all-in rates generated in step 5.

The civil engineer now has a detailed estimate for the project.

Detailed estimates from the client's design team may now be used to compare, contrast and choose from a range of builders' tender submissions, and then choose where each independent bidder has submitted their price to complete the work tasks using the standard format given in step 2.

The prospective contractors' detailed estimate, in this summarised standard BQ format, may now be used to submit accurate tender-bids, with a clear and objective item-by-item indication of profitability, if successful.

A scenario might involve the detailed estimation of some typical work tasks such as the general excavation of 25 m³, the supply and placement of 0.12 tonnes of reinforcement bar to *in situ* concrete and the supply and placement of steel fabric reinforcement to slabs.

The following step-by-step guide illustrates detailed estimate preparation.

Steps 1 and 2
Subelement items in standardised CESMM BQ format are as follows:

Ref		Unit	Quantities	Rate	Total
	Earthworks				
	General excavation			$	$
E423	Material other than topsoil, rock or artificial hard material Maximum depth: 0.5–1 m	m³	25		
	Concrete ancillaries **Reinforcement for *in situ* concrete**				
	Reinforcement				
G516.1	Plain round steel bars Nominal size: 20 mm diameter *To isolated beams, straight and bent, supply and place*	tonne	0.12		
G562.1	Steel fabric Nominal mass: 2–3 kg/m² *Welded, nominal side and end laps, supply and place*	m²	25		

Step 3
Work method, gang-rate efficiency information must be gathered. This must be validated by industry suppliers and vendors, as well as time-and-motion studies to update task performance times.

A data set allows subsequent detailed estimates of BQ task items.

Table 2.5 summarises the type of information essential to allow progression towards a detailed estimate of a project's (method statement) itemisation of jobs.

Step 4
A tabulated method statement is prepared. A method statement is prepared (in preparation for the inclusion of costs) for (i) excavation, (ii) rebar reinforcement and (iii) mesh reinforcement and is represented in Table 2.6.

Step 5
Taking the work-breakdown structure identified in step 4, and inputting the values generated in step 3, an updated table of task item costs follows (Table 2.7).

Step 6
An updated item description tabulation that now inputs costs (calculated in step 5) into the previously unpriced BQ (prepared in step 2) is presented as follows:

Ref		Unit	Quantities	Rate	Total
	Earthworks				
	General excavation			$	$
E423	Material other than topsoil, rock or artificial hard material Maximum depth: 0.5–1 m	m³	25	7.77	194.25
	Concrete ancillaries **Reinforcement for *in situ* concrete**				
	Reinforcement				
G516.1	Plain round steel bars Nominal size: 20 mm diameter *To isolated beams, straight and bent, supply and place*	Tonne	0.12	2516.35	301.97
G562.1	Steel fabric Nominal mass: 2–3 kg/m² *Welded, nominal side and end laps, supply and place*	m²	25	20.36	509.00
					$1005.22

Table 2.5 Efficiencies

Work task efficiencies	Unit	Labour rate	Plant/ machine rate	Material rate
Excavate by machine to reduced levels where depths don't exceed thicknesses of				
0.25 m (precise excavation)	h/m³	0.15 h/m³	0.15 h/m³	—
1.00 m (less precise digging)	h/m³	0.12 h/m³	0.12 h/m³	—
2.00 m (less precise digging)	h/m³	0.12 h/m³	0.12 h/m³	—
Cost/hour of general labourer	$	$21.50/h	—	—
Cost/hour of excavator + driver	$	included	$36.00/h	—
Extra for semiskilled labour for reinforcement steel fixing	$	+$3.50/h	included	—
Reinforcement				
12 mm hot rolled bars	$/tonne	—	—	1350.00
20 mm high yield bars	$/tonne	—	—	1450.00
Fabric sheet type	$/m²	—	—	11.56
Reinforcement waste				
Bar	%	—	—	2.5%
Fabric and lapping	%	—	—	11%
Steel fixing outputs				
12 mm bar	h/tonne	50	—	—
20 mm bar	h/tonne	30	—	—
Fabric sheet	h/m²	0.05	—	—

(continued)

Table 2.5 Efficiencies (Continued)

Work task efficiencies	Unit	Labour rate	Plant/ machine rate	Material rate
Tie Wire				
Tie wire cost	$/kg	—	—	5.00
Tie wire for 12 mm bar	kg/tonne	—	—	0.09
Tie wire for 20 mm bar	kg/tonne	—	—	0.10
Fabric sheet snap-spacers	$/m²	—	—	4.00
Overheads and (market) profit	%	—	—	12.5%

Table 2.6 Method statement

	Method statement description	Labour	Plant	Material	Running total	Total ($/unit)
i.	**Excavate to reduced levels <1 m deep**					
	Labour: ... h/m³ output @ $...	...				
	Plant: ... h/m³ output @ $...		...			
	Material:			—		
	Safety: nominal temporary earth retention				...	
	Overheads/profit: 12.5% of...				...	
						...
ii.	**Reinforcement, 20 mm mild steel**					
	Labour: ... + skill... =... x... h/t	...				
	Plant:		—			
	Material: (bar...) + (waste...% =...) Tie wire...kg @... (nominal waste)			... ...		
	Safety:				...	
	Overheads/profit: ...% of...				...	
						...
iii.	**Reinforcement, fabric with side laps**					
	Labour: fix mesh output... @ $...	...				
	Plant:		—			
	Material: fabric... + (waste...% =...) spacers $.../m² (nominal waste)			... ...		
	Safety:				...	
	Overheads/profit: ... % of...				...	
						...

Table 2.7 Method statement costs

	Method statement description	Labour	Plant	Material	Running total	Total ($/unit)
i.	**Excavate to reduced levels <1 m deep**					
	Labour: 0.12 h/m³ output @ $21.50	2.58				
	Plant: 0.12 h/m³ output @ $36.00		4.32			
	Material + safety extras:			—		
					6.90	
	Overheads/profit: 12.5% of 6.90				0.87	
						7.77
ii.	**Reinforcement, 20 mm mild steel**					
	Labour: 21.50 + skill 3.50 = 25.00 × 30 h/t	750.00				
	Plant + safety extras:			—		
	Material: (bar 1450) + (waste 2.5% = 36.25) Tie wire 0.10 kg @5.00 (nominal waste)			1486.25 0.50		
					2236.75	
	Overheads/profit: 12.5% of 2236.75				279.60	
						2516.35
iii.	**Reinforcement, fabric with side laps**					
	Labour: fix mesh output 0.05 @ $25.00	1.25				
	Plant + safety extras:			—		
	Material: fabric 11.56 + (waste 11% = 1.28) spacers $4.00/m² (nominal waste)			12.84 4.00		
					18.09	
	Overheads/profit: 12.5% of 18.09				2.27	
						20.36

The detailed estimate example described above addresses the on-site operational costs (direct costs) of the various work tasks involved in the realisation of a project (more examples of detailed estimating are presented in Chapter 7).

The steps above touch briefly on percentage additions for overheads. Several other additional cost plan inclusions (such as overheads) require consideration to allow a more all-encompassing prediction of project cost. Discussion of these other (indirect cost) items is presented in Section 2.2.4.

2.2.4 Cost plan inclusions

In addition to measured operations related to work on-site, cost plans (and final BQs) require the consideration, review and inclusions of other areas liable to contribute to any expenditure set aside for project participation.

As well as direct costs, estimates need to include (in addition to Section 2.2.3.1)

- *Specific contingencies*: Costs related to particular activities to cover the cost of extra work not foreseen at the design stage for rock, flooding and the like
- *Indirect costs*: Percentage spread of nonproject overheads such as accounting services, running the main office with indirect labour (receptionists and the like), advertising, telephones, travel and indirect resources (company cars), as well as indirect cost such as recruitment and human resources services
- *Markup*: Assessment of extent to which unit rates reflect both profit percentage established by senior management and direct overheads (of fencing, site utility electricity and water, along with site cleaning and debris removal)
- *Goods and Services Tax (GST)*: Local tax (≈10%) for items sold/consumed
  - GST is a broad-based tax of 10% on goods, services and other items sold or consumed in Australia. Generally, businesses registered for GST will include GST in the price of sales to their customers; this is applicable to all construction and civil engineering work.[17]

Other cost inclusions (over and above direct on-site expenditure) for a project may be grouped under several general headings that include:

- Preliminary items
- Preamble items
- Prime costs
- Provisional sums
- Contingency sums
- Profitability mark up

Discussion of these additional cost considerations are presented Sections 2.2.4.1 to 2.2.4.6.

2.2.4.1 Preliminary items

Preliminaries set out the practical and contractual conditions of the project. Respective environmental conditions and location-specific variables described within the preliminaries are an important consideration in an estimating process in which external factors contribute greatly to task method statements and related cost estimations.

Preliminary information such as urban site constraints, rural isolated conditions, project staging and seasonal inclement weather, availability of a skilled workforce, supply chain logistics and the like must be factored into estimates.

To allow appropriate tender submissions, civil engineering design teams must ensure that the project is adequately described in terms of the following preliminary items:

- Location, site description, nature and extent of work, type of contract, identifiable factors that may affect the effective execution of works and project overheads that may result from specifics such as
- Supervision requirements, temporary works expectations for false-work and scaffolding, protection of the site generally, services, facilities, security and any additional health and welfare concerns

Coupled with the preliminary items, tender documentation (usually BQs as part of the contract documents) might seek to include general preamble information.

2.2.4.2 Preamble items

Preambles sections (in a BQ) give a brief overview of a specification document and provide a summary of key aspects such as

- The quality and description of materials generally
- Standards of workmanship
- Any relevant location specific information, such as national testing needs

2.2.4.3 Prime cost sums

Prime costs sums are approximate sums of money, included by a civil engineering estimator, for works and services to be carried out by specialist subcontractors. These specialists are either nominated directly by the client's representatives or chosen (by a main contractor who has successfully won the tender) from a listing of domestic firms.

Prime cost sums represent approximate estimates that cannot be given detailed estimates because they represent work that is outside the remit of a civil engineering practitioner. Work of a specialist nature, in which only an approximate prime cost sum is applicable, includes areas such as

- Specialist architectural works incidental to civil engineering works
- Mechanical and electrical services (lift installation and the like)
- Materials and goods obtained from a (client) nominated material supplier

2.2.4.4 Provisional sums

Provisional sums are amounts of money set aside in a cost plan (or a BQ) by a client for unforeseen work. Estimates for work not known at the time of tendering might include provision of new mains electricity/gas supply to a site, permanent access footpaths and the like in which

- *Provisional sum 'defined work'* is where a contractor makes allowances in programming, planning and pricing preliminaries.
- *Provisional sum 'undefined work'* has no such allowance and is thus a justifiable future percentage adjustment to a contract period final account.

2.2.4.5 Contingency and daywork amounts

A somewhat overused provision that should be used sparingly is the contingency amount, which is placed into the contract sum to cover the cost of extra work not foreseen at the design stage. Similarly, daywork rates, which are largely a pro rata cost buildup technique for unknown labour/plant/materials, again somewhat flies in the face of a detailed estimate of a finalised all-inclusive design specification because it seeks to provide a means to cost unexpected/unknown work outwith the known scope of works.

2.2.4.6 Profit markup

The detailed cost estimate (of on-site operations) prepared by main contractors for their tender-bid (Tender timeline shown below) now requires a profit markup to reflect intent and market competiveness.

It is usual for firms to incorporate profit in an item-by-item fashion, as main contractors experience cash deficits for most of the contract. Firms want to receive payment as early as possible. This can be done by putting a higher profit markup onto the early activities on-site, in lieu of nominal profit percentage markups for latter on-site activities; reasonable weighting applies.

Once profit markup is incorporated, the tender-bid is submitted. The detailed project cost estimate is retained and used to predict cash flow; cash flow is discussed in Section 2.3.

<small>TENDER TIMELINE</small>

- Tender-bid prepared by the builder of on-site and other costs + profit markup
- Tender-bid submitted (without further solicitation) to the client's representatives
- Check all bid submissions received for
  - Arithmetic errors
  - Change in tender documentation
  - Appreciation of workmanship
  - Past experience
  - Subcontractor suggestions
  - Methodology
  - Inappropriate profit loading
- Representatives nominate a builder
- The client and builder sign a contract
- Work starts; cost monitoring begins

2.3 CASH FLOW PREDICTION AND INCOME/REVENUE MONITORING

Once the civil engineering project starts on-site, all parties involved seek to ensure that detailed, preprepared cost predictions are maintained and that a project cost blowout is avoided. This requires both the client's representatives, as well as the construction firm engaged to build the project, to measure and monitor their respective profit and loss expectations at regular intervals throughout the job.

Ultimately, civil engineers seek a comparison between actual performances with predetermined expected inputs. This can be done by periodically measuring and comparing differences between the job's earnings and it's expenses in which earnings related to a contract might be expected to include

- Predominately, the amount paid by the client for carrying out the contract
- To a much lesser extent, the sale of surplus or scrap materials or the plant and the renting out of the plant and the loan of labour to other projects

Expenses related to a contract might be itemised as a contractor's:

- Direct labour cost of wages, piece-work payments, labour-only subcontracts
- Plant charges with depreciation, maintenance costs and operating expenses
- Invoices for materials either for permanent works or temporary works
- Site (direct) overheads for access roads, office facilities, insurances and bonds
- Subcontractors' accounts based on agreed terms

Collecting and keeping records of (a client's) project payments centres on progressive interim payments' information for the remuneration of all measured work on-site, less a client's retention amount (stipulated as a single-figure percentage in the contract) to safeguard any default by the builder. Interim payment records will itemise dates alongside the percentage completion of phases or project components.

Interim valuations are usually conducted monthly. Once agreed upon by both the client's team and the builder, they represent a client's payment of work done thus far; an interim payment is the quantity completed on-site multiplied by the BQ rate, towards a maximum payment previously agreed as the contract sum (notwithstanding any claims extra-over this authorised and contractually binding project sum).

Collecting and maintaining a record of the builder's expenditure on-site requires a system to capture labour, plant, materials and overheads; outgoing costs such as labour costs are recorded by time sheets (for all locations/durations) sent to the head office. This is then itemised into basic pay, overtime, transport, accommodation and others. These actual labour costs are then compared with projected estimates for component construction, whereas plant costs are recorded by location and working hours plus idle time as well as any delays or breakdowns as a function of either the ownership cost or the cost to hire. Materials costs are recorded by delivery notes, purchase orders, on-site store then location installation; overhead costs are recorded by direct invoices related to site-fencing and utility payments, with indirect nonproject invoices retained from main office invoices.

A civil engineering contractor's cost control mechanism might seek to measure profit from the value of an activity (what a client will pay at specific time), with a deduction for the cost to carry out that activity on-site. This is often termed a *cost and value reconciliation*. A tool to assist in the visualisation of a cost and value reconciliation is an 'S-curve' cash flow,[18–21] the preparation and monitoring of which is discussed in Section 2.3.1.

2.3.1 S-curve cash flow

Cash flow prediction gives a basic forecast of projected money flow over time and requires a range of information including: payment periods, amounts to be retained by the client, the monthly value of the work (or a valuation of the work) on-site, the contractual defects' liability period after which retained money is released to the builder, as well as the value of the any delayed payment at specific stages of time. A projected income for a point in time (or more specifically, the income for a building activity at a specific period) stems from total expenditure deducted from total income.

The preparation of an S-curve illustrates cash flow over time. The x-axis of a prepared graphic represents the period (and by implication a work activity occurring at a specific period in the project schedule), although the y-axis represents the cumulative project contract sum (and again by implication, a contract sum payment for a work activity at a specific time).

The cumulative S-curve cash flow diagram (Figure 2.2) typically illustrates the following:

- In the first-third of the project duration, 25% of the project's total cost is spent on initial activities such as site set-up, site clearance and excavation.
- In the second-third of the project's duration, 50% of the project's total costs are spent on the major activities of superstructure and the conclusion of all substructure activities.
- In the final-third of the project duration, the remaining 25% of the project's costs are spent on finishings, fixings and services.

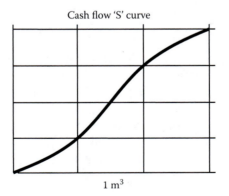

Cash flow 'S' curve

1 m^3

Figure 2.2 S-curve.

Clients might be advised by their representatives to expect to pay for work done on-site (payment outgoings) in a regularised periodic fashion, based on agreed interim/periodic valuations for activity completions at specific times. Building contractors might also seek to chart income against outgoings over time. A practice example for S-curves, to allow visualisation of cash flow over time, is discussed in Section 2.3.1.1.

2.3.1.1 S-curve cash flow: Practice example

An S-curve cash flow diagram links, on the one hand, a preprepared time-schedule of tasks, and on the other hand, the preprepared detailed cost estimate(s) of the complete series of project work tasks.

The following five steps guide the civil engineer in charting an S-curve cash flow diagram:

Step 1: Access and review the civil engineering project's preprepared project network/project schedule/critical path describing the timeline of total activities on-site; this usually takes the form of a (software package) Gantt chart/bar chart of interrelated task start-and-end dates (as discussed in Chapter 3).

Step 2: Access and review the various method statements, detailed cost estimate(s) of all the respective tasks that make up the construction project (as discussed in Section 2.2).

Step 3: From the data sets prepared for steps 1 and 2 above, tabulate a payment prediction schedule (of interim monthly valuations of work done); this usually takes the form of a table of (potential) monthly/fortnightly/weekly cumulative payments for work done (to be done) on-site.

Step 4: Plot: tasks occurring at specific months/weeks along an x-axis against cumulative task costs along a y-axis; usually, the plot of points from project beginning (nought weeks, $0) to the final project end point (100% weeks, $100% contract sum) produces a squeezed/skewed 'S' shape.

Step 5: The building contractor can extend the S-curve yet further by superimposing (i) an S-curve of the project value (the client's expected interim payments) on (ii) an S-curve of the project cost (the contractors' expected cost outgoings); usually, this allows the building contractor an opportunity to predict specific times/months when they expect to have a negative cash flow. In other words, when they (the building contractor) may need to borrow to cover their on-site expenditures.

A scenario might involve a civil engineering project conducted over an agreed contract period of 40+ weeks that has a contract sum of approximately Aus$1.3M. In this case, several interim monthly payments are expected from the client to the building contractor. The following step-by-step guide illustrates S-curve preparation.

Steps 1 and 2
Data sets representing the project schedule and the detailed cost estimates of tasks are accessed and combined.

Step 3
Table 2.8 illustrates a predictive cash flow estimate in which a client's team (in future liaison with the building contractor), values the work (to be) done on-site.

Predictive payments relate to the project's preprepared schedule and the detailed unit rates within the contract (usually the agreed BQ).

In Table 2.8, column three indicates an expectation for seven monthly interim valuation payments, followed by the release of retention money

Table 2.8 Interims

Project progress (week)	Interim payments	
	Cumulative (retention) payments ($)	Interim periodic valuation payment
1	—	
4	—	Interim valuation 1
5	140,000	Payment 140K
8	140,000	Interim valuation 2
9	280,000	Payment 140K
13	280,000	Interim valuation 3
14	460,000	Payment 180K
17	460,000	Interim valuation 4
18	640,000	Payment 180K
22	640,000	Interim valuation 5
23	900,000	Payment 260K
26	900,000	Interim valuation 6
27	1,100,000	Payment 200K
30	1,100,000	Interim valuation 7
31	1,240,000	Payment 140K
34	1,240,000	Practical completion
40+	1,270,000	Retention release
42+	1,300,000	Final payment

at practical completion. Column two shows the amount of monthly payments related to, column one of specific periods related to scheduled work tasks.

Step 4

The S-curve in Figure 2.3 may be plotted by the civil engineer to guide their client in the project's expected outgoings.

The client's cash flow payments, in this case being $1.3M, is paid over ≈42 weeks.

The client's S-curve payments represent the cumulative value of the civil engineering project, stipulated in the contract documents as the contract sum, itemised specifically in the contract documents' BQ, and scheduled within a previously agreed upon time frame as the contract period.

Any future deviations from these parameters need full justification in a contractual claim to be submitted by the builder to the client.

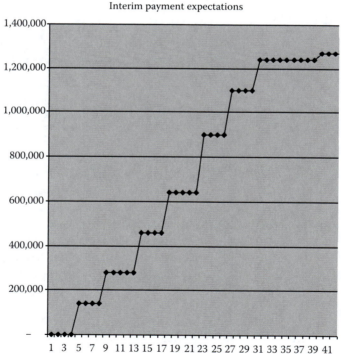

Figure 2.3 S-curve interims.

Step 5

Client-orientated S-curves and builder-orientated S-curves are somewhat different in the way they represent the costs and value of a civil engineering project, respectively.

A contractor's in-house S-curve is likely to be based on their own specific weekly forecast. The contractor's estimated rates are linked to activities on the project programme, and the total costs for each work section is identified by the time allocated/allowed to complete the task.

A contractor will seek to superimpose interim monthly payments expected from the client, upon their own estimate of the weekly costs to carry out the tasks, to allow an assessment of where they might expect to be out-of-pocket at any time during the duration of the project. The main contractors seek to predict and minimise any negative cash flow. This might be presented graphically as follows:

In this practice scenario, Figure 2.4 represents a client's valuation payments, superimposed on a contractor's periodic costs. Generally, the building contractor seeks to keep their cost outgoings below the payment incomings. Any negative cash flow is identified (and circled as shown in

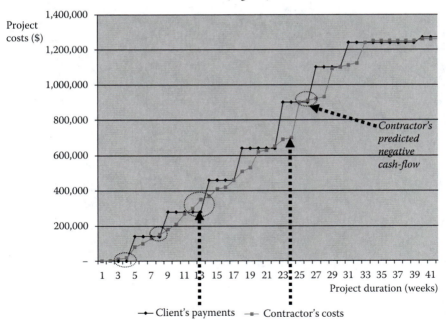

Figure 2.4 Cash flows.

Figure 2.4) and flagged by the building contractor as periods/months in which they may need to borrow, to cover on-site outgoing costs.

Having estimated the cost in detail, and predicted, and planned the project's cash flow, civil engineers must now monitor progress on-site to ensure that the parties do indeed maintain and implement the cost plan. Cost income and monitoring is discussed in Section 2.3.2.

2.3.2 Cost and income monitoring

Cost monitoring allows civil engineering building contractors to maintain and update their preprepared cost-estimate plans.[18]

Cost and value reconciliation for ongoing projects requires a periodic comparison of the original cost estimate with actual outgoing expenditure towards an analysis of discrepancies and the taking of appropriate action.

Cost accounts for specific work items (such as 'excavation' in Table 2.9) provide civil engineers variances from estimates and flag areas of inefficiency.

Variance analyses of cumulative (monthly) work done, charts progress and cost-efficiency to date, and allows an on-going assessment of profitability break-down, or where certain work items (such as excavation to foundations; Table 2.10) will incur a loss unless action is taken to rectify current inefficiencies.

Variance analysis generally requires a month-on-month updating of cost areas such as budgeted value of work to be completed, budgeted value of work *actually* completed, total budget of the work *as planned*, *actual* cost of work completed, revised total estimate cost of the project, and overall budget performance level.

Cost and income monitoring updates go towards a civil engineering building contractor's final account reporting. Final account reports

Table 2.9 Cost account variances

Excavation (weeks 1–4; status: complete)	
Resources	Cost (Aus$)
Labour	5K
Plant	10K
Materials	—
Safety[a]/false-work	1K
Indirect overheads	1K
Total cost	17K
Value (valuation payment)	15K
Work-item profit	—%
Work-item loss	(−2K)

[a] http://www.commerce.wa.gov.au/worksafe/.

Table 2.10 Variance analyses: Cumulative work done

Excavation to foundation	
Total value of work	$18,750
Budgeted value of work done	$7350
Value of work done	$6250
Actual cost of work done	$6950
% complete	33%
Variance: cost–value	–$700
Variance percentage	–11%
Forecast variance	–$2100

are distributed to a range of stakeholders. These financial reports summarise not only the condition of specific projects (and the efficiency of the trades involved in that project) but also give an indication of the building-contracting company as a whole. Reports and distribution might include

- Company balance sheets of assets and liabilities for directors and shareholders
- Site accounts for interested parties such as insurers, financiers and banks
- Project-specific cost accounts as a general overview for senior management
- Budget cost reports of variance analyses for construction management teams

Although the names and formats of these reports may vary from organisation to organisation, their ultimate purpose, as an indicator of the financial efficiency of the civil engineering building contractor, remains ever important.

The preceding discussion charts the need for civil engineering projects to accurately address capital construction cost control. Section 2.4 shall extend this discussion towards complementary methods to address time–value of money.

2.4 THE TIME–VALUE OF MONEY AND (CIVIL) ENGINEERING ECONOMICS

Cost monitoring and the generation and charting of revenue over time for civil engineering works, as discussed in Section 2.3, requires further contextualisation in terms of the time–value of money where, generally speaking, cash received today is worth more than the same amount received in the future. This is perhaps a fundamental concept of what has commonly come to be known as *engineering economics*.[22–24]

For civil engineers and contractors, the time–value of money is used in project appraisals to compare cash flow (cash-out versus revenue-in) longitudinally, such that if $90.91K is available today, and a bank's investment rate is quoted at 10% for this large sum, then in 1 year's time, the updated bank account will read $100K (90.91K + 9.09K). Contractors then will attempt to maximise revenue/income as early as possible on a big job in recognition that $90.91K received now is better than $90.91K received next year, and that $90.91K received now is as good as $100K received next year.

Resources appraisals are also applicable, where opportunity cost might be used to determine whether an experienced civil engineering contractor that works and earns $150M annually in the (somewhat uncertain) resources industry, should seek instead to get involved in the (somewhat more stable) public sector with potential fee generation of $50M. Opportunity cost for this decision would be $200M ($150M on income foregone, plus $50M), notwithstanding any future tender list inclusions/employability.

The rate of return that can be paid on existing funds is also important in project appraisal. In essence, the cost of capital is the rate of return in which capital stemming from money or property is used to produce wealth, such that equity capital is one's own investment and borrowed capital is obtained from lenders with a (bank's) interest rate determining a project's worthiness. A basic illustration of rate of return appraisal follows.

RATE OF RETURN FOR CIVIL ENGINEERING

- An engineering manufacturer designs/makes/sells precast concrete lintels (the firm uses their own capital).
- Current market value finds the firm generating a 7% rate of return (profit).
- Banks offer an 8% interest rate; logically then, the cost of capital is at least 8%.
- Perhaps this firm should simply just invest with the bank and do nothing?

Inflation must similarly be taken into account in contracting to facilitate good cost-planning and control because inflation reduces the buying power of money, resulting from increases in the average price paid for goods and services.

Inflation might be illustrated in terms of a firm that invested $100K in a bank 1 year ago at an interest rate of 8% and whose account, although now showing $108K, is subject to inflation (an average increase in prices) of 2%. The inflation factor here can be stated as (102/100) 1.02, resulting in the actual worth of today's $108K as (108K/1.02) $105.88K. Conversely, a decrease in goods and services prices increases buying power, but can

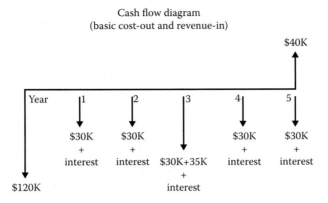

Figure 2.5 Cash flow diagram.

cause deflation, which is apt to result in holding-out for further decreases, resulting in large stockpiles and knock-on unemployment to avoid creating more unused stock.

Given that engineers benefit generally from graphical representation, illustrations and cash flow diagrams (Figure 2.5) of costs-out and revenue(s)-in, clarify cash flow.

A basic cash flow diagram to illustrate cost-out and revenue-in is shown in Figure 2.5 to represent a firm that buys a back-actor/digger for $120K (using borrowed finance at $x\%$).

Yearly maintenance and operational costs are $30K, alongside a major overhaul of $35K midterm, with second hand resale of $40K in 5 years (notwithstanding interim payments received for project participation).

2.4.1 Present, annual and future worth project assessment techniques

Time value of money analyses seek to assist in determining project financial viability where net present value (NPV) is today's value of a project's future expected cash flow:

NPV = (Present-value-revenue) – (Present-value-cost). Thus,
NPV > 0 is financially viable
NPV < 1 is potentially rejected (for private sector projects), and
NPV = 0 is equal to an internal rate of return and largely represents a project's breakeven point.

It is perhaps worthy of note, however, that public sector organisations such as the Public Transport Authority of Western Australia[25] are much less likely to determine project viability and go-ahead based simply on NPV calculations.

Indeed, approximately 65% of all Public Transport Authority of Western Australia projects predict negative NPV, but still get senior stakeholder 'go-ahead'.

Table 2.11 Economic forecasting

			Project		
		Economic forecast assessment techniques			
Factor name	Find x given y	Notation	Standard notation equation	Factor to multiply the given figure	Excel functions
Single-payment, future compound amount	Future/present	$(F/P, i, n)$	$F = P\,(F/P, i, n)$	$(1+i)^N$	FV $(i\%, n, P)$
Single-payment, present worth	Present/future	$(P/F, i, n)$	$P = F\,(P/F, i, n)$	$\dfrac{1}{(1+i)^n}$	PV $(i\%, n, F)$
Uniform series payment, present worth	Present/annual	$(P/A, i, n)$	$P = A\,(P/A, i, n)$	$\dfrac{(1+i)^n - 1}{i(1+i)^n}$	PV $(i\%, n, A)$
Capital recovery	Annual/present	$(A/P, i, n)$	$A = P\,(A/P, i, n)$	$\dfrac{i(1+i)^n}{(1+i)^n - 1}$	PMT $(i\%, n, P)$
Uniform series, compound amount	Future/annual	$(F/A, i, n)$	$F = A\,(F/A, i, n)$	$\dfrac{(1+i)^n - 1}{i}$	FV $(i\%, n, A)$
Sinking fund	Annual/future	$(A/F, i, n)$	$A = F\,(A/F, i, n)$	$\dfrac{i}{(1+i)^n - 1}$	PMT $(i\%, n, F)$

Excel function to find the interest rate (or rate of return) for any cash flow series entered into a series of contiguous spreadsheet cells	IRR (first_cell:last_cell)
Excel function to display the compound interest rate when the annual cash flows are the same	RATE (number_years, A, P, F)
Excel function to find the number of years n for a given A, P and F	NPER $(i\%, A, P, F)$
Benefit–cost (B/C) analysis expects investment P to bring benefit Q (revenue + public good intangible benefits)	

It might be argued that 'intangible', public good, environmental impact reduction, social benefit as part of a benefits/cost analysis are perhaps much more appropriate for public sector projects and perhaps, by extension, a majority of civil engineering infrastructure development.

Projects undertaken must consider the return profit on the investment and the extent to which project costs can be recovered by revenue (benefits) + profit for risk. Some methods to assess if a project gives a good return include assessment of present worth, annual worth, future worth and internal and external rates of return, all at a given interest rate i, for a number of years n.

Several economic forecast techniques, as well as Excel spreadsheet financial formula, are shown in Table 2.11.

Generally, a (private sector) project's minimum attractive rate of return must be higher than the interest rate from a bank savings account, where profitability needs to be considered across the entire life cycle of the built asset.

Further discussion of the whole-life of a project, and life-cycle analysis is tackled in Section 2.5.

2.5 LIFE CYCLE COST ANALYSIS: CIVIL ENGINEERING APPLICATIONS

Asset management and the need for decision support tools, both at the planning phase and throughout the life cycle of a civil infrastructure, is a key issue for the construction industry; it remains a challenge for civil engineers to reduce the costs of a resource or asset over the duration of its useful design life.

Comparison of the available civil engineering design options may find one clearly cheaper in initial capital construction costs; however, costs induced over the lifetime of the component may well find maintenance and repair costs outweighing costs initially saved at construction. Life cycle cost analysis (LCCA) provides a method that helps determine which specification option gives the most cost benefit over the long term, and is reviewed in many texts, not least by Australia/New Zealand Standards, AS/NZS 4536.[3,26,27]

LCCA requires an analysis period and the determination of a net present value, discount rate as a measure of the time–value of money applied as a percentage per annum, so that an effective comparison of fit-for-purpose design alternatives may be made.

Discount rates and sensitivity analyses are important in life cycle cost outputs. Among different agencies in Australia, the discount rate used in civil engineering works can be seen to vary, from the Eastern state Victoria 'VicRoads' opting for a value of 7%, to the National Road Transport Commission of Australia recommending 5%.

A Royal Institution of Chartered Surveyors[28] rule-of-thumb toward addressing discount rate is perhaps somewhat more empirical, and results from an input of relevant available data, such that

i. No-risk return = Treasury bond rate of return – Inflation
ii. Average-risk premium discount rate = Average equity return – Treasury bond
iii. Discount rate = No-risk return + 0.5 × Average-risk premium discount rate

An LCCA spreadsheet may be developed (albeit, off-the-shelf LCCA packages are increasingly available) to facilitate the input of data related to a design alternative's capital cost, plus operation-and-maintenance costs, taking into account salvage/resale. Although some suppliers/vendors provide LCCA applications,[29] such software must be used, as always, with an awareness of the wide range of 'default' assumptions that may be somewhat too indistinct for specific in-depth study, not least due to the inability of existing software to calculate, from first principles in line with the Royal Institution of Chartered Surveyors (RICS) rule of thumb above, an objective discount rate.

Information sources for quantitative data may be generated not only by industry standards and building cost information services (BCIS) but also perhaps specifically from the various state and local authorities (i.e. roads/main roads and traffic authorities) concerned with the generation of typical costs and maintenance periods. In the case of road pavements, for example, data related to alternatives of, on the one hand, an asphalt surface (thin asphalt-surfaced unbound granular pavement; AC) and, on the other hand, a concrete pavement (continually reinforced concrete pavement; CRCP) might be generated, input into a developed spreadsheet and compared empirically to provide an objective measure of whole costs, and subsequently enhance objective specification decision making.

2.5.1 LCCA case study: Review of road alternatives

The discussion in this section describes the processes involved in developing an LCCA model for an infrastructure pavement design project.*

Design options considered involve CRCP and AC. Three lifespan periods of 30, 60 and 100 years, with a discount rate calculated at 8% (incorporating sensitivity analyses), finds that the asphalt pavement alternative is the less expensive option, recommending asphalt pavement over concrete pavement in this local West Australian environment. The following discussion explains how these findings were generated.

Roads make up a large proportion of total civil infrastructure in Australia, with highways being major schemes that are owned by the State Government

---

* Input from Anna Pham,[30–45] received with thanks.

in perpetuity; sole ownership indicates that whole-of-life costs studies are extremely relevant for asset management.

Generally, asphalt pavements are favoured due to less expensive acquisition costs albeit conversely in-use maintenance might be expected to be substantial; a truism exists in the industry that although concrete roadways have a higher capital cost, upkeep costs might be expected to be lower. An LCCA was conducted to empirically assess the cheaper option.

As mentioned previously, LCCA takes into account all costs that occur over the effective life of the resource. Discounting is applied so that all costs induced over several periods are converted into present-day monetary values, summated into an NPV. Early application yields the best outcomes and allows for design specification changes to occur during planning. Figure 2.6 shows how the cost of making changes increases while the opportunity for savings decreases as time goes on.

The determination of a suitable discount rate for the analysis can lead to differing results, and civil engineers and analysts seldom reach a consensus figure; as mentioned previously, the Eastern Australian State of Victoria (VicRoads) opts for 7% whereas the National Road Transport Commission pegs the discount rate at 5%. The rate adopted here sought a factor of treasury bond rate of return, inflation rate, and average equity return rate to place long-term serviceability into context in which concrete pavements include methods such as full-depth repair, slab stabilisation or joint and crack resealing, whereas rehabilitation of asphalt pavement involves levelling and resurfacing, performed in conjunction with widening. Overall, the classic sawtooth effect of spend and deterioration occurs: spend money to build/record a gradual deterioration, spend money on maintenance/record a further deterioration, spend money to refurbish or replace/record a further deterioration until the end of life, scrap and seek residual return/then spend money for replacement.

A project-specific LCCA model/spreadsheet was developed using Excel, which enabled each component of the analysis to be input and compared;

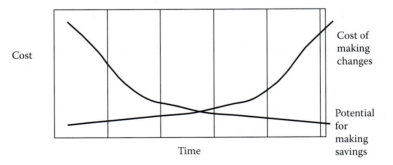

Potential savings and cost relationship from the NSW treasury

Figure 2.6   NSW treasury data.

different specification options and cost components over the useful life of the pavement were entered, namely, asphalt-surfaced (AC) and concrete pavement (CRCP) designs. In addition, stakeholder consultation with MainRoads Western Australia clarified the specification(s), cost and maintenance data generated.

The discount rate was calculated by taking into account factors such as inflation and interest, a 5.44% treasury bond rate of return based on a reasonable 10-year yield, an inflation rate attained from the Reserve Bank of Australia's website of 1.5%, and an average equity return rate assessed as 13.6% was taken from Investment-Wise. These values allowed the discount rate to be calculated as 8.0% (sensitivity analysis applied).

NPVs of asphalt and concrete pavement alternatives (AC and CRCP) over periods of 30, 60 and 100 years showed that the asphalt alternative (somewhat surprisingly, given the expectation for high maintenance and refurbishment) was the cheaper alternative, as shown in Table 2.12.

The initial cost for the concrete pavement accounted for half of the total costs, whereas the initial costs were only 20% of the entire life cycle costs for asphalt. Maintenance for asphalt conversely was double that of concrete. The higher costs at the construction phase of the pavement lead to fewer maintenance costs over the life of the specification in question (in the case of the concrete pavement, higher costs lead to fewer maintenance works over the life of the pavement).

Reconstruction was more expensive for asphalt compared with concrete reconstruction, perhaps reflecting the higher involvements in the removal and replacement of the more extensive layers associated with the asphalt alternative (Figure 2.7).

Table 2.12 LCCA comparison: Blacktop versus concrete

Option/type of pavement	Construction cost ($/m²)	Maintenance costs ($/m²)	Rehabilitation/ reconstruction costs ($/m²)	Salvage value ($/m²)	NPV ($/m²)
30-year analysis period					
1 AC	74.20	14.52	0	0	88.72
2 CRCP	133.00	2.45	0	0	135.45
60-year analysis period					
1 AC	74.20	15.33	6.04	2.65	98.22
2 CRCP	133.00	2.87	1.48	−0.74	136.61
100-year analysis period					
1 AC	74.20	15.46	6.31	2.77	98.74
2 CRCP	133.00	2.91	1.51	−0.75	136.67

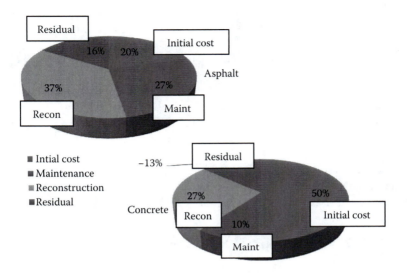

Asphalt vs. concrete LCCA 60 years

Figure 2.7 LCCA breakdown.

The initial costs for the asphalt and concrete alternatives showed an anticipated large variation. The costs for the concrete alternative were seen to be significantly greater (by 44%). The costs of concrete's initial construction in Western Australia is almost double that of asphalt and is a major factor that deters pavement agencies from using concrete (Table 2.13).

Table 2.13 Capital cost comparison

Initial capital costs			
Blacktop		Concrete	
Description	*Rate ($/m²)*	*Description*	*Rate ($/m²)*
175 limestone sub-base	13.77	150 mm sub-base LMC including finishing, curing/provision of typical quantities of subgrade beams for accesses and intersection	35
230 base course	30.84		
Two-coat emulsion seal	6.76		
Tack coat	0.34		
30 dense-graded asphalt	13.59	240 mm continuously reinforced concrete pavement: supply and place concrete, and longitudinal joints and slab anchors	60
30 open-graded asphalt	8.90		
		Finish cure and texture base	3
		Supply and place steel reinforcement	35
Total	74.2	Total	133

The maintenance costs formed a bulk of the overall costs over the life of each alternative; asphalt pavement options historically require more routine repairs to fix deterioration such as potholes and other structural failures. Concrete, on the other hand, with its higher durability and strength characteristics, can withstand the elements and the same amount of traffic loading without extensive restoration. The reconstruction costs for concrete pavement occur at year 50, whereas the reconstruction costs for asphalt pavement carries a much higher rate (as a result of more regular refurbishment). The costs of reconstruction were significantly higher than the initial costs of construction because, in Western Australia, the opportunity or the ability (in real terms) to recycle and reuse materials beyond concrete as a sub-base fill is limited, thus nominal residual amount(s) are generated and used to offset new builds. Overall, in Western Australia, asphalt is recommended over concrete.

2.5.2 LCCA case study: Pavements

The following discussion extends the previous concrete and asphalt comparison example, and considers further pavement application options in Western Australia.*

Recycled concrete, recycled demolition-arisings and recycled asphalt materials are compared with raw material equivalents at three life cycle periods of up to 90 years to assess environmental/economic considerations for local pavements, namely,

- Footpath alternatives compared over a lifetime toward user benefit
  - Four alternative footpath specifications were considered: (i) asphalt, (ii) *in situ* concrete, (iii) brick paving and (iv) limestone paving options.
  - Paths were designed to the same loading and performance parameters for a new housing development requiring 1000 m of a 1-m-wide path.
- Road base recycled materials were compared against virgin raw materials.
  - Structurally fit-for-use recycled concrete road base (CCRB) versus recycled demolition road base (CDRB) materials were matched against virgin limestone.
- Road wearing course recycled asphalt pavement benefits, maintenance and whole-life costs were considered.
  - Three methods are in use locally for heavily trafficked roads: (i) standard hot mix asphalt, (ii) warm mix asphalt and (iii) warm mix asphalt containing recycled asphalt pavement; considering environmental and social merits or otherwise, these options are analysed over road life span.

---

* Input from Liam Gayner[26,46–58] received with thanks.

A discount rate, this time calculated at just over 6% and testing over various life spans, finds that sustainable (recycled) options are superior to (virgin) specifications when considering the range of alternative pavement specification design options across the triple bottom-line criteria of economic, environmental and social factors. An Excel spreadsheet framework was programmed and all cost information was entered, formatted and interpreted. Once values had been ascertained, they were entered. Data plots were then made and alternatives compared for best life cycle benefit (Table 2.14).

Four footpath options (1000 m × 1 m, 1000 m³) were considered for 60 (and 30) years at a discount rate of 6.25% (11% and 3%):

- Asphalt footpath: 100-mm-thick cement-treated crushed rock base course with 25 mm bituminous surface
- Concrete footpath: 100-mm-thick unreinforced 32 Mpa concrete
- Brick footpath: 75-mm-thick, 110 mm × 230 mm standard clay brick
- Limestone footpath: 75-mm-thick rectangular stone

In this case, although concrete has a higher initial cost compared with asphalt, it is the cheapest option over a life cycle.

After 10 years, a concrete pavement, if chosen, will save the taxpayer money despite the cheaper initial cost of asphalt.

Concrete ends up being $7337 cheaper over 60 years; a reasonable saving.

The other two options' initial costs are so high that no amount of life span saving is likely to attract clients or their design representatives.

The limestone paving stone option scored best (after review of user preferences for aesthetics, surface comfort and enhanced property value perception);

Table 2.14 Footpaths

Item cost	Footpath pavement options ($/1000 m³)			
	Asphalt	Concrete	Brick	Limestone
Capital	28,650	33,330	59,400	105,000
Replace	17,858	5715	9784	17,181
Maintenance	21,905	21,716	22,315	21,905
Salvage	61	407	−254	77
Life cycle	68,475	61,138	91,245	144,164
Benefit	70	103	93	118
B/C	1.0	1.7	1.0	0.8
Rank	2	1	2	4

Note: Footpath options: 60 years at 6.25% DR: $/1000 m³.

however, it's restrictively high (doubled) cost excludes it from any real practicable consideration.

Brick pavers scored well, with an overall benefit–cost (B/C) ratio of 1.0. Over the pavement life cycle, brick (although $22,770 more expensive than the bituminous option) makes gains against the B/C score of (the users apparent dislike of) asphalt.

Overall, for footpaths, with a confidence index of 1.53 (in which 0.25+ is considered positive), the concrete option is deemed the most practicable pavement option here.

The road base case study concentrates on initial and salvage costs (maintenance road base costs are nominal) and haulage. Three road base options are available:

- Limestone virgin material
- Recycled concrete (grade 1)
- Recycled demolition material (grade 2)

These are assessed over the life of a heavily trafficked road (a bus lane construction of 1000 m³), for a practicable period of 51 years (Table 2.15).

- New limestone is initially cheaper than the recycled options (assuming that haulage distances are equitable).
- The two recycled options gain superior nonmonetary benefits. Grade 2 is more workable so it receives a slightly higher score than grade 1.
- Considering the B/C ratio, virgin limestone would become a more attractive option (compared with the recycled grade 2 alternative).
- Currently, Western Australia's recycling facilities are much less abundant than quarry facilities.

Table 2.15 Road base options

	Road base options		
Item cost	Virgin limestone	Recycled concrete (Grade 1)	Recycled demolitions (Grade 2)
Capital	9783	15,600	12,000
Replace	7667	10,725	11,000
Maintenance	13,423	15,000	13,500
Salvage	1931	854	876
Life cycle	32,783	42,179	37,376
Benefit	115	141	149
B/C	3.5	3.3	4.0
Rank	2	3	1

Note: Bus lane of road base options: 51 years at 6.25% DR: $/1000 m³.

Design engineers involved in road base specification decision making might be justified in a choice of virgin limestone in this scenario.

Wearing course specification options include

- Hot mix asphalt, which is a traditional method using high temperatures (160 degrees) to achieve maximum workability and bonding.
- Warm mix asphalt, which is a newer and increasingly popular innovation that lays 'warm' (100 degrees) asphalt.
- These two methods can be combined with a third sustainable option of using recycled asphalt pavement with virgin materials.

These three road wearing course options are deemed most practicable for 250 m² material coverage (at a depth of up to 400 mm; Table 2.16).

- The options are very close in terms of cost and social/environmental virtues.
- Initially, the more environmentally friendly options are slightly cheaper.
- Nonmonetary analysis scores benefit the two sustainable options, although hot mix asphalt has better constructability and durability.

In general, the wearing course option using warm mix asphalt with recycled asphalt pavement is deemed the best option for this scenario.

Overall, it might be suggested that in this case study

- *In situ* concrete should be used for the construction of residential footpaths.
- Recycled road base materials should be used, where haulage does not preclude application.

Table 2.16 Wearing course options

Cost	Hot-mix asphalt	Warm-mix asphalt	Recycled asphalt
Capital	40,876	39,658	38,242
Replace	8713	11,008	10,615
Maintenance	37,286	39,150	40,574
Life cycle	86,875	89,816	89,431
Benefit	94	98	99
B/C	1.08	1.09	1.10
Rank	2	3	1

Note: Wearing course options: 51 years at 6.25% DR, $100 m³ (250 m² × 400 mm depth).

- Warm mix asphalt, which incorporates recycled asphalt pavement (at a proportion of 10%), should be used for the wearing course of Western Australia's highly trafficked arterial roads.
- LCCA techniques should be integrated into design decision making in Western Australian construction and civil engineering projects towards the prediction of potential whole-life cost savings

2.5.3 LCCA case study: Floating structures

The following discussion compares three alternative substructure options (pile-on-water vs. floating substructure vs. traditional land options) on Swan River, Perth, Western Australia.*

The following case study finds that, following an LCCA of alternatives, the lowest NPV whole-cost is a pile-on-water substructure option in lieu of other specification solutions.

Given that water occupies 70% of the world's surface, the push towards building on water is increasing; floating substructures present an alternative design option for offshore construction. Indeed, in Western Australia, a $300 million waterfront development is ongoing in Perth's city centre. Currently, there is limited costing information that takes into account floating substructure specification options related to supporting superstructures. Preliminary designs of substructure alternatives were conducted, followed by a detailed design of the three options to incorporate respective component specifications.

Capital costs for each of the substructure options (pile-on-water, floating and traditional) are tabulated in Table 2.17; purchase of land requirements for the traditional (default/control) land substructure option meant that its (default net present) value is high (Table 2.17).

Maintenance costs for each of the substructure options sought to determine the timings and intervals of maintenance works for each component of the substructure options, the service life of each component was determined by using a method developed by the Commonwealth Scientific and Industrial Research Organisation. A summary of the main maintenance and inspection works and their respective associated costs for each of the substructure alternatives can be seen in Table 2.18 (followed by disposal costs in Table 2.19).

Table 2.17 Floating options

Substructure option	Pile-on-water	Floating	Traditional land
Capital cost	$348,314	$444,664	$1,301,286

Note: Capital costs substructure options, supporting 400 m² restaurant superstructure.

---

* Input from Philip Gajda,[59–66] received with thanks.

Table 2.18 Maintenance

Option cost details	Pile-on-water substructure		Floating substructure		Traditional land substructure	
	Maintenance intervals (years)	Total cost per interval	Maintenance intervals (years)	Total cost per interval	Maintenance intervals (years)	Total cost per interval
Grout injection for longitudinal cracks	5	$5000	5	$16,000	5	$10,000
Patch jobs—minor works	20	$51,660	—	—	—	—
Patch work—major works	25	$158,200	20	$140,000	20	$70,000
Inspection by engineer (visual and sounding hammer)	1	$740	1	$1110	740	$740
Diver services (inspect piles, report to engineer)	1	$560	1	$840	—	—

Note: Maintenance/inspection costs substructure options, supporting 400 m^2 restaurant superstructure.

Table 2.19 Disposal costs for substructure options

Substructure option	Pile-on-water	Floating	Traditional land
Disposal cost (unfactored future cost)	$56,393	$61,386	$67,617

Table 2.20 NPV whole-cost for substructure options

Substructure option	Pile-on-water	Floating	Traditional land
Initial costs	$348,314	$444,664	$1,301,286
Replacement cost/salvage value	$63,663	$55,027	$34,052
Annual costs	$696,856	$736,340	$37,960
Total costs (NPV)	$1,108,834	$1,236,032	$1,373,300

The discount rate for the LCCA was calculated using a methodology developed by the RICS as a function of treasury bonds rate of return, inflation rate, average equity return and construction risk. The resultant discount rate for the LCCA conducted here was calculated as 6%.

The results of the LCCA for the pile-on-water, floating substructure and traditional land substructure is shown in Table 2.20. Results indicate corresponding NPVs, at a discount rate of 6%, for initial costs, replacement cost/salvage value, annual costs and total costs for each substructure option.

From the results of the LCCA, it is noted that the pile-on-water substructure option has a lower NPV whole-cost and cost-in-use ($1.1 million) than both the floating substructure alternative specification design option as well as the (default/control) traditional land-based option (sensitivity tests largely endorsed the initial findings); the pile-on-water option is the most economically attractive over the whole-life of the project.

The difference in NPVs between the pile-on-water and floating substructure options was an initial/capital cost of $96,000. The majority of (additional) costs for the floating substructure came from the concrete works involved with the floating substructure. Approximately $334,000 or 75% of the initial/capital costs for the floating substructure came from *in situ* concrete and concrete ancillaries. The need for low-density concrete, relatively complex formwork and reinforcing bar installation essentially meant that the floating substructure option had a higher NPV.

2.5.4 LCCA framework towards roof covering comparison

This section discusses a framework for a life cycle costing approach towards user-friendly means to compare alternative roof elements.[67–70]

A format was developed allowing a user to progress through a series of drop-down menus allowing inclusion and sharing of as little or as much detail as is required. This simple spreadsheet format, incorporating Visual Basic, assesses alternative assets (based on a given number of alternative component specifications) that describe whole costs in terms of the building elements of substructure, finishes, fittings, services and builder's work. Alternative specification whole-cost differences are presented explicitly to describe the capital, maintenance, operation and residual costs in terms of the time–value of money to assist in decision making.

Alternative specifications are identified (via updated design changes shared by an overarching holistic system where possible). Designers specify materials from any number of alternative systems that make up subelements, elements and group elements (as shown in Figure 2.8). Design alternatives are then examined in terms of life span and deterioration (ideally utilising fuzzy system/artificial neural networks, probability and statistics to better assess cost data for availability, tangibility and certainty), allowing comparison of alternatives (as shown in the LCCA of roof-cladding options; Figure 2.9).

Section 2.5.3 and the previous case studies illustrate the various structured techniques for (life cycle) cost control, from initial approximate estimating (Section 2.2.1), preliminary estimating (Section 2.2.2), then detailed estimating of the final design (Section 2.2.3), toward cash flow and income monitoring on-site (Section 2.3), project variability and NPV (Section 2.4) and finally whole-cycle cost analyses (Section 2.5).

Cost estimation must go hand-in-hand with project timelines. Chapter 3 discusses structured methods to assess and present project timelines.

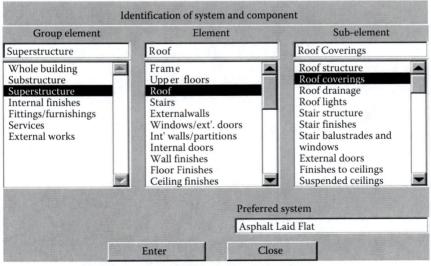

Screen capture: LCCA subelement identification

Figure 2.8 LCCA elements.

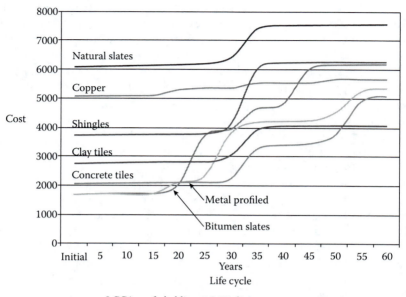

LCCA roof-cladding at 2.5% discount rate over 60 years

Figure 2.9 LCCA roofs.

Chapter 3

Timelines and scheduling civil engineering projects

A civil engineering development project involves a number of phases and stages in its design timeline. At the initial concept phase, a client briefs a designer on an overall need, assesses whether the construction of an asset is feasible and confirms a project's maximum budget. An outline proposal and scheme design then allows final development of the brief. Full design, materials specification and construction methodology review commences, necessitating the preparation of a project timeline and a schedule of when the various on-site tasks might be expected to start and end.

This section details timelines for civil engineering projects in terms of

- Scheduling: Charts, networks, critical path
  and evaluation   (Section 3.1)
- Rescheduling: Programme crash-costing
  and resources aggregation   (Section 3.2)
- Scheduling risk   (Section 3.3)
- Linear time-and-distance scheduling   (Section 3.4)
- Method statements   (Section 3.5)
- Value management   (Section 3.6)
- Critical path and critical chain   (Section 3.7)
- Agile management   (Section 3.8)
- Delay and fluctuations in projects   (Section 3.9)

An assessment of the time frame presupposes that working drawings of the detailed project design are available. Detailed design allows planning and scheduling the proposals for on-site civil engineering construction works and the determination of a project timeline and the establishment of a contract period.

Civil engineering construction projects, clarified by comprehensive design drawings and detailed specifications of both materials and workmanship, need to be broken down into major constituent tasks. A job is principally described in terms of an elemental division of substructure/superstructure, finishes, fittings and services/external works. Further subdivisions describe subelements such as foundations, frames, floors, roofs and the like.

The standardisation of elements and components in construction projects is somewhat variable, with different bodies and national standards organisations defining and categorising in slightly different ways;[1,2] the Civil Engineering Standard Method of Measurement by the Institution of Civil Engineers provides a benchmark.[3]

Civil engineering subelement work classification packages provide a detailed assessment of a job, where on-site activities might be described (and measured) as items such as demolition, earthworks, *in situ* concrete (and associated ancillaries of formwork and steel reinforcement), precast concrete, pipework, structural (and miscellaneous) metalwork, timber, piles (and ancillary work such as casting and interlocking), roads and paving, rail track, tunnelling, masonry work, waterproofing (and associated painting), sewer and water main work and simple building work incidental to civil works.

In scheduling a project, the civil engineer must identify a logic flow of the critical on-site tasks and determine a sequence for the various activities to be carried out; in other words, clarify precedence relationships and critical activities and organise the available resources accordingly.

3.1 SCHEDULING TECHNIQUES

Several tools and techniques (typified somewhat currently by off-the-peg software applications offered by Microsoft Project, Primavera and Bentley[4]) assist in the preparation of a project's schedule. The theoretical foundations[5] laid by Henry Gantt and other historical leaders in the field have had a great effect on scientific management, with the implementation of Gantt/bar charts, network diagrams, critical path methods and programme evaluation and review techniques. The discussion in Section 3.1.1 reviews the common scheduling methods.

3.1.1 Gantt charts

The Gantt or bar chart remains one of the simplest yet most effective scheduling techniques available to order construction task periods. Subelement work activities are identified and entered (preferably chronologically) into the left-hand column of a bar chart. Duration, start dates and end dates are determined and entered with reference to a civil engineering method statement (see Section 2.2.3.1); resultant blocks of time are then charted on the right-hand side.

In Figure 3.1, a hypothetical substructure element consisting of subelements earthwork (excavations), formwork (as part of concrete foundation ancillary work) and the supply and installation of *in situ* concrete

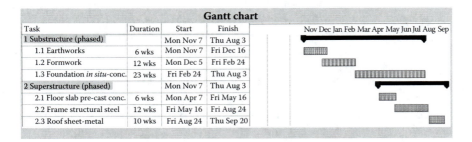

		Gantt chart		
Task	Duration	Start	Finish	Nov Dec Jan Feb Mar Apr May Jun Jul Aug Sep
1 Substructure (phased)		Mon Nov 7	Thu Aug 3	
1.1 Earthworks	6 wks	Mon Nov 7	Fri Dec 16	
1.2 Formwork	12 wks	Mon Dec 5	Fri Feb 24	
1.3 Foundation *in situ*-conc.	23 wks	Fri Feb 24	Thu Aug 3	
2 Superstructure (phased)		Mon Nov 7	Thu Aug 3	
2.1 Floor slab pre-cast conc.	6 wks	Mon Apr 7	Fri May 16	
2.2 Frame structural steel	12 wks	Fri May 16	Fri Aug 24	
2.3 Roof sheet-metal	10 wks	Fri Aug 24	Thu Sep 20	

Figure 3.1 Schedule example.

foundations is deemed to take 6, 12 and 23 weeks, respectively; formwork activities begin as excavation is completed. Similarly, some superstructure steelwork may begin with the substructure.

Although this graphical representation certainly helps to visualise the timing of the activities, it is somewhat impossible to produce without first developing

   i. A detailed description of the project via specifications and working drawings

  ii. Confirmation of the detailed quantities required for each subelement (Section 2.2.3.1, step 2)

 iii. A relevant work method gang-rate efficiency data set (Section 2.2.3.1, step 3)

 iv. A method statement that builds on worker rates and quantities to determine activity times and resource requirements (Section 2.2.3.1, step 4)

  v. An assessment of task precedence: Determination of any tasks that may be carried out concurrently (if resources are available), as well as an identification of those tasks that cannot begin until another task has been initiated

The requirement for an understanding of task precedence (item [v] above) leads to the discussion of network and precedence diagrams in Section 3.1.2.

3.1.2 Network and precedence diagrams

Graphical (network) models take lists of project activities and draw their logical interrelationships to estimate activity duration and calculate start and end times. Off-the-shelf software packages toggle bar charts with networks of task activities and highlight the key (critical) activities involved (discussed in the following). The ultimate goal of network diagrams is to identify which (critical) tasks link together and follow one another, and which on-site activities may be allowed to have a more flexible (floating) start or end time, relative to an overall project duration.

A basic network diagram draws arrows and nodes to represent activity precedence: concreting after formwork after excavation (Figure 3.2).

Precedence in activities can occur as finish-to-start or indeed can be determined in multiple relationships such as start-to-start, start-to-finish or finish-to-finish.

Concurrent tasks can be represented such that (Figure 3.3) prior to the final installation of, for example, a gabion wall, excavation of rockfill, the assembly of the steel cage and the preparation of the sub-base can be conducted in parallel.

Precedence diagrams (Figure 3.4) are similar to arrow diagrams but provide more information (on the node) about durations, start and end times and total float (flexibility in days related to an activities' start or end date relative to the overall programme).

By examining activity float (flexibility or inflexibility in start and end times), civil engineers determine the overall timescale of a project and those critical and key activities that must start and finish in sequence to ensure

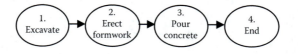

Figure 3.2 Arrow diagram: Activity-on-node.

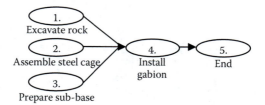

Figure 3.3 Arrow diagram: Concurrent tasks.

Earliest start	Duration	Earliest finish
Latest start		Latest finish
Activity/task description		
Total float	Activity ref. number	

Figure 3.4 Precedence diagram node data.

timely completion. Discussion of critical, key activities and critical path methodologies is presented in Section 3.1.3.

3.1.3 Critical path method

Having assigned duration and cost estimates to each activity (via civil engineering method statements) and having drawn a logic flow network of interrelated connections and precedence(s), it becomes possible to now determine the longest (cumulative) time through the network of project tasks; this calculation of total project duration is termed the *critical path method* (CPM).

Precedence node information for each task is input from progression through a full forward-run through the network and then a backward-run.

Forward-run, first to last activity gives

- *Earliest start* (ES): Before an activity can start all predecessors must finish
- *Earliest finish* (EF): The sum of earliest start time + Duration

Backward-run, last to first activity gives

- *Latest finish and start* (LF and LS) times

Flexible start and finish time task floats may then be assessed (see Precedence node information input).

PRECEDENCE NODE INFORMATION INPUT

A forward-run through a network of tasks gives

ES = Maximum earliest finish of *all predecessors*
EF = ES + Activity duration time

A backward run through a network of tasks gives

LF = Minimum LS of *all* following activities
LS = LF – Activity duration time
Float = LS – ES ...or... LF – EF
    If zero, the task is termed a critical task.

The flexible 'float' time for each activity (in which any delay in that one particular on-site work task will not affect overall duration) may be determined by subtracting the earliest possible start time from the latest possible start time (LS – ES) or indeed from subtracting the earliest possible finish time from the latest possible finish time (LF – EF).

Total float is flexible time that is shared amongst more than one activity (in which delay in one task with float will affect and reduce a neighbouring task's flexibility). *Free-float* time on the other hand, is where the flexible start or finish time for a single activity does not affect any other activity. Activities with no float are termed *critical activities* and are linked together to form a *critical path*. Any delay to a task on the critical path results in an overall delay in project duration. Cumulative critical path durations represent a project's contract period.

It is advantageous to be able to review a project's overall time plan programme. It gives a civil engineer the opportunity to determine the likelihood that a project can be completed within the time frame originally envisaged. Techniques to assist with programme evaluation and review follow in Section 3.1.4.

3.1.4 Programme evaluation and review technique

Civil engineers who wish to be as sure as they can be that they can complete a project within a certain timeline might be inclined to use a programme evaluation and review technique (PERT). Although off-the-shelf software scheduling packages seldom show explicit PERT calculations, these packages do allow schedulers an option to input worst, pessimistic time and also best, optimistic times, alongside what the civil engineer considers to be the most likely task duration time, and then view the related best, worst and most likely bar chart schedules of the project.

Traditional[6,7] PERTs are based on probability distributions for task duration—a bell curve of possible times, usually across three potential inputs.

Three possible time estimates might be used in a PERT assessment of task duration and overall project time.

The best (most optimistic) time assumes that all conditions are optimum. The worst (most pessimistic) time assumes that anything that can go wrong, does go wrong. The most likely task duration is somewhat in the middle of these two extremes (see Excavation; Figure 3.5).

An evaluation of potential time variance(s) across the entire project requires calculating (risk and) task completion times. The most likely time (LT) is assumed to be three standard deviations from both the best time (BT) and also the worst time (WT).

The variance of times allows the assessment of standard deviation and ultimately the assessment of different potential completion times, such that a civil engineer might assess the expected variance of times (V), using standard deviations, as the WT minus the best time, divided by six standard deviations, all squared.

Therefore, variance of times (V) = $[(WT - BT)/6]^2$ (see Probability of timely [on time] project completion).

The most likely time (LT) might be suggested as being four times more likely to occur than both the WT and the best time (BT), such that a civil engineer might have an expected duration (T) for excavation (as a factor of

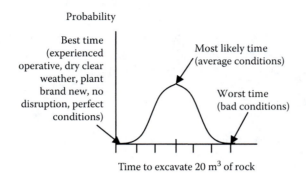

Figure 3.5 PERT.

the LT, WT and BT) as best time plus four multiplied by a likely time plus worst time, divided by six likelihoods.

Thus, task duration time $(T) = (BT + 4{*}LT + WT)/6$ (see Probability of timely [on time] project completion).

PROBABILITY OF TIMELY (ON TIME) PROJECT COMPLETION

Variance of times $(V) = [(WT - BT)/6]^2$
Task time $(T) = (BT + 4{*}LT + WT)/6$
Project variance $(PV) = \Sigma V$ critical path items
Project standard deviation $(PSD) = \sqrt{PV}$
Chance of completing $= (New - Expected)/PSD$

The expected time and related variance can be (tabulated and) assessed, to give the civil engineer an indication of the expected overall variance of the project, by taking into account respective *critical path* task variances. Thus, the project variance of the critical path activities (and the project standard deviation) shows the chances of completing a project on-time (in which the project standard deviation is the square root of the project variances, as above).

The probability of an on-time project completion requires the civil engineer to assess/recall the critical path and then add up the cumulative project variances and standard deviations such that, once calculated, a more 'empirical' statement of whether or not a project can meet its deadline is forwarded to the client.

Where a client revises a due date request, recalculation involves subtracting the original estimate of time from the revised expected date for completion, then dividing by the project's standard deviation. This chance of completing the revised time frame is given as a percentage.

The above PERT analysis can be somewhat overly time-consuming for a civil engineer who simply wants their on-site activities to follow originally planned durations and estimated task completions. PERT remains available, however, as a tool for those who seek detailed clarification of the rationale for project delays and the chances of meeting newly imposed deadlines.

A practice example for scheduling and planning project task times and an indication of how to calculate the chances of achieving revised project time frames (based on calculated standard deviations) follow in Section 3.1.5.

3.1.5 Project scheduling and plan review: Practice example

The following five-step guide allows the civil engineer to schedule a project (and interpret the output of time management software packages).

Graphical representations of a timeline of projects and tasks adopt the following steps:

Step 1: Access and review the preprepared method statements (of the respective element, subelement and civil engineering work classification items) of the tasks that make up the project (as discussed in Chapter 2).

Step 2: Confirm task precedence(s) and establish respective earliest and latest start and finish times.

Assess opportunities for flexible float in the start and end times of individual activities, without affecting overall project duration.

Determine the critical path of activities based on an absence of float.

Step 3: Access and review an off-the-shelf software package able to represent a project's scheduled time frame.

Prepare a Gantt chart of project duration(s).

Step 4: Review method statement (likely task time) durations estimated in step 1 towards acknowledgment of risks, to assess the worst, pessimistic time if all goes terribly, as well as the best, optimistic time if all goes perfectly.

Use this range of potential durations to assess standard deviations from an overall expected time.

Step 5: Use standard deviations prepared in step 4 towards assessing the chances of being able to meet revised time frames imposed in the future by stakeholders.

A scenario might involve a civil engineering project for the supply and installation of 25 standard galvanised mesh rock-filled box gabions of 1 m^3

Gantt chart

Task	Duration	Start	Finish	Nov Dec Jan Feb Mar Apr May Jun Jul Aug Sep
1 Substructure (phased)		Mon Nov 7	Thu Aug 3	◀━━━━━━━━━━━━━━━━━━━━━▶
1.1 earthworks	6 wks	Mon Nov 7	Fri Dec 16	▨▨▨

Figure 3.6 Gantt example.

each (with a CESMM4 reference[3] of 'X410' for miscellaneous work associated with earthworks) that requires localised rock excavation, cage assembly sub-base preparation and then subsequent preparation and installation of gabion(s).

Step 1
Acknowledge phased substructure works in the general scheme of the project (Figure 3.6).

Break down a method statement of task resources (and respective durations) into

- Substructure element
  - Earthworks subelements
- Method statement items related to 25, 1-m³ rock-filled gabions
  - Excavation of rock fill, approximately 3 weeks
  - Cage assembly, estimated at 2 weeks
  - Sub-base preparation estimated at 1 week
  - Gabion installation estimated at 1 week, and 1 week to test

Step 2
Establish earthworks' activity precedence(s) and the subsequent earliest and latest start and finish times for rock excavation, cage assembly, sub-base preparation and gabion installation and testing based on task duration times determined from step 1.

Determine and complete activity node information (see Section 3.1.3), where T = task times and durations generated previously from step 1 method statement(s).

A 'forward-run' through a network of tasks helps to fill-in the blanks for

ES = Maximum earliest finish of all predecessors
EF = ES + Activity duration time

A 'backward run' through a network of tasks gives

LF = Minimum LS of all following activities
LS = LF – Activity duration time, with
Float (TF) = LS – ES or LF – EF; if 0, the task is deemed critical and is highlighted

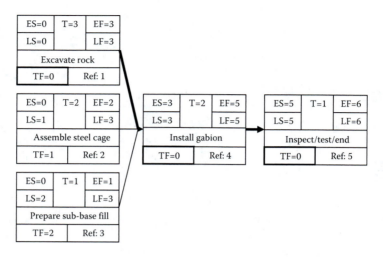

Figure 3.7 **Precedence.**

Figure 3.7 shows a precedence diagram drawn for interrelated gabion installation tasks.

Step 3
After cumulative task precedence(s), input the information into a relevant bar chart spreadsheet or scheduling software package. Figure 3.8 applies to gabions.

Step 4
Using task duration times (step 1), calculate the potential for deviation from timelines by assessing potential expected duration variance(s) related to the civil engineer's updated (risk) analysis of an on-site activity's worst possible and alternatively best possible time to complete a task.

Action by highlighting critical tasks and tabulating is found in Table 3.1.

Item	Duration	Start	Finish	Nov	Dec
1 Substructure (phased)					
1.1 earthworks	5 wk	Nov 1	Dec 8		
1.1 excavate rock	3 wk	Nov 1	Nov 22		
1.2 assemble cage	2 wk	Nov 1	Nov 15		
1.3 prepare sub-base	1 wk	Nov 1	Nov 8		
1.4 install gabion	2 wk	Nov 22	Dec 9		

Figure 3.8 **Gabion.**

Table 3.1 Critical tasks

Task	Best time (BT), in weeks	Likely time (LT), in weeks	Worst time (WT), in weeks	Expected time (in weeks), T = (BT + 4LT + WT)/6	Potential variance (in weeks), V = [(WT − BT/6)]²
Excavate rock	2	3	5	(2 + 12 + 5)/6 = 3.17, ≈3.5	[(5 − 2)/6]² = 0.25, 0.25
Make cage	1	2	3	(1 + 8 + 3)/6 = 2, 2	[(3 − 1)/6]² = 0.11, 0.11
Dig base	0.5	1	2	(0.5 + 4 + 2)/6 = 1.08, ≈1.5	[(2 − 0.5)/6]² = 0.06, 0.06
Install gabion	1	2	4	(1 + 8 + 4)/6 = 2.17, 2.5	[(4 − 1)/6]² = 0.25, 0.25

Step 5

Use standard deviation and expected times in step 4 to assess if a new maximum deadline imposed by client-stakeholders is possible. In this case, assess the chances of meeting a new, imposed maximum 6-week limit for the installation of gabions.

- Use critical path earthworks items (excavate rock and install gabion) and calculate project variance by adding critical path item variances together.
- Project variance = 0.25 + 0.25 = 0.5.
- Calculate project standard deviation such that.
- Project standard deviation = $\sqrt{0.5}$ = 0.707 or 0.7 of a week (or 4 working days out of a 5-day working week).

Recall that in a normal curve for gabion installation (Figure 3.9), the most likely time to install gabions is 5 weeks, but that there is a 50%

Normal curve
for gabion completion

Standard deviation, 0.707 weeks

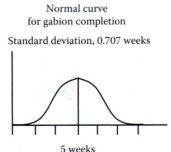

5 weeks
Expected time
to install 20 m³ of rock gabions

Figure 3.9 Normal curve.

chance of completing installation in less than 5 weeks and a 50% chance of completing installation in more than 5 weeks.

If a new maximum of 6 weeks is imposed by the client, rather than the expected 5 weeks, the number of standard deviations from the mean will give the chances of completing by this new maximum deadline. Therefore, to calculate the chance of success:

> Chances = (New due date – Expected date of completion)/Project standard deviation
> = (6 – 5)/0.707
> = 1.41 standard deviations

The civil engineer should retain the value of 1.41 standard deviations and then recall and seek to apply normal curve area tables (with reference to engineering financial first principals calculation reminders and normal curve area tables[8]) to assess the chances of meeting the new 6-week deadline imposed by the client for the work on-site.

If the civil engineer recalls and consults their normal curve areas tables, then the figure of 1.41 standard deviations corresponds to a value of 92% of the area under the normal curve (see Figure 3.10) and, thus, it can be stated that there is a 92% chance of completing installation in 6 weeks.

If the client demands a 99% assurance of completion, then, once again, consultation with the normal curve areas table is required, which will indicate that 2.33 standard deviations, beyond the mean of 5 weeks, allows a 99% certainty of gabion installation.

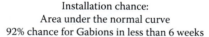

Installation chance:
Area under the normal curve
92% chance for Gabions in less than 6 weeks

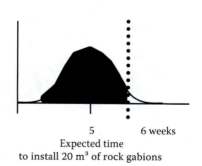

5   6 weeks
Expected time
to install 20 m³ of rock gabions

Figure 3.10 Gabion normal curve.

In other words

$$99\% \text{ chance of installation} = \text{Likely date} + (\text{Tables} - \text{Value} \times \text{Standard deviation})$$

$$99\% \text{ chance of installation} = 5 \text{ weeks} + (2.33 \times 1.41) \text{ weeks}$$

$$99\% \text{ chance of gabion installation} = \text{In just over 8 weeks}$$

The civil engineer can then communicate (with relative arithmetical confidence if nothing else) that gabion installation is expected to take 5 weeks and that there is a 92% chance to install the gabions before 6 weeks, and also a 99% chance that the gabions will be installed before 8 weeks.

The scenario above shows how civil engineers can use their knowledge of task resourcing requirements to plan and schedule a timeline for a project and also attempt to predict the chances of meeting specific deadlines.

The discussion in Section 3.2 builds on the scheduling techniques presented, to show how the civil engineer might use task resourcing (requirements) data to speed up a programme of works or to address project delays and bring schedules back on track.

3.2 RESCHEDULING TECHNIQUES TO IMPROVE AND ADAPT PROJECT TIMELINES

All projects must maintain a balance between the parameters of quality, cost and time. The civil engineering design team ensures the quality of their products and services through specific detailed drawings and specifications issued to describe a preferred design solution to a client's problems and needs.

Although the quality constraints of a project may be reviewed at a preliminary estimating stage through value management reviews of alternative materials specification (see Section 2.2.2), traditionally, quality considerations should not/cannot change once work begins on-site. If stakeholders do seek to alter a project's overall time and its constituent activity timelines, it is the cost (and not the design quality) aspects of a project that must be addressed once work has begun.

If a project falls behind schedule or the client's team of representatives are instructed to enact measures (within the contract between the client and builder) to bring forward a proposed date of practical completion, extra resourcing (cost) is one way to speed-up (or condense and crash[9,10]) project programmes. This technique is discussed in Section 3.2.1.

3.2.1 Project programme condensing and crashing

Where a need arises to condense a project's programme duration, it is somewhat logical to reduce activity time by adding extra resources. Time savings in specific tasks may then result in a knock-on reduction in the overall project programme, if the activity is a key, critical task on the project's critical path.

A civil engineer may save time by increasing resources/costs but must recognise which tasks are suited to extra resourcing and which tasks are not: on the one hand, the curing time of green concrete remains unaffected no matter how many extra workers or plants are deployed; on the other hand, excavation time might be halved by having two backhoes/back actors to dig a required strip foundation, rather than using one machine.

Introducing a third machine to dig foundations; however, might be completely unfeasible in a restricted site, thus the civil engineer must also recognise that there are limits to extra resourcing and certainly a cap on a shortest possible time to complete certain activities. There are limits on condensing an overall programme of site works that stems from respective critical path task crash/reduction limits.

If stakeholders seek to speed-up a project's timeline by adding resources, they must

- Identify and review all critical path tasks
- Calculate the periodic (weekly or monthly; crash/reduction) cost of deploying extra manpower and extra machines for each respective key activity
- Select the key task with the least expensive (crash/reduction) cost per period
- Assume deployment and payment for extra resources to achieve time saving
- Update the overall programme, acknowledging the edited (quicker) period
  - Update new task durations
  - Review any knock-on changes to the critical path

This process is repeated if another time period saving is sought.

Reductions in the timeline as a result of increased (crash) costing for extra resources requires a reexamination of the task's method statement and detailed cost estimates of the direct costs for manpower, machines and materials (Section 2.2.3), as well as an incorporation of (a percentage of) the project's entire indirect costs (Section 2.2.4).

The brief practice example in Section 3.2.1.1 illustrates the basic premise of condensing a project's time frame through extra resources costs.

3.2.1.1 Project programme condensing and (cost) crashing: Practice example

The five steps to reducing/crashing the duration of a project through extra resourcing are

Step 1:   Access and review task method statements (Section 2.2.3), detailed cost breakdowns (Section 2.2.3), task precedences (Section 3.1.2) and the critical path activities (Section 3.1.3), and tabulate these values.

Step 2:   Calculate, from step 1 data, then tabulate pro rata extra-costs (crash cost) to deploy additional resources so that each activity is completed 1 week earlier.

Step 3:   Identify one activity on the critical path with the least expensive crash cost (calculated in step 2); revise its duration accordingly.

Step 4:   Reassess the task precedences (task network node information) and, where appropriate, identify revised critical activities and redraw the critical path.

Step 5:   Repeat this process to further reduce the programme by 1 more week.

A scenario might find stakeholders, already on-site, seeking to reduce the installation time of gabions from an originally scheduled 5 weeks (in Section 3.1.5) to a reduced maximum installation time of 4 weeks.

Steps 1 and 2
Table 3.2 shows steps 1 and 2 towards time reduction through extra resourcing.

Steps 3 and 4
Cage manufacture and sub-base preparation are not applicable because they are not critical items on the critical path. Rock excavation and gabion installation are critical activities (see Section 3.1.5); thus, the civil engineer

Table 3.2 Crash costs

Task	Time (in weeks)		Cost ($)		Critical path
	Likely time	Crash time	Usual cost	Crash cost/week	
Excavate rock	3	1	1800	800	Critical path task
Make cage			Not applicable for this round		Not critical path
Dig sub-base			Not applicable for this round		Not critical path
Install gabion	2	1	4500	1800	Critical path task

extends their preprepared detailed estimates to determine any potential (crash) time savings of 1 week (via the deployment of extra resources), reassessed/recalculated at $800 and $1800 per week, respectively.

An obvious choice might be to target rock excavation activities at an extra cost to the project of $800 per week (in lieu of installation activities, which would cost an extra $1800 per week), to reduce the overall time frame from 6 weeks to 5 weeks.

Step 5

If further programme time frame reductions are sought, this process is repeated by addressing (newly updated) critical path activities. Although the cost-crashing methodology above seeks to reduce a project's duration after commencement on-site by adding more resources to a task, often the reverse situation (of adding to a project period) arises at an early planning stage. A contractor's resources are finite and cannot all be used at the same time, no matter how much time would be saved if this was possible, thus resource allocation becomes key.

The discussion in Section 3.2.2 highlights methods, at the planning stage, for the best on-site allocation of limited resources, and builds upon general approaches to addressing resource constraints and aggregation.[11]

3.2.2 Resources and time frame

Adding extra resources and thus increasing the project sum to reduce project timelines is a somewhat extreme measure and can be avoided by an earnest preparation of early-stage method statements and detailed plans that acknowledge, in advance, workforce gang efficiency (Section 2.2.3.1, step 3) and the finite nature of available plant.

In planning a civil engineering venture, the constructor must seek best use of their existing/accessible resources. The availability of float (Section 3.1.3) and flexibility in task start-times and finish-times within the overall scheme can help with best deployment of a contractor's resources; sometimes, however, the planner has no such float flexibility and must add to programme duration to avoid resources being double-booked. Civil engineers must review resource availability when programming the timeline for a project and determine an efficient distribution of their available workforce and plant across several on-site activities.

Charting a preliminary measure of week-by-week resources (to be) used on-site is a key way to assess which task uses what resource, at what time. Method statements (Section 2.2.3.1, steps 3 and 4) are a starting point to determine the type and quantity of women/men and machines required to supply, install and complete a particular activity.

The basic practice example in Section 3.2.2.1 describes how to aggregate available resources to determine advantageous deployment for a project in its entirety.

3.2.2.1 Project resources deployment: Practice example

Five steps towards best use of resources are

Step 1: Access and review task method statements (Section 2.2.3), task precedence (Section 3.1.2) and the floats of all noncritical path activities (Section 3.1.3).

Step 2: Tabulate week-by-week total number of workers on-site, alongside respective task start, end and float times, based on a prepared precedence diagram (Section 3.1.5).

Step 3: Chart week-by-week total number of workers on-site in a Gantt chart format.

Step 4: Seek equitable week-by-week aggregate of total number of workers on-site, by reviewing then retiming worker allocation, based on flexible float start and finish times, then tabulate the revision.

Step 5: Chart the revised week-by-week total number of workers on-site in a Gantt chart format.

A scenario might involve best use of the resources required for the installation of gabions (in Section 3.1.5), where a builder seeks to schedule the most practicable timeline without disrupting existing employee deployment. The builder seeks to (i) avoid employing extra workers above the firm's existing four skilled workers and (ii) avoid sending the existing four skilled workers off-site at periods of low activity.

Step 1
Gabion installation start and finish and float calculations are as previously identified (in Section 3.1.5, step 2) and are deemed to require: rock excavation, three operatives for 3 weeks; cage assembly, one worker for 2 weeks; base, one worker for 1 week; gabion installation, four workers for 2 weeks; and ongoing quality compliance (all activities), four workers for 1 week.

Step 2
Tabulation of week-by-week total number of workers on-site, alongside task start, end and float times, identify an uneven spread of on-site workers' needs (Figure 3.11).

Task	Weeks				
	wk1	wk2	wk3	wk4	wk5
a. Excavate rock	3	3	3		
b. Assemble cage	1	1	Float		
c. Prepare sub-base	1	Float			
d. Install gabion				4	4
e. Inspect and test and compliance check					4
Workers on-site	5	4	3	4	8

Figure 3.11 Worker weeks.

Step 3
The charting of the week-by-week total number of workers on-site (by task blocks *a–e*) in a Gantt chart format, clarifies an uneven resource distribution and the need for aggregation and smoothing, using available task floats (Figure 3.12).

Step 4
Aggregation of workers finds float to start activity 'b' in week 2 and moving 'e' to week 6. This results in a total of four workers on-site for a 6-week project (Figure 3.13).

Step 5
Charting the revised week-by-week total of workers on site (Figure 3.14) shows the extent to which the original schedule in step 3 above has been smoothed to allow a more equitable use of the four skilled employees of the firm by extending by 1 week.

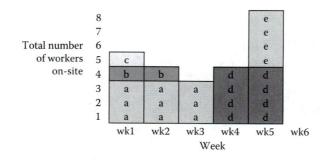

Figure 3.12 Week-by-week total number of workers on-site (by task blocks a–e) in a Gantt chart format.

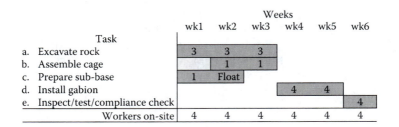

Figure 3.13 Worker week floats.

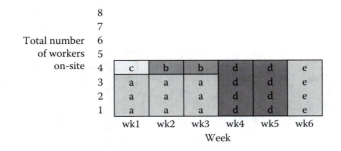

Figure 3.14 Revised week-by-week total number of workers on-site in a Gantt chart format.

Following review of task resourcing, projects must also address task risk, as discussed in Section 3.3.

3.3 RISK: STRUCTURED REPORTING

Civil engineers require an appreciation of risk in specific project tasks and the extent to which identification and subsequent mitigation might affect both time and cost (Sections 2.2 and 3.1) in the completion of on-site activities.

Risk management must address the chances of an incident's effect on project progress and requires measurement in terms of both consequence and likelihood; the civil engineer must be in a position to assess potential loss, injury or disadvantage, and provide an objective description of an incident's probability or frequency.

A structured approach is needed to ensure that all potential risks and hazards are identified and quantified through the preparation of a plan (summarised in Risk management in civil engineering) that develops responses and seeks to control such response options.

National and international standards organisations assist risk management by the provision of risk principles and guidelines.

RISK MANAGEMENT IN CIVIL ENGINEERING

- Identify all potential risk incidents and their knock-on consequences.
- Assess objectively the chances of the incident actually happening.
- Develop structured risk treatment approaches and a risk mitigation plan by stakeholder consultation to identify, contextualise, analyse, evaluate, treat and then monitor and review all tasks' risks.

Addressing risk in a structured way is enhanced by a working knowledge of

- Risk management—Principles and guidelines AS/NZS ISO 31000: 2009; ISO 31000:2009(E)
- Risk Management Guidelines Companion HB 4360:2004 to ISO Guide 73:2009
- Risk Register (Tables 10.1 and 10.2) and Risk Treatment Schedule Plan (Tables 10.3 and 10.4) of AS/NZS 4360:2004 and related updates

A process to align risk categories with risk components might be suggested as a way to checklist and identify a project's threats. A structured approach to categorising the components extends from the national/internationals standards above.

Categories of risk include

- Organisational, governmental, commercial or financial funding approvals issues as well as risks related to development locations, zones and environments
- Risk arising from uncertainties related to project design approval, design technology, site and site conditions
- Risks related to the category of building and the operational process itself, which requires discussion in terms of respective risk components, where

Components of risk for the civil engineer centre on construction operations on-site and might be argued to include

- The source of the threat such as the dangerous activity of materials' placement
- Event or incident where something happens such as plant failure during placement

- Consequence in which damage is caused such as damage from improper placement
- Cause of what and why the incident occurred such as human error due to a lack of plant operation or old plant with nominal maintenance and upkeep
- When in the on-site work schedule and where on the actual site incidents happen

A measurement scale for the categorisation of a project's respective tasks' risks, in terms of a percentage calculated relative to the overall contract sum, provides one way to prioritise all identifiable threats; any task that carries a risk 'calculated' at 10% of the project sum is given a high priority of attention. In addition, high-risk construction activities and related safe work method statements (risk/hazard assessment statements) are a legal imperative before undertaking many on-site activities and certainly must acknowledge the risks incumbent in demolition work and structural steel erection.

An engineer's (safe work[12]) method statement for particular on-site tasks seeks to address and mitigate risk at the early planning stage.

Risk repeat reporting (beyond task method statements) gives additional structured opportunities for monitoring, review and mitigation measures.

A likelihood/consequence risk matrix allows ease of reference (see Risk likelihood/consequence matrix).

Risk likelihood/consequence matrix

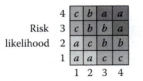

Once again, perhaps it is pertinent to repeat Operational Health and Safety requirements:[12]

- Operational Health and Safety requires compliance with National Standard for Construction Work (Commerce 2011).
- All staff should undergo construction induction training in accordance with Occupational Safety and Health Regulations 1996 (Worksafe 2009); whereby clients (Protection 2007a), designers (Protection 2007b) and main contractors (Commerce 2007) are responsible for exercising the regulations as required.

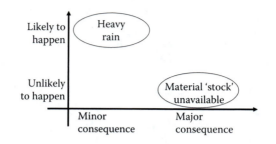

Figure 3.15 Risk summary graphic.

Risk reporting takes the form of a project's risk matrix summary and a compilation table of a project's collective risk ranking.

By way of example (Figure 3.15), inclement weather conditions deemed within meteorological norms,[13] might be classified as likely to happen but with nominal consequences.

On the other hand, material unavailability because of supply restrictions from a trusted long-term (international) supplier may be very unlikely, but have major consequences if it did occur.

It is worth noting that risk is increasingly being prioritised in terms of 'consequence' over 'likelihood'; project risk mitigation must have addressed all extreme consequences prior to site commencement. The BP oil spill in the Gulf of Mexico in 2010 is somewhat reflective of the trend/shift in risk weighting towards mitigation of consequence over likelihood.[14]

Tabulation of all the risks related to the project, alongside the task(s) that may be affected, with an indication of the risk ranking of importance if the incident did happen, as well as the cost in both time and money of the identified mitigating measure (cross-referenced to the method statement where appropriate) can be summarised in the example given in Table 3.3.

In this scenario, the potential for 'heavy rain' in the winter/monsoon season has already been included in the detailed estimate of foundation excavation with the mitigating measure of a suitable water pump on-site, included in the direct costs of the project. *Material no-show* on the other hand, has potentially a much more severe mitigating cost and would require major head office negations to identify, contact and confirm a new supply chain, somewhat beyond the remit of the civil engineer on-site (Table 3.3).

It is only by evaluating mitigating measures, that risk is truly addressed in timelines; be they bar charts or the alternative linear techniques described in Section 3.4.

Table 3.3 Risks

Risk	Related tasks	Risk ranking (% of the project sum)					Mitigating cost, labour/plant/ material/safety (method statement cross-ref) and monitoring period
		1	2	3	4	5	
		≤0.1%, $	0.1%–1%, $	1%–5%, $	5%–10%, $	>10%, $	
Heavy rain	Excavate foundation		X <1%, $5K				Water pump availability (method statement task 1), monitor: monthly
Material no-show	Superstructure steel installation					X 25%, $250K	Call/confirm new supplier (method statement task 2), monitor: as-is, where-is

3.4 ALTERNATIVE SCHEDULING TECHNIQUES FOR CIVIL ENGINEERING PROJECTS

Industry and most software scheduling packages emphasise bar chart presentation tools for project task timelines (Section 3.1). Alternative methods are, however, also worthy of note for civil engineering projects and include linear (time and distance) scheduling techniques.[6,15,16]

Many civil engineering infrastructure tasks are somewhat cyclical; activities may repeat over kilometres or tens of kilometres. Substructure earthworks and general clearing activities followed by sub-base, peripheral and fundamental drainage works, then asphalt black topping, concrete pavement works or rolling stock surfaces, culminating in finishes and signage, are a feature of road, rail and pipework projects. Similarly, multistorey structural works may repeat floor upon floor upon floor.

In repetitive civil engineering projects, the rate of progress becomes the dominant ongoing measure.

A repetitive activity scheduling or line of time balance technique may be applied to repetitive/phased element construction (see Figure 3.16), where activity timelines are shown for phases or floors alongside weekly progression.

Similarly, distance-based linear (time and distance) scheduling techniques may be used to represent construction timelines for transportation infrastructure projects and the building of pipework routes.

The practice example in Section 3.4.1 illustrates a basic linear scheduling application.

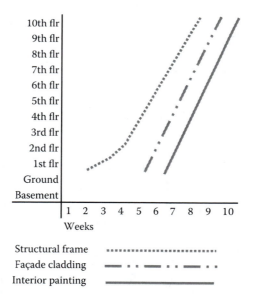

Figure 3.16 Repetitive activity.

3.4.1 Linear scheduling: Practice example

The three steps for the preparation of a linear schedule, sometimes called a speed diagram, are as follows:

Step 1: Identify a project as a development involving repetitive tasks and calculate
  • The gang efficiency rates for the various tasks
  • Flexibility and float opportunities in start or finish times of the tasks

Step 2: Tabulate precedence and chart all (repetitive) tasks along an 'x' axis of time and a 'y' axis of distance. Input one-of tasks as blocks of time as necessary.

Step 3: If appropriate, seek opportunities to speed up the overall project duration by
  • Identification of the activity with slowest, least efficient rate of progress (effectively, the task with the flattest slope on the timeline)
  • Add additional resources as appropriate/deemed necessary

A scenario might involve the wish to develop a linear schedule (time/distance, speed diagram) for a 1.8-km road and a desire to use the linear schedule to assess project duration to supply, install and compact the sub-base, the base course/emulsion seal, painter work and signage and an animal access culvert; and seek to determine the task best placed to receive extra resources if stakeholders wish to speed-up the project.

Step 1

Road task efficiencies sourced (from past experience or text) are deemed to involve

  • Supply/install/compact 350 mm sub-base
    • Gang/worker efficiency of 180 m/d: 1.8 km in 10 days
  • Supply and install 175 mm base course with emulsion seal
    • Gang/worker efficiency of 300 m/d: 1.8 km in 6 days
  • 1.8 km in 6 days painterwork and signage
    • Gang/worker efficiency of 1000 m/d: 1.8 km in 2 days
  • Minimum float is 2 days for all tasks

Step 2

Tabulated tasks for the road project in terms of their respective precedence and worker efficiency rates and float (Figure 3.17).

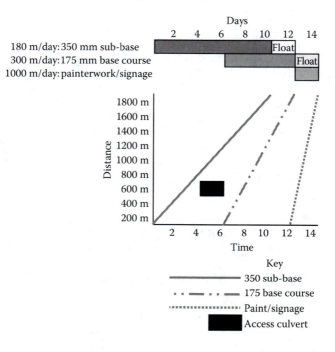

Figure 3.17 Linear schedule.

The linear schedule depicting distance by time (project speed) for each task is charted. The overall project duration is noted as 14 days. Insert one-of task 'boxes'.

Step 3
To speed-up the project, the task with the flattest slope is chosen because this represents the slowest task rate. Assess if extra resource deployment to the sub-base task will speed-up overall project duration sufficiently.

Adaptations of this linear scheduling technique involve insertion, into the graphic, of blocks/boxes of other finite, one-of tasks. In the example above, the installation of an underground animal access culvert in week 5, at the 600 m marker peg along the route, is inserted into the linear schedule as a 'black box/block' activity.

Inherent in time schedules, is the need for timing within work breakdown structure(s) and method statements. Method statements are discussed in more detail in Section 3.5.

3.5 METHOD STATEMENTS

The following discussion of method statements seeks to complement previous discussions on construction programme schedules and cash flow diagram S-curves.*

Method statements can be categorised as follows:

- Prequalification submissions of techniques used to complete previous projects, which might be applicable to a current project proposal to assist with tender list inclusions, or tender review, which gives specific general statements or proposals to carry out the tasks at hand towards potentially saving the prospective client time and money.
- A detailed (contractor's own) internal document that is not shared with external parties towards full description of the method statement of task realisation for assisting in the preparation of a tender-bid.
- Alternatively, a method statement that specifically sets out to discuss occupational health and safety considerations for a range of potentially dangerous and hazardous activities.

Generally, a method statement produced for internal use (to assist with tender-bids) is perhaps most worthy of note because it provides a fully detailed identification of the supply and installation of the elements and subelements of a project.

Method statement details depend on the information available to the main contractor at the time of preparation and will seek to identify methods for the main components of the work, scheduling of work (labour, plant and equipment), details of any subcontracted work, bulk quantities of materials required and any necessary temporary works, alongside environmental protection and safety measures, transport requirements and management of traffic on-site.

A range of stakeholders, including the prospective site manager, might be expected to be involved in determining construction methods that, coupled with a project schedule, enables the estimation of a relatively accurate project value and, not least, provides the means to monitor actual costs on-site to ensure that costs are kept within the tendered amounts. Anecdotally, on-site construction operations detailed within method statements are less likely to be followed by personnel on-site and, perhaps as a result, are less able to assist complex projects or those that have strict completion deadlines to maintain cost and time predictions and planned parameters.

3.5.1 Method statements and risk

A first step in risk management is identification; one possible approach to this is to include an assessment of specific risks related to explicitly

---

* Input from Colleen Smythe,[17-39] received with thanks.

itemised work tasks. After the identification of risk, mitigating measures might then be addressed directly within method statement breakdowns. Such a method might be argued to provide a more detailed and accurate reflection of the costs (both financial and time) to mitigate identified risk, rather than the somewhat more commonplace approach (often as a result of intense tender selection deadlines) to deal with risk in an arbitrary way by simply adding a contingency of 10% or so onto the overall project cost.

Such quantitative risk analysis, as direct risk mitigation measures placed directly with specific method statements, might come closer to estimating the project cost contingencies as accurately as possible for identifiable risks at particular times. Currently, linking method statement breakdown tasks specifically to a quantitative risk analysis is rare.

Qualitative risk analysis, if applied at all, is somewhat more of an academic exercise that incorporates a range of (statistically based) methods such as simple assessment, probabilistic analysis, Monte Carlo stimulation, Latin hypercube sampling, sensitivity analysis and multiple estimating using risk analysis (MERA), outlined briefly as follows

- *Simple assessment*: A method for estimating risk by calculating the expected effect of each significant risk and then adding the values together to give an estimate of the overall risk allowance required via a relatively simple mathematical technique best suited for small projects
- *Probabilistic analysis*: Applying (often subjective if historic cost spreads are unavailable) probabilities to estimates, using a range of probabilities and associated values to determine an expected value applicable to a relatively small number of on-site risks
- *Monte Carlo simulation*: A (roulette wheel, computer software) probabilistic analysis technique to address a large number of risk factors by simulating elements and combining them statistically to quantify project risks and provide risk parameters
- *Latin hypercube sampling*: Based on the Monte Carlo technique, takes less time because it uses a systematic rather than random sampling
- *Sensitivity analysis*: Demonstrates the effect that risk has on a project by varying the values of small key separate variables and measuring results to determine effect; albeit perhaps unrealistic because other variables are likely to be effected too
- *Decision tree diagrams*: Can evaluate alternative approaches to a risky proposal
- *MERA*: A quantitative risk analysis method to find a risk estimate by identifying and costing risk events associated with a project, which, rather subjective, may provide a meaningful estimate

3.5.2 Method statement continuity

Within both industry and individual companies, there is significant variation amongst the variables included in method statements for actual projects. Overall costs are a useful indicator of the effects of variations from predicted method statement values; however, contractors cannot determine the specific stage of the work breakdown structure task that contributed most to a cost overrun.

Contractors therefore might seek to track the costs at the different stages of the work breakdown structure utilising an ongoing S-curve. If records were kept in enough detail (as a progressive method statement continuity process) then this may determine where most cost overruns occur. Such information may then be presented in a diagram superimposing both schedule and cash flow; this combined diagram, prepared for a wharf berthing dolphin construction (Figure 3.18) is shown as follows:

Method statements itemise the labour, plant and material input(s) required for particular work tasks. Once allocated at the early planning stage, stakeholders might perhaps seek to reflect on whether the methods and materials chosen represent a best distribution of all available resources. Section 3.6 structures such a reflection and describes specific ways to conduct a value management exercise to assess opportunities for potentially improved resource utilisation.

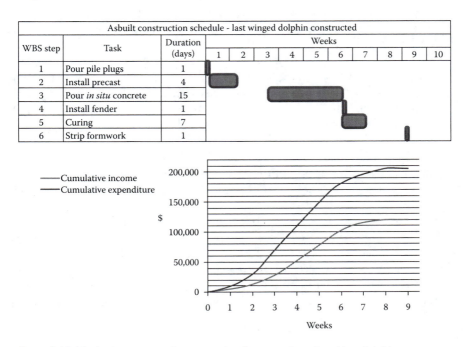

Asbuilt construction schedule - last winged dolphin constructed												
WBS step	Task	Duration (days)	Weeks									
			1	2	3	4	5	6	7	8	9	10
1	Pour pile plugs	1										
2	Install precast	4										
3	Pour *in situ* concrete	15										
4	Install fender	1										
5	Curing	7										
6	Strip formwork	1										

——Cumulative income
——Cumulative expenditure

Figure 3.18 Method statement incorporating S-curves for a berthing dolphin.

3.6 VALUE MANAGEMENT

This section discusses value management (VM) in infrastructure projects and reviews the techniques and appropriate stages to implement and introduce VM.*

VM has long been regarded as an effective means to eliminate unnecessary capital and life cycle costs (it might be noted that perhaps life cycle cost analysis, discussed previously, might be deemed to fall under a VM umbrella); however, although many are familiar with the theory, its use in industry is sporadic.

The discussion here and in Section 3.6.1 documents the potential benefits to be achieved by using value management techniques and the attitudes of industry professionals towards the feasibility or need of establishing a somewhat more obligatory value management procedure for all (civil engineering) developments. Generally, findings suggest that although the Western Australian industry is well aware of both the concept of value management and the accrued benefits of addressing the life cycle of a project, there is resistance to any legislative requirement for VM to become part of the contractual documentation that is used to bring the design solution to fruition.

Value management grew from a post–World War II need to address the significant shortages of available resources, skilled labour and raw materials, where the need to substitute one material specification with another similarly fit-for-purpose alternative material specification was found to often result in benefits of overall project cost savings or time savings in the production process.

The term *value management* is somewhat interchangeable with labels that include *value analysis* and *value engineering*, and although Australian engineering currently displays a localised preference for the more North American branding of value engineering, the term value management (VM), with the establishment of the Institute of Value Management Australia in 1977, might be suggested to remain relevant.

Although VM implementation costs approximately 1% of a project's value (as a result of a practitioner's time to review a range of design solution specification alternatives in totality), current moves by civil engineering towards staged/fast track/design-bid-build/turnkey delivery methods, means that there is practically no time to allow any real review of the range of potential design solution options. Particularly for Western Australian mining clients and their representatives who seek quick installation of civil engineering components, to the exclusion of infrastructure cost-in-use variables, as a means to simply get them up and running, to allow mine resource extraction and transportation (and resultant profit generation of that resource) to commence as soon as possible. Albeit that perhaps this

---

* Input from Carlo Cammarano,[40–58] received with thanks.

historical attitude is changing, however, with senior (client) personal and design consultants recognising that most support structures and facilities put in place two or three decades ago to facilitate mining operations have now reached and overrun their expected life cycle, and require essential expansion refurbishment and retrofitting.

3.6.1 Value management techniques and tools

Value management techniques to address infrastructure design solution considerations of alternative like-for-like material specifications, might be suggested to include

VALUE MANAGEMENT APPROACHES

- Brainstorming via open, noncritical discussion of alternative design solutions.
- Review of alternative materials' 'strengths, weaknesses, opportunities and threats' (SWOT) to help identify all factors that might apply fit-for-purpose options.
- Function analysis system technique (FAST) assessment of the interplay between (sub)components towards determining what elements are principally required to do and what functions are superfluous.
- Risk analysis provides a structured approach to identify potential risks associated with project elements and attribute costs (financial and time) to mitigate (through technical installation processes or supply chain issues of) such risk.
- Cost–benefit analysis (B/C) comparing tangible as well as intangible benefits and costs for a particular project's overall value, not only in terms of ultimate project go-ahead but also perhaps more relevantly assessment of the (need for the) contributory components or subcomponent alternative specification inputs.
  - In the hugely profitable mining industry, infrastructure investment is usually worthwhile, thus this method is relevant when considering alternative overall solutions for the same projects.
- Stakeholder analysis technique, where a range of key (client and the like) stakeholders with influence or authority over a project are identified to assist in focussing scope.
  - Although perhaps it might be somewhat glibly stated that such a stakeholder analysis is largely a client's briefing procedure, this more structured, defined approach might well identify, and allow consultation with, any individuals who may otherwise have slipped through the client briefing 'net'.

- Substitute, combine, adapt, modify, put to another use, eliminate, reverse (SCAMPER) similarly provides a structure under which alternative design solutions might be assessed for their respective fit-for-purpose attributes.
- All party consultations with the full range of contractors and subcontractors.
- Life cycle cost analysis (LCCA), as discussed previously, is part of a VM process that allows consideration of element-by-element alternative design specifications over the facilities' useable life, from design through construction and installation through operation and maintenance until decommissioning; all in terms of a consideration of the time–value of money (discount rate) where one dollar today is worth less in the future.
- Life cycle assessment (LCA), incorporated into a VM procedure applied to the design process at the early stages, allows an analysis of alternative materials (and alternative installations) in terms of their environmental effect related to option-by-option comparisons of pollution-in versus pollution-out.

Towards the assessment of the importance of value management and the techniques above, a pilot study targeted 10 experienced professional civil engineers working in and around Western Australia and found that nine of the interviewees had an in-house structured management policy, which included appropriate value management procedures largely concerned with the review of principal fit-for-use component specification alternatives.

- It was suggested by the respondents that early contractor/subcontractor involvement is an important variable because construction knowledge can be successfully integrated into the design procedure to create an effective solution in terms of constructability.
- Many suggest a combination of techniques in the development phases of the value management study, stating that stakeholder concerns should be explicitly considered along with the potential risks of each alternative.
  - These issues are further considered by a cost–benefit analysis to compare qualitative and quantitative factors that affect the cost and benefit.
- Time factors play a part, and therefore a number of the techniques were not considered as feasible within a VM approach, most notably LCCA and LCA.

Table 3.4 highlights the most common techniques considered applicable to a value management study of civil engineering projects in Western Australia.

Table 3.4 Value management techniques utilisation

VM technique	Application in projects (%)
Risk analysis	80
B/C analysis	8
Stakeholder analysis	70
Brainstorming	70
Issues generation and analysis	50
Value analysis	20
SWOT analysis	20
FAST analysis	30
SCAMPER	0

The timing of value management implementation is worthy of note; although (unsurprisingly) all 10 of the design and construction companies interviewed indicated that they used (or would wish to use) value management at the conceptual design phase, 5 of the 10 noted that value management can be used during feasibility and detailed design; it has been suggested that the greatest possible savings can be achieved during the planning and definition phases, which will ensure that appropriate investment decisions are made. Four of the 10 respondents mentioned that they incorporate value management in the post construction and maintenance phase, commenting that it is *still* possible for cost reductions/effective construction methods through VM/NE applications by subcontractors (Table 3.5).

Several interviewees commented that company policy or design procedures do not allow sufficient time to implement VM, particularly in the traditional 'design-bid-build' contracts where there is added pressure to fast-track the construction phase. Indeed, given the need for civil engineering consultants and contractors to realise their respective infrastructure projects early, enabling early mine operation resource production, this is unlikely to change. Time limitations were found to be at the top of the list of issues identified as affecting (negatively) VM uptake (Table 3.6).

Table 3.5 Timing and stage of value management implementation

VM stage	Application in projects (%)
Conceptual design	100
Feasibility	50
Detailed design	50
Postconstruction/maintenance	40
Construction	10

Table 3.6 Mean (importance) score of issues affecting VM uptake

VM issues	Mean (importance) score
Time limitations	2.6
Lack of understanding	2.5
Ambiguous specification	2.3
Faulty ambiguous drawings	2.1
Lack of commitment	2.1
Lack of support	1.9
Unforeseen constraints	1.9
Budget limitations	1.7
Confrontational relationships	1.6
Nonstandard drawings	1.5

Although factors such as time limitations and ambiguous design solutions were found (Table 3.5) to detract from VM uptake, if value management techniques are adopted as part of a materials specification/installation review, then benefits are argued to accrue, including perceived improved project value for the client, cost savings and improved cooperation. Indeed, some respondents feel that the benefits would be greater if all participants were briefed about value management procedures so that they are conscious of the importance of their individual participation. Table 3.7 places perceived benefit in order of (mean) importance.

Respondents subsequently identified scenarios in which having the time to conduct, at least some form of, VM allowed opportunities to reflect on planned specification choices for civil engineering projects. By way of example, three retaining walls at three locations were identified in various city projects. The retaining walls were constructed using the T-Block wall system. These walls are manufactured from precast concrete sections with baled rubber tyres placed in between. Before the decision was made to use the T-Block walls, the manufacturer was contacted to provide the

Table 3.7 Mean (importance) score of benefits of value management

VM benefits	Mean (importance) score
Improved project value for client	2.9
Improved effectiveness	2.7
Cost savings improved profit	2.6
Improved relationship/coordination	2.5
Documentation of issues/opportunities	2.5
Improved reputation	2.1
Reduced claims or disputes	1.8

Table 3.8 Case study specification comparison: Percentage comparison

	Total cost	Cost of T-Block	Cost savings	Capital cost saving
Concrete	$358,000	$229,011	$129,600	36%
Limestone	$220,400	$154,687	$65,713	30%

comparative cost of construction for similar walls and provide data towards a value management assessment. The alternative materials considered were T-Block wall options of concrete or limestone.

A capital cost reduction potential was highlighted (Table 3.8). This was presented as the initial step in an overall value management exercise that sought to factor in life cycle cost analysis, supply chain variables, tradesman expertise and client/(user) aesthetic considerations, as well as work breakdown structure activity scheduling for the wider aspects of the project. Respondents suggested that the initial impetus for value management resulted in a significant cost saving accumulation due to a compounding of factors related to the repetitive nature of the projects' multiple retaining walls.

Generally speaking, although the cost to provide and maintain (civil engineering) facilities is but a drop in the (profitability) ocean for the huge West Australian resources industry, suggestions that companies can't afford to *not* use VM rings as true for mining infrastructure as it does for more traditional building design projects because value management has the potential to provide increased worth and effectiveness at all stages of a project's life cycle.

Certainly, the quality of the final decision is influenced by the level of information provided initially, compounded by the pressures to fast-track design projects to on-site construction. It might be suggested that the types of projects in which value management has the most potential for high returns on investment include projects that are complex and unique, or alternatively, are repetitive in nature. The benefit of value management across all types of civil projects depends in large part on the commitment and initiative of each individual member making up a design (and value management) team.

The discussion above on value management is somewhat complemented by Section 3.7, which reviews the critical path items and opportunities for buffers to accommodate delay. Section 3.7 discusses critical chain scheduling.

3.7 CRITICAL CHAIN PROJECT MANAGEMENT SCHEDULING

This section provides a critique on critical chain project management (CCPM) scheduling as a means to facilitate meeting construction project duration targets.*

---

* Input from Andrew Crew,[59-66] received with thanks.

Infrastructure projects and civil engineering developments generally are notoriously bad timekeepers despite the range of methods available to structure, plan and monitor timelines. Indeed, an Australian Constructors Association Report entitled 'Scope for Improvement', which surveyed more than 200 industry professionals a year or so ago, found that one out of every three infrastructure projects (in which survey participants were associated with) were completed in overtime. Indeed, in the Western Australian market, professionals identified significant liquidated damages associated with projects conducted in their unique home state. Potentially then, there remains a need to reexamine the range of project management techniques available and reassess their practical worth, and go beyond industry's most frequently used time-scheduling technique, CPM (critical path method, an approach discussed previously).

CCPM is perhaps a way to address unsuccessful, overtime delivery of projects for the local construction industry. CCPM is an application of the theory of constraints, which seeks to establish procedures to better manage worker behaviour and more efficiently allocate resources required to complete the key tasks related to the work breakdown structure of a civil engineering venture. CCPM goes beyond the CPM by seeking to incorporate risk assessment more explicitly.

Although industry is somewhat reluctant to embrace CCPM in more standard civil engineering developments (due to either simply being unaware of any potential benefit to be gained from its use or due to its call for 'daily' reporting), critical chain does remain an option worth exploring; particularly in disaster mitigation and community asset reinstatement, as discussed in more detail towards the end of this section.

It might be suggested that project blowout failure rates are somewhat a function of traditional CPM's ambiguities in dealing with three key factors, namely

- Fundamental project uncertainties causing delays
- Resource dependencies between activities
- Human behavioural factors

Such that project uncertainties are accounted less accurately than might be expected in CPM schedules through the (mis)allocation of contingency time to activities' duration estimates, based on the planners' subjective perception of risk to which the activity is exposed. Planners will estimate according to their worst past experience of that type of activity, leading to excessive safety time being allocated, which in turn seldom motivates the performance of a respective project team.

In addition, it might be argued that the insertion of respective contingency time(s) into each activity's scheduled duration makes it very difficult for the project team to monitor project progress along the schedule. It can be argued that this prevents the project team from the accurate estimation of a work path or expected completion date, which prevents it from effectively intervening to accelerate progress and protect committed completion dates.

Efficient allocation of a project's committed resources is an important aspect of successful project management; however, traditional CPM schedules do not largely require activity resource allocation. Omission of some resource dependencies is likely to result in unexpected delays, as activities await the resources required.

It is common within the Western Australian construction industry for payments to be paid based on the achievement of milestone progress points. This form of progress measurement can be argued to result in an environment of somewhat interim payment focussed project management in which the project team, in an effort to appear successful, pushes to achieve these milestone payments at the expense of other works. Although this form of progress measurement is not specifically required by CPM, CPM seldom provides any relevant guidance.

Given these one or two anomalies in the traditional CPM, it is now worth considering if the CCPM system might be able to progress matters.

CRITICAL CHAIN PROJECT MANAGEMENT

- Applies the theory of constraints to the project environment
- Addresses 'contingency' time to deal with uncertainty in the form of buffer periods installed into the programme
- Regularly monitors project and feeding buffers' consumptions towards more efficiently allocating resources required to complete tasks

It might be argued to be a structured way to increase project flexibility, allowing advances in the programme to be capitalised upon and providing additional feedback loops for management to base intervention decisions on.

3.7.1 Implementation of critical path project management

A review of the CCPM method is presented in the following. The discussion attempts to summarise the key points involved in its implementation.

3.7.1.1 Cutting estimated task durations

The first stage of CCPM addresses (effectively cuts) the safety/contingency time from activity durations estimated within the project schedule:

- The amount of time to be trimmed from each activity duration estimate should reflect the amount of uncertainty embedded in its tasks.

- Significant cuts to activity durations provide opportunities for the halving of a reasonable task duration estimated by an experienced scheduler using CPM.
- Thus, a proposed new expected duration is subsequently deemed to reflect the median duration of works, with no inbuilt (extreme worst-case scenario) time to accommodate delays that *might* occur.

However, a degree of contingency time will remain to deal with uncertainty. In CCPM, this time is installed into the programme in the form of buffer periods.

3.7.1.2 Buffer period installation and management

There are three types of buffer periods used in CCPM, as follows:

- Project buffer
  - Half of the time cut from critical activities are compounded (summed together) and inserted into the schedule at the end of the critical chain.
  - Thus, the estimated activity duration is to be cut by 50%, and half of the time that has been removed is placed in the project buffer.
  - Effectively meaning that the overall activity duration estimation calculation, including safety time, is approximately (up to) 75% of the original projected duration using traditional CPM scheduling methods.
- Feeding buffers
  - Feeding buffers are calculated using the same techniques as the project buffer above.
  - The difference between feeding and project buffers is that
    - Feeding buffers contain safety time for each noncritical work path opposed to the critical chain and are inserted into the project schedule.
    - Whereas, the noncritical work path feeds into the critical chain and as such there is more than one feeding buffer within the project schedule, although there is only one project buffer.
- Resource buffer
  - Unlike the project and feeding buffers, a resource buffer is not a summation of safety time to account for delays.
  - Rather it is a rouse (wake-up call) in the form of a notice, given to the relevant subcontractor 1 week and again (perhaps 1 day) prior to a particular plant or personnel being required to undertake a critical activity.

- These resource buffers are inserted into the project network in the form of resource dependency arrows between activities.
- A requirement of CCPM project network development is that each activity is allocated the resources required for its completion, allowing resource allocation priorities to be developed.

Monitoring of the project and feeding buffer's consumption is to be done on a regular basis (perhaps, if time allows, even on a daily basis) by routinely updating the project schedule, where delays in activity completion are subtracted from their corresponding buffer. Once delays have been subtracted, a buffer's so-called *burn rate* must be calculated.

A burn rate can be suggested to represent the following:

- It is the proportion of the percentage of buffer consumed to the percentage of work completed along a particular activity task.
- It acts as a feedback mechanism for the project team.
  - Once a work path produces a burn rate equivalent to or more than 1, intervention is required from the project team and the site manager of the project.
  - Such intervention (in the form of additional/reallocation of resources) brings the work task on a path back on schedule.
  - This site manager's action then protects the overall expected completion date and delivery of the civil engineering project to the client.

3.7.2 Updating project schedules in CCPM

For buffers to be managed effectively, CCPM advocates a significant increase in project progress reporting. Regular (ideally daily) on-site subcontractor supervisors will be required to submit a regular/daily progress report to the main contractor or the site manager to update them on progress made on activities throughout the day and predict expected activity completion dates.

Based on these regular (daily) reports, each activity and task that is under way is updated on the overall project schedule programme to reflect the situation across the entire on-site works, namely,

- Once each schedule has been reviewed, the site manager combines these into a summary schedule, showing percentage progression of (or off) each work path.
- Based on the projected dates, corrective action can be taken by the project management team to protect committed completion dates.

CCPM also calls for an increased openness between the contractor and client, brought about by allowing the client stakeholders to have access to this working schedule. Therefore, once the schedule is updated, it needs to be posted on a preagreed upon (online) forum to which the client and client's representative have access.

Somewhat by definition, no matter how quickly you complete noncritical activities, the project cannot be completed any sooner than the time it takes to complete those which are critical. Therefore, by measuring the progress of a project by its progression along the critical chain, a true estimate of progress can be made and payments made accordingly. CCPM calls for progress payments to be paid as a percentage of the total quoted price of the project based on the percentage of the critical chain completed. Therefore, the progress and cost report must include a detailed breakdown of critical chain activities and the progress that has been completed along it thus far. The information required to ensure the accuracy of these monthly reports is to be taken from the (constantly) updated schedules.

3.7.3 CCPM application opportunities in civil engineering

Given the rationale above, opportunities for, and perceptions of, utilisation of CCPM (in Western Australia) for civil engineering project applications, was sought. Pilot study data collection occurred in the form of semistructured interviews with very senior civil engineering project management professionals locally. The following highlight the concerns and issues related to the potential to implement CCPM.

Liquidated damages factors
Respondents mentioned that it was very common locally for the application of harmful liquidated damages because of time blow-outs, resulting in the contractor losing money. The high level of harm caused by liquidated damages within the Western Australian construction industry indicates a greater level of uncertainty than was contractually expected from the builder/contractor. Somewhat ironically perhaps, CCPM advocates contractors and subcontractors sign up to even larger liquidated damages contractually, therefore, exposing them to more risk in exchange for substantially larger profit margins. That being said, a CCPM argues strongly to enable the project team to better understand and deal with delays by providing additional feedback loops, minimising uncertainty and making harmful liquidated damages less likely, as well as the greater potential for on-time (ahead of time) bonuses.

Cost competitiveness issues
The structured interviews identified that there is a general trend of cost competitiveness that is increasing within the construction industry, especially in the lower tier building sector. For larger contracts, the sample

identified that competition is not placing significant downward pressure on profit margins (within Western Australia). Because CCPM calls for large financial incentives for its implementation, large infrastructure projects would be (most feasibly perhaps) able to offer these incentives.

Milestone payments

The senior managers in the sample interviewed identify the truism that pushing for milestone payments at the expense of other works is common within the construction industry. This practice increases the risk of delayed works being neglected until they become critical and potentially result in delaying (overall) project completion.

Subcontractor issues

It is known that subcontractors within the construction and civil engineering sector independently allocate resources to their activities and job sites. This means that daily progress is not necessarily consistent and, generally, subcontractors allocate resources only once an activity has become critical, largely confirming the CCPM assumption that adding safety time to activity durations does not help in delivering projects, but rather that an activity expands to fill the time. Without consistent progress (or subcontractors being able to accurately predict the finishing dates of activities) the updated schedule will not help in planning future works.

Payments and cash flow

The senior respondents in this study suggest that, if payments are made based on the progress along the critical path, the cost of works to the contractor will not correspond to the value of work. Meaning, in some cases, that the contractor will need to pay more than they are expected to earn for the works, especially in a 'services heavy' project and that these costs could bankrupt smaller contractors. If, however, finance was obtained by the contractor, all additional costs due to interest owed would most likely be passed on to the client and result in an increase in the cost of the project. It was also confirmed that contractors aim to make additional funds by remaining cash-positive throughout the project and reinvesting this additional money.

It was identified by both senior project management and contracts administrator respondents that any cash flow problems experienced by the contractor due to insufficient payments from the client would result in delayed or withheld payments to subcontractors and suppliers, which could very negatively affect cash flows or indeed prevent them from agreeing to participate in CCPM projects. Indeed, another problem related to systems of payment was the ability to incorporate any change in scope or change in critical chain/path (notwithstanding contractual claims mechanisms); it was argued that this leads to a delay in payments to the contractor, which will be passed on to subcontractors and suppliers.

Duration estimation

Accurately estimating durations to reflect the safety free median activity durations was described by sample project managers interviewed as a 'major resourcing issue'. Yet, all participants also recognise that target programmes—developed by contractor representatives and used on-site to drive subcontractor performance—estimate 'unrealistic activity durations', which assume 'a best case scenario' (much like the activity durations called for in CCPM); however, it was also recognised that if CCPM installs larger liquidated damages into subcontracts, the high level of subcontractor interdependence and interfacing will significantly increase the risk to subcontractors, making them less likely to contractually commit to CCPM projects. Further insight is required into such liquidated damages within CCPM contracts.

Staff and plant idleness

The senior panel consulted suggested that 'starting noncritical activities earlier than required to keep resources on-site and absorb excess capacity' to be common practice in the Western Australian construction industry. However, CCPM calls for a structured progression through the schedule in an effort to complete works sequentially and prevent multitasking. Discipline will be required by the contractor to not push ahead with future works if the required management resources are not available to the contractor on-site, requiring an understanding and belief in the ideals of CCPM brought about through the continual training of the contractor's project team.

Training requirements

For a contractor to implement CCPM, significant continual investment will have to be made in the training of staff members and auditing procedures. This investment in people poses significant problems due to the relatively high staff turnover experienced in the Western Australian construction industry, as identified through the structured interviews. In the Australian construction industry, during recent years, there has been a 13% mobility rate within the industry and 4% exit rate from the industry, in which a mobility rate refers to employees leaving one company for another within the construction industry.

3.7.4 CCPM variables already in use in the construction industry

Although few examples can be found of CCPM in standard civil engineering projects in Australia, a number of its principal themes do seem to be in existence albeit less explicitly under the umbrella of CCPM, two such examples being target programmes and approximate subcontractor start dates.

Target programmes

The project programme generally will have much contingency time installed into it because programmers often have a less than clear understanding of the project constraints, site conditions, activity methodologies and other details specific to the project being undertaken. Therefore, contractors develop their own programmes, which take into account many of these previously unknown variables. These more accurate target programmes assume a best-case scenario, as expressed in the structured interviews, meaning that contingency time is not installed into the (updated) activity duration estimates.

This means that the target programme is the contractor's version of a CCPM programme without the installation of buffers. The purpose of these target programmes is to accelerate subcontractor performance by a perceived 'setting' of 'unrealistic activity durations'; therefore, subcontractors constantly *believe* they are running late. However, the development of these target programmes is individual to each project manager with the time cut, dependent on their unique perception and experiences of uncertainty.

Instead of buffers being installed and managed in a structured way, contingency time is cut from the schedule and it is hoped that the difference between the target programme and project programme will account for uncertainties and delays. The problem with this method of management is that the criticality of each delay is not identified by the project team and, thus, the effect that each has on the project completion is not fully understood or appreciated by stakeholders.

Subcontractors' approximate start dates

CCPM calls for subcontractors to commence activities directly after the completion of the preceding activity and as such become a relay race of resourcing. This requires subcontractors to commit to durations for works rather than set dates and also requires them to commence works upon instruction.

The practice of giving subcontractors rough starting dates to begin works was identified, through the interview panel of senior managers, as common within the Western Australian construction and civil engineering industry whereby the project programme is recognised as a guide only and dates are generally confirmed during meetings closer to the commencement date.

3.7.5 Potential CCPM uptake factors

Having assessed the extent to which some variables that contribute to the approach of CCPM are already being implemented, the following discussion looks at the extent to which further uptake of CCPM might occur in civil engineering.

Reductions in project duration(s)

Given perceived logistical problems with CCPM implementation (and, not least, the less competitive environment that results from civil engineering support projects that traditionally service the Western Australian resources and mining industry, that in itself seems to buck every trend, not least the global financial crisis of recent years), the investment required to successfully implement CCPM would significantly outweigh the potential benefits provided by CCPM in securing tenders and increasing profit margins for construction contractors already active in the sector.

However, partial implementation of CCPM techniques in the development and management of target programmes, as outlined above, will increase the control of these programmes by providing a framework in which to establish activity durations, monitor progress and manage buffer periods. Additionally, in the case that the economic climate within the Western Australian resources and mining sector, which currently feeds the civil engineering and construction industry worsens significantly beyond its current levelling out, CCPM implementation would become more feasible, providing an alternative to the client in a highly competitive environment.

Multiple projects

CCPM seeks to provide the client with an up-to-date working schedule as discussed above, which shows the effect of delays in responding to contractor questions on the project's overall duration through a 'colour-coordinated' buffer management system. In the case of a client with multiple construction projects running simultaneously, as generally is the case for the government's public works department, this ability of CCPM to give the client an easily recognisable feedback loop in which to prioritise decisions could provide significant benefits in reducing delays experienced by all projects.

Known resources

Inconsistent resource allocation (as is argued generally by a panel of seniors to be the case within the Western Australian construction industry) means regular monitoring of progress is unlikely to give an accurate feedback mechanism to judge progress and estimate completion dates. Monitoring *daily* might be argued to elicit preemptive action on behalf of the project management team, potentially causing delays and straining relations between the two parties.

Therefore, projects with a known amount of resources, such as infrastructure projects or on building contracts where no sections of work are subcontracted, would be two project types most able to feasibly utilise CCPM monitoring and buffer management techniques.

Large liquidated damages

The changes necessary for a contractor to undertake/implement CCPM requires a lot of time, investment and understanding to be successful.

To make this large level of investment feasible, sufficient financial incentives need to be provided contractually by the client based on the real cost to the client whilst the infrastructure is not operational. Therefore, projects with potentially relatively large liquidated damages are perhaps best suited to CCPM because they are better able to provide the large financial incentives and bonuses required to make CCPM implementation financially viable.

High levels of competition

By implementing CCPM and advertising its potential benefits in reducing project completion durations, contractors offer an alternative to the client, with an estimated project completion date significantly shorter than that offered by traditional CPM. It might be argued that this shorter duration is likely to be balanced by a somewhat higher project price. In the case of a significant increase in industry competition for projects (private sector market troughs perhaps), this alternative scheduling tool and the investment required for it might prove viable for consideration/implementation.

Emergency relief

In an emergency situation, shortening project durations becomes the primary project success criteria to minimise impact by restoring vital infrastructure. It can also be assumed that the maximum amount of resources available will be utilised, therefore, the resource pool is known.

> Essential requirements of short durations and finite resources make *emergency relief* construction works (and perhaps more pertinently, the somewhat longer-term reestablishment of essential residential and community amenities) well-suited perhaps to CCPM techniques (notwithstanding a project manager's understanding of, and ability to apply proficiently, the underlying principals).

Higher contractor profit margins might realistically perhaps be balanced with political and humanitarian incentives in delivering and completing projects as quickly as possible. It might be argued that any organisation predisposed to, and experienced in, emergency relief construction works could potentially benefit most from the implementation and offer of works scheduled by CCPM, if experienced practitioners were able to hit the ground running with such an approach.

Overall, perhaps full implementation of the CCPM scheduling tool cannot really be recommended for the construction and civil engineering industry in a buoyant market; however, target programmes on the other hand, might well be used on-site and benefit from managing according to CCPM principals.

- Development of such target programmes, including activity duration estimations and resource dependencies in accordance with CCPM techniques, although it requires additional reporting of progress through regular/(daily) subcontractor progress reports and complementary contractor's schedule management, can be argued as being potentially advantageous; programme scheduling might be addressed explicitly, towards mutual benefit, within local standard forms of construction contracts, such as the 'programming' clause 32 of Australian Standards 4000.
- Similarly, opportunities may exist for stakeholders associated with disaster mitigation and related community reestablishment projects, to gain from an implementation of CCPM techniques (and its incorporation into selection criteria for awarding such disaster reconstruction contracts), to address/reduce project duration/delivery times.

Having discussed scheduling opportunities that may address delay, it is perhaps pertinent to now discuss scope management approaches that embrace agile systems.

3.8 AGILE MANAGEMENT

The following discussion reviews opportunities for traditional engineering management in construction projects, towards alignment with agile management approaches.*

> Stated somewhat glibly, *agile management* is a method that embraces flexible scope. Perhaps the 'trick' to agile management is to benefit from having a 'flexible' scope, but to avoid the disadvantages of 'scope creep' and any knock-on expensive variations.

Civil engineering projects, like project realisation in all other sectors globally, can be successfully characterised by variables of flexibility and speed of delivery; in other words, agility in addressing and realising respective project brief(s) and the subsequent completion of built assets, alongside their effective operation and maintenance over a usable life cycle. Projects that do not, at the outset, embrace responsiveness in their approaches are often those that result in being overbudget and overtime.

As mentioned in previous chapters, media reports note $7 billion blowouts and 4-year delays for high-profile projects statewide (never mind across the nation and beyond); suggesting that there is a need to address, once again, the tools and techniques of project management and their

---

* Input from Aziz Albishri,[67–82] received with thanks.

implementation as design teams strive to deliver a quality product at a predicted cost within a predicted timescale for construction; in other words, a need for a more agile approach to the planning and running of a job.

Agile management concepts and approaches used predominately in the manufacturing industry seem to offer effective methodological system(s) for companies seeking to win tenders and secure procurement routes. Given the attractiveness of this approach, there is perhaps an opportunity to transfer this set of agile skills towards efficiency gains, from the manufacturing sector to the construction and civil engineering projects; a potential then, for agile management approaches to address concerns, locally and internationally, regarding civil engineering project time and budget overruns.

The level of big (mega) project instigation and implementation between parties (bound nationally or internationally) has accelerated dramatically in the last 10 years. These large-scale projects can represent infrastructure ventures such as dams, transport facilities, nuclear power plants, mining and oil and gas exploration. An example of the scale and technical requirement of such 'mega' projects can be illustrated locally by the recent consortium of six multinational companies engaged to realise the $43 billion liquefied natural gas project in Gorgon rural Western Australia, whereas other more international examples of mega projects are exemplified by endeavours such as the Germany–Italy rail route across the Alpine Mountains, as well as the $50 billion project between the United States and Russia to improve access across the Bering Strait.

These projects may differ from each other physically, but are similar methodologically by seeking to balance the three principles of quality, cost and time factors of a traditional project management approach. The measurement and monitoring of quality, cost and time is made increasingly difficult by seemingly constant changes in the general scope of works and related deviations from the initial brief; mega project variation tracking creates huge complications despite a raft of clauses of contract seeking to explicitly regulate extensions related to a scope change.

To address efficiency levels in the realisation and delivery of big projects, a number of project management reviews have been undertaken. In 1990, project management lean production and agile thinking were adopted and adapted from the successful Japanese experience (and continues to evolve in studies related to the Korean experience).

The evolution of the relatively new agile project management (APM) technique can be traced from Deming (discussed in Chapter 4) through the so-called 'Toyota' development successes, to the agile movement. Although popular in manufacturing industries, there is no explicit definition for agile project management that relates to the construction and civil engineering fields. Agile application, however, remains worthy of consideration, particularly to assess its relationship with 'scope'.

3.8.1 Agile management and project scope

Continuous change of project scope is common in the initiation of very large-scale multidisciplinary projects from planning to execution, due to the high degree of specialist input and need for expert consultations over time, and this leads to the requirement for coordination and control of the process of technical input across all project stakeholders at various stages and phases of development; agile approaches may at least be considered towards providing an ethos to work within.

Large-scale engineering projects have high capital cost and long time-scales. Consortiums or partnerships are a first step towards such projects; a multitude of parties making the final decision. Procurement routes (discussed in full in later chapters) then require consideration: partnering, alliance, lump sum and turnkey routes of engineering procurement still require scope clarification. Similarly, joint venture agreements take place based on the nature of (specific) tasks undertaken. Therefore, scope of work clarification is the core of project briefing starting from the initiation phase onwards.

Scope and the resultant schedule, quality and investment factors are key to project go-ahead. The importance of the scope of work lies as the crucial factor for project design and, to meet its needs, project management's basic processes include initiation, planning and design, execution, monitoring, control and finally close-out, which may be explained further:

- In the traditional initiation phase, the scope of work comes directly after the project idea; if scope is not specified, project success is unlikely.
- The second phase of traditional project management is planning and design.
- Traditional controlling of projects monitors scope compliance.

Agility approaches to management try to balance fixed duration with variable scope on a package-by-package basis. It is for the construction design team to assess the extent to which a 'package' may lend itself to 'flexible scope' opportunities. Perhaps fittings and fixtures (at the poststructural integrity compliance design stage) might adopt an agile approach.

Agile methods characterise a shortening of project schedule concerns with 'unknown' requirements that are not (needed to be) specified at initiation and planning stage of project process. One of the major advantages that distinguish agile methods from other methods is the ability to change the requirements at any stage at lower cost or seek flexibility to change the scope. The success of such methods depends on senior management commitment and the creation of an appropriate environment. Agility method adaption and adoption in civil engineering and building may depend on learning some success factors in information technology and software engineering.

In general, the software industry has been adopting two methodologies for projects: sequential (linear) and iterative approaches to achieve the highest degree of flexibility and adaptability to uncertainty, with iterative approaches represented by frequent checkpoints for review. Agile methodologies have daily activities starting with a team review, development of activities, testing and the daily outcome to integrate with the final project.

Large-scale (mega) projects usually involve firms from different countries and engineering fields including contractors, suppliers and fabricators. Given the above discussion, it might be argued that improvement of project performance in construction and engineering fields might begin at initiation phases and specifically scope-passing between detailed plans that incorporate flexibility or agility in implementation.

Civil engineering opportunities for agile management may be summarised at this stage as requiring further case study exploration of scope flexibility. Sections 3.7 and 3.8 have sought to deal with time blowouts in construction. Section 3.9 continues this theme and further reviews the factors that cause delay.

3.9 DELAY AND (OIL PRICE) FLUCTUATIONS IN CIVIL ENGINEERING PROJECTS

This section identifies the range of factors (such as oil price) apt to influence delay in construction projects across a range of different environments.*

Projects around the world experience delay, that is to say, they go beyond initial time and cost estimates. There are a multitude of causes of delay that are argued as being beyond the design/site civil engineer's control and somewhat beyond organisation and monitoring of the three M's (materials, women/men and machines). Delay factors for both local and overseas projects often include socioeconomic, sociopolitical and environmental factors; oil prices typify such factors.

The extent to which geographic location and socioeconomic variables affect construction delay generally is worthy of analysis and, in particular, delay in construction projects (and an assessment of the means to address and mitigate such delay) in both oil-producing and oil-dependent countries. Key factors of delay can relate to reliance of construction budget(s) on oil price in locations where 85% of the national construction and infrastructure budget is derived from the oil income, and where fluctuations in dollar per barrel costs have a knock-on effect for construction project interim payments and final accounting and practical completion. Indeed, energy (and

* Input from Maryam Alavitoussi,[83–92] received with thanks.

global financial) crises, both decades ago and in recent years, highlight the importance of seeking measures to address delays that arise not only from on-site factors but also from variables in a wider international context.

Fundamentally perhaps, delay issues on a more global scale can be brought back within the remit of the civil engineer through the somewhat simplistic application of fluctuations (and escalation) clauses in the general conditions of the contract. An accurate means to predict and then allocate risk and fluctuation allowances at both the precontract feasibility stage as well as in on-site activities can be directed and controlled through contractual obligations and responsibilities that seek to balance risk.

Delays create a host of problems in projects: time overruns, cost overruns, negative effects on related financing arrangements, interest on loan factors, client–contractor confidence reductions and designer/client concerns over final quality and structural integrity. There are several factors that contribute to construction costs from the estimating stage to project completion. These factors can be divided into two main categories: intrinsic and extrinsic factors. Intrinsic factors are related to construction organisations, whereas extrinsic factors are those related to sociocultural, economic, technological and political environments within which these organisations operate. Unlike extrinsic causes, intrinsic causes of delay in construction projects can be overcome by efficient project management. Extrinsic causes are aligned to international risk.

Global risk factors differ from region to region. It has been suggested that each country, and very often, specific regions or districts commonly identify location-specific key variables that contribute to delays on-site.

Often, objective practical methods to mitigate the negative effect of delays on projects through improved predicative models and the assessment of potential multiplier models (as part of fluctuation or escalation analyses) may cover estimation inconsistencies, particularly in regions where fluctuations in variables such as the oil price can be a significant triggering factor in delays in civil engineering construction projects, where such fluctuations often cause projects to go well beyond their initial predicted time and cost estimations.

Because the price of oil is determined in an international market and does not remain constant at all times, fluctuations in oil prices are argued to have a potentially negative effect on construction projects; however, civil engineers can still maintain a degree of control if, at the outset of project feasibility, they are *au fait* with all relevant fluctuation and escalation contractual clauses (that can be applied retrospectively in an unforeseen situation) to help take speculation out of the built-up unit rates to estimate a job.

Reliance on oil export as well as import leads to vulnerability as oil prices are proved to be extremely unstable. Often, high proportions of construction projects are reliant on oil revenue. Certain countries, the United Arab Emirates for instance, have been successful in decreasing their dependency on oil income by building new cities, real estate and tourism projects. These reforms have resulted in the diversification of their export structure. Statistics show that in the United Arab Emirates, construction constitutes more than 80% of the gross domestic product, whereas the share of oil and gas is less than 10%; this is in contrast with neighbouring countries in which the share of oil and gas in gross domestic product dominates.

Economic and political instability globally contributes not only to delay in on-going construction projects across the world, but also to affect 'go-ahead' or 'abort' decisions for new proposals. Often, there is a negative effect from supply chains that are traditionally sourced overseas. Economic instability leads to fluctuations in localised exchange rates. Because many international currencies are not stable, the price of equipment (or indeed the salary and exchange rate expectations of the pool of professional civil engineers) will fluctuate significantly from the price expectations and estimates at the feasibility stages of projects. Indeed, a case in point might be the consideration of fuel within a unit rate (discussed in previous chapters), which can contribute up to one-third of the plant rate breakdown. It is perhaps unreasonable to expect contractors alone to absorb such escalation risks of fuel in more uncertain and volatile markets.

3.9.1 Fluctuation and escalation clauses to address oil prices

Fluctuation clauses contained in the general conditions of a contract are perhaps key to allow civil engineers to retain control and avoid delays, even in situations that appear outwith their design/site remit(s).

Traditional or lump-sum contracts generally include (alongside provisions for advance payment, a retention bond, a bond for the payment of off-site materials and for third-party rights or collateral warranties) fluctuation options in a schedule; albeit in a limited way, such that fluctuations (namely, levy and tax changes) as well as escalation clauses, are aligned within sub-contracts that include provisions for fluctuations by formula adjustment allowing any such amounts to be adjusted under the main contract.

Therefore, escalation clauses—which are provisions in a contract that help adjustments in prices in the event of an increase in fundamental costs—are important to address specifically unexpected costs, such as those resulting from fluctuations in prices from not only raw materials (such as perhaps the doubling of the real cost of steel over the past decade or so) but also as (perhaps more directly) fuel costs increase over the course of a construction project; such escalation clauses (formula calculations) can begin to take speculation out of explicit unit rate work task calculations and allow concentration on the main estimation variables of the three M's (women/men, machines and materials).

Escalation clauses in contracts establish (the potentiality of) change in contract price in which factors such as the price of oil, which is beyond the immediate control of the project stakeholders, results in changes to the builders' costs on-site. Effectively, escalation can be applied to construction projects using building cost information service formula adjustment methods and, indeed, engineering contracts such as NEC3's Engineering and Construction Contract, contain such special clauses.

This chapter and its various illustrative case studies have sought to delineate a timeline. Previous sections have highlighted time management variables including Scheduling networks and bar chart techniques (Section 3.1), Rescheduling: programme crash-costing and resources aggregation (Section 3.2), Scheduling risk (Section 3.3), Linear time-and-distance scheduling (Section 3.4), Method statements (Section 3.5), Value management (Section 3.6), Critical path and critical chain (Section 3.7), Agile management (Section 3.8) and Delay and fluctuations in projects (Section 3.9).

Cost analysis in Chapter 2, and time assessment in this chapter, now give way to a discussion of quality control in Chapter 4.

Chapter 4

Quality control in civil engineering projects

Cost and time factors in civil engineering projects are somewhat easily quantifiable as dollars and days. Similarly, controlling and monitoring money and months on the job requires comparisons of value estimates versus actual outgoings and progress.

Quality control in civil engineering work requires similar levels of measurable clarity.

In civil engineering projects, quality can be guaranteed by a design team's detailed drawings and comprehensive materials specifications, which adhere fully with current building codes as well as a builder's full and complete compliance with these contractual documents.

Quality systems in civil engineering

Quality systems might be summarised as

- Detailed checklists for civil engineering drawings and specifications compliance
- Checklists ensuring that prescriptive or performance requirements are tested and met

Quality control and proof of structured quality systems must increasingly be embedded explicitly within the documentation to progress a civil engineering project. Often, a client's tender-list of potential builders are required to have current quality system certification before even being allowed to submit a tender-bid for a project.

Indeed, upon awarding a project to a preferred building firm, the standard form of contract that is signed by the parties requires full compliance with clauses that stipulate that a contractor must have a conforming quality system in place and adhere to it specifically throughout the works.

This chapter reviews quality control in civil engineering projects in specific terms:

Civil engineers, whether representing a client or charged to build a proposed solution, must address quality requirements at all stages of design, supply and installation. Quality control systems help parties to ensure compliance with national standards in the pursuit of fit-for-purpose construction.

Section 4.1 discusses quality management systems in more detail.

4.1 QUALITY SYSTEMS AND QUALITY STANDARDS

Structured approaches to quality systems and quality compliance requirements are available from international and national standards organisations,[1] which prepare, publish and disseminate a wide range of principles and guidelines; these include

Systems:
- Australian Standards/New Zealand Standards (AS/NZS) ISO 10005:2006—Quality Management Systems—Guidelines for Quality Plans
- Originated as AS/NZS ISO 9004.5 (Int.):1995

Management guidelines:
- ISO 9000-1 Quality Management and Quality Assurance Standards—Guidelines for Selection and Use
- ISO 9004-1 Quality Management and Quality System Elements—Part 1: Guidelines for Management

Three quality standards:
- ISO 9001—Quality Systems—Model for Quality Assurance in Design, Development, Production, Installation and Servicing
- ISO 9002—Quality Systems—Model for Quality Assurance in Production, Installation and Servicing
- ISO 9003—Quality Systems—Model for Quality Assurance in Final Inspections and Testing

Additional guidelines:
- ISO 9000-2—Quality Management and Quality System Elements—Generic System Guidelines
- ISO 9000-3—Quality Management and Quality System Elements—Guidelines for Software
- ISO 9000-4—Quality Management/System Elements—Guidelines for Program Management
- ISO 9004-2—Quality Management/System Elements—Part 2: Guidelines for Services
- ISO 9004-3—Quality Management/System Elements—Part 3: Guidelines for Processed Materials
- ISO 9004-4—Quality Management/System Elements—Part 4: Guidelines for Quality Improvement
- ISO 9004-5—Quality Management/System Elements—Part 5: Guidelines for Quality Plans
- ISO 9004-6—Quality Management/Elements—Part 6: Guidelines on Configuration Management
- ISO 10013—Guidelines for Developing Quality Manuals

Accreditation:
- JAS-ANZ is the government-appointed accreditation body for Australia and New Zealand that is responsible for providing accreditation of conformity assessment bodies in the fields of certification and inspection
- Certifying bodies are called Conforming Assessment Bodies (CABs) in Australia

The International Organisation for Standardisation,[1] with country-specific input, provides quality control guidance documents (see Selected quality standards) that seek to provide the minimum rules for creating a complete management system of quality assurance. National adaptations are approved by more than 90 countries and give a starting point for addressing quality in civil engineering projects.

SELECTED QUALITY STANDARDS

- Quality management systems, guidelines for quality plans: AS/NZS ISO10005
- Quality management systems: ISO9004
- Final inspection and testing: ISO9003
- Development, production, installation and servicing: ISO9002
- Design, development, production installation and servicing: ISO9001
- Quality management: ISO9000

After an individual company's initial request for review of their procedures, a nominated accreditation body will be invited to assess the firm's practices and procedures. If deemed in-line with ISO expectations, that company is certified as ISO-compliant. This procedure is repeated every 3 years or so to retain ISO certification.

Generally, the Joint Accreditation Systems of Australia and New Zealand (JAS-ANZ),[2] alongside proprietary material certification bodies (such as Code-Mark[3]), are somewhat typical of the national government–appointed accreditation bodies (signatories to bilateral international agreements) responsible for endorsing a firm's procedures. CABs are given licence (by JAS-ANZ and the like) to inspect and certify companies as having suitable quality systems in place.

Thus, accreditation of a firm's practices and procedures towards ISO quality certification is available in Australasia from the JAS-ANZ, which accredits 106 CABs, which in turn certifies approximately 80,000 organisations. Including accreditations and technical assistance projects, JAS-ANZ provides services in more than 20 countries.[2]

Accreditation systems also might be deemed to include the 'Code-Mark Scheme', which seeks to provide a mark of conformity that proprietary building products are based on technical documentation, alongside regular review of manufacturing and quality control processes that monitor compliance with the codes such as the Building Code of Australia.[4]

Clients look for these quality accreditation certificates in both civil engineering design consultancies and civil engineering construction firms as qualifications that confirm an organisation's credentials and suitability to tender and bid for, and subsequently participate in, civil engineering projects.

The different departments in a firm, whether a small design consultancy or a large contractor, must each acknowledge their role in maintaining quality in each and every stage of a project's procedures and deliverables: from conception, through feasibility and design drawings, to construction on-site and decommissioning where appropriate.

Quality systems for all design services and drawings outputs, and all on-site activities, might ideally be deemed to be based on a process of stage-by-stage 'plan–do–see' (Figure 4.1).

Famously termed the *Deming cycle*,[5] quality compliance is planned, monitored on-site and actioned to address any noncompliances.

Influential authors such as Deming remain pivotal to theoretical discussions of quality.[5]

Perhaps more specifically for civil engineering works, common rules require explicit clarification to coordinate operations and achieve principal quality objectives for installation compliancy and testing on-site. These quality compliance rules may be in printed format, but are more likely to be available in a constantly updated soft copy medium for quick and easy

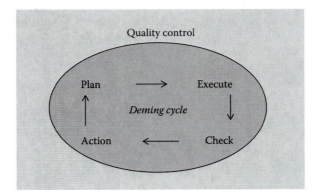

Figure 4.1 Deming cycle.

cross-reference with national standards and respective building codes for inspection and related ongoing quality check record-keeping.

Progression from senior management policy dictates through to items of more relevance for the civil engineer's help checklist the fit-for-purpose supply and installation of materials and project elements. On-site work instructions and expectations for testing and compliance checking, as well as confirmation that these tests have been done, are a major means to ensure that a quality system remains relevant.

The quality systems might be expected to involve the following:

- Quality policy provides an overarching head office short policy statement related to a firm's certification of, and commitment to, their in-house quality system(s).
- Quality (soft copy) manuals and specific guidelines to staff set out the rules and cross-reference all relevant standards for substructure and superstructure work.
- Quality procedures discuss the sequences, actions, roles and responsibilities on-site for preinstallation and postinstallation tests, inspections and record-keeping.
  - Working instructions, again in soft copy format, detail information about the particular tasks to be done to ensure fit-for-purpose installation.
- Quality verification documentation and compliance confirmation(s) reports, prepared periodically during site works, must be forwarded ultimately to the client for future asset and facility management and life cycle upkeep.

Of the eight specific items that make up the ISO 9000/9001 quality system[6] (see Quality management system ISO 9000/1), items addressing product realisation and measurement analysis and improvement are perhaps most directly applicable to the supply and installation of building components. Thus, items (vii) and (viii) are worthy of note for civil engineers and materials specifications that require extensive compliance testing.

QUALITY MANAGEMENT SYSTEM ISO 9000/1

Eight specific items make up the ISO 9000/9001 quality system, namely,

    i. Scope
    ii. References
    iii. Definitions
    iv. Requirements
    v. Management responsibility
    vi. Resource management
    vii. Product realisation
    viii. Measurement analysis and improvement

Product realisation and measurement items are specifically referenced in quality management systems guidelines and requirements such as AS/NZS ISO 90001:2008 and ISO 9000-1 Quality management and quality assurance standards—Guidelines for selection and use.

The practice example in Section 4.1.1 illustrates a basic application of a quality system; this example links with the 'Specification decisions: Practice example' in Chapter 5.

4.1.1 Quality system checklist: Practice example

An understanding and application of the tenets of quality control systems[7] helps address concerns related to potentially noncompliant products.

Three steps in quality control, with particular regard to AS/NZS ISO 90001:2008, items (vii) and (viii), 'product realisation' and 'measurement analysis and improvement' are

Step 1: Identify a project as requiring a quality system ISO 9001 checklist for the following:
      • Product realisation factors of design and development, purchasing, production and product provision
      • Quality assurance certification of (proprietary) products by World Technical Assessment affiliates[8]; thus, product certification quality assurance include national and international organisations such as
        • World Federation of Technical Assessment Organisations (WFTAO), a worldwide network for coordinating and facilitating the technical assessment of innovations in the construction field

- CSIRO (Australia's Commonwealth Scientific and Industrial Research Organisation)
- Australian Building Codes Board guidelines
- Code-Mark
- British Board of Agrément certification of building product 'fit-for-use' quality standards, recognised internationally
- SIRIM, Malaysian quality certification and the like

Step 2: Identify a project as requiring a quality system ISO 9001 checklist for measurement analysis and improvement factors of monitoring and measurement, control of nonconforming products, analysis of data and improvement

Step 3: Implement quality checklist during design, planning and construction

A scenario might involve working in a location where material quality is less than assured with poor adherence to available standards and specifications; specifically, a new apartment block is required in a location known for multistorey residency collapse resulting from construction faults and poor quality concrete.

Step 1

Use quality systems to checklist 'product realisation'. Specifically, target 'design' by consideration of options for concrete design and supply. Recognise that, in this location, the civil engineer has different methods of specifying concrete mix design (relevant to standards such as AS-1379 and AS-3600), such that designated concrete requires a concrete producer to hold conformity certification based on product testing and surveillance to quality system ISO 9000, and where deemed appropriate, conformity certificates from international and national industrywide quality assurance organisations that certify 'fit-for-purpose' suitability of (construction) elements and products.[9]

Note that designated mix design, in this location, is preferred compared with options such as

- Designed concrete, in which a builder is responsible for specifying strength
- Prescribed concrete, in which a builder specifies proportions and performance
- Standardised concrete, in which prescribed concrete is selected from lists
- Proprietary concrete, in which producers are not required to declare compositions

Step 2

Use quality systems to checklist 'measurement and analyses' of the designated mix. Specifically, target 'monitoring, measurement and control of nonconforming products', in which nonconforming concrete is identified in tests measured against requirements in specifications, drawings and procedures to detect discrepancies between actual installed versus intended fit-for-purpose, such that interim stage-by-stage tests *condemn* work if

- Concrete samples sent to an independent testing laboratory are deemed lacking and fail to reach a required strength after a required period of curing.
- Nonconforming elements are condemned and removed, necessitating rework.

Step 3
Remix/reinstall element to conform to specifications; repeat step 2 as required.

4.2 QUALITY AND CONTRACTUAL REQUIREMENTS

Quality control measures apply to installation processes as well as materials supply. Adequate workmanship is very much part of the responsibilities and obligations set out in the contractual agreement between the client and the builder.

Fit-for-purpose material(s) supply is expected in a construction contract (reinforced by the supply of goods acts and consumer laws). Similarly, workmanship is a dominant constructor obligation, not only in the contract but also in overarching common law(s).[10] Common English law (as it relates to common law, consumer law/supply of goods, obligations and responsibilities) forms the basis of jurisprudence in Australia, the United States, Canada, New Zealand, Malaysia and India, and is somewhat distinct from civil (Roman) law systems in Europe (Scotland), South America and Japan. In general terms then, fit-for-purpose working applies to expectations for a builder's application of suitable skill and judgement.

Materials and workmanship of a suitable quality and standard are specified in contract documents. Specifically, the general conditions of contract contained within standard forms of contract between a client and a builder requires adherence to quality clauses. The Australian Standards 4000 (AS4000)[11] standard form of contract is typical in containing specific clauses that emphasise fit-for-purpose quality; clause 29 of AS4000 details such expectations for quality in a contract.

The general conditions of contract between a client and a builder for construction work in Australia are typified by two Australian standards (and their respective variations), namely, AS4000-1997 General conditions of contract and AS2124-1992 General conditions of contract.

GENERAL CONDITIONS OF CONTRACT QUALITY
CLAUSE RESPONSIBILITIES AND OBLIGATIONS

(AS4000 clause 29)

- Supply/install 'suitable' materials.
- Proper 'tradesman-like workmanship'.
- Establishment and maintenance of 'confirming quality system' to build.
- If nonconforming work is identified, then remove, replace, demolish, rework.
- 'Rework within 8 days' or done by other and paid by contractor.
- Potential for noncompliant work as 'variation' to contract specification.
- Expectation of good quality up to and 'until defect liability' period expiry.

Quality systems in engineering projects are applied stepwise:

- The production of drawings and specifications alongside checking mechanisms
- Work method statements to address elementally the work under the project
- Inspections and test plans, cross-referenced to contractual quality systems
- Recording of results for all tests and inspections for work breakdown structures
- Forwarding of data logbooks of all materials specifications compliance tests to the client for asset management, maintenance, operation and refurbishment purposes

Although quality systems (encapsulated within by AS4000 clause 29) seek to ensure compliance with drawings and specifications, any nonconforming installation and damage that does occur on a project needs to be put right and 'reinstated' under the direction of related separate clauses (such as AS4000 clause 14).

It might be argued that, to avoid future quality noncompliance, explicit improvement must be enacted. Thus, engineers need to measure the efficiency and effectiveness of the current production process, analyse the information generated and then seek to address any inconsistencies or deviations from compliance dictates. This type of continuous improvement procedure is discussed briefly in Section 4.3.

4.3 QUALITY AND CONTINUOUS IMPROVEMENT

Quality systems (ISO 9001) have a requirement for 'measurement analysis and improvement' and specifically directs the civil engineer in 'continuous improvement',[7] not only of the quality system checklist itself but also, by implication, the continuous improvement of the building element and component or product to be designed, specified, supplied and installed on-site. To avoid future nonconformity and enhance existing products, structured approaches to continuous improvement need enactment.

Continuous improvement related to ensuring quality for civil engineering projects may be reviewed in standards such as AS/NZS ISO 90001:2008 item 8.5.1 (Continual improvement), item 8.5.2 (Corrective action) and item 8.5.3 (Preventive action to determine nonconformity), evaluate the need for action to prevent future nonconformance and, by implication, improve the current product.

Continuous improvement (total quality) management principals are traditionally deemed to include customer focus, employee empowerment, benchmarking, just-in-time approaches, as well as Taguchi concepts related to targeting specifications as accurately as possible during the production process.

Structured approaches to continuous improvement repackage somewhat total quality management (TQM) techniques; the continuous improvement label is favoured over TQM. Despite this relabeling, original TQM tools remain available to those seeking to objectively measure production processes and a supplied product's compliance with design specifications (see Continuous improvement approaches, measuring planned versus actual).

CONTINUOUS IMPROVEMENT APPROACHES,
MEASURING PLANNED VERSUS ACTUAL

- Customer feedback ascertained by charting complaints
- Employee feedback resulting from regular on-site 'tool box' meetings that discuss efficiency of production and installation
- Benchmarking what one company does, against transparent procedures and products of an independent market leader

Continuous improvement, in essence, seeks to measure what exists and compare it with expectations. One approach that seeks to better align 'planned quality' with 'actual product delivery' targets supply chains to reduce on-site material storage inventories through supplying materials just-in-time for installation on-site.

The on-site approach of just-in-time supply of materials and components (where and when needed), seeks to avoid inventory issues but more specifically address on-site storage space limitations and reduce the potential for spoil and waste. Although not applicable in every situation, just-in-time affords the opportunity to avoid storage (and any knock-on problems resulting in late identification of defects in large quantities of inventory) and instead, install immediately upon arrival.

This is somewhat in line with current trends towards lean construction (what has previously been termed modularisation and historically was known as prefabrication).

Clearly, a just-in-time supply of materials is less useful for civil engineers who have the luxury of large rural building sites, or who simply want to avoid the 'nervousness'[12] of hoping that a just-in-time delivery will indeed arrive on-time, not be late and not delay a project's overall timeline and critical path. It is perhaps worthy of note that just-in-time materials supply has been termed the 'nervous' system of lean production by organisations such as the U.S. Environmental Protection Agency.[12]

Civil engineers who seek further continuous improvement of products or deliverables are directed towards available (TQM) tools and techniques such as

- Taguchi methods
- Kaizen applications
- Zero defects approaches
- Six Sigma applications
- Ishikawa fishbone approaches to defects identification and rectification
- Statistical process control techniques

as well as other methods to chart 'existing' versus 'intended' production rates via

- Check sheets, scatter diagrams
- Cause-and-effect diagrams
- Pareto charts
- Flow/process charts and production histograms

A range of continuous improvement approaches might be gleaned from agencies and private sector organisations such as the U.S. Environmental Protection Agency and dominant industry practitioners such as Bechtel.[12,13]

Although efforts to continuously improve processes and products have previously been discussed, it is perhaps pertinent to return to the ongoing need to frame every activity within safe work practices. Indeed, early chapters discussing cost and time continually made reference to mitigating measures towards addressing risk and ensuring health and safety on-site.

Section 4.4 discusses in more detail occupational health and safety in civil engineering and in the construction industry in general.

4.4 OCCUPATIONAL HEALTH AND SAFETY IN CONSTRUCTION

Civil engineering and the construction industry in general is a dangerous place to work. Annually, Australia might expect close to 30 fatalities industrywide. Safe-Work Australia monitors work injuries and regularly benchmarks the various industry sectors.[14] Deaths as a result of working in the industry have changed little over the past decade or so. Table 4.1 identifies the construction industry's placement as second most dangerous sector nationwide.[15]

Accidents in the construction industry cause distress to all involved; resultant levels of loss and damage to victims can be extreme. The industry must continue to structure and implement health and safety policy and practice towards making the sector a better place to work. This must occur at all stages in the realisation of a project, from planning and design through on-site construction and beyond to the certificate of practical completion and defects liability stages.

During project planning, dedicated roles must oversee safety culture. At the design stage, work methods and work breakdown structure must seek to incorporate safe-work practices into activity schedules and, not least, factor into feasibility cost estimates and time schedules suitable risk mitigation to address all safety concerns.

During all on-site construction work, safety plans must be adhered to and monitored strictly, with particular reference to safe site locations for the full range of resources, incident reporting procedures, inspections, worker and line manager consultation procedures, adequate levels of training and policing of training certificates, suitable means to monitor major

Table 4.1 Safe-Work Australia findings for worker fatalities by industry (2003–2011)

Industry of workplace	Year							
	2003– 2004	2004– 2005	2005– 2006	2006– 2007	2007– 2008	2008– 2009	2009– 2010	2010– 2011
Agriculture/ forestry/fishing	39	39	33	27	27	45	28	33
Construction	25	18	24	30	38	27	21	27
Manufacturing	8	8	23	17	16	13	16	22

Source: Work-Safe-Australia. 2011. Available at http://www.safeworkaustralia.gov.au/sites/swa/about/publications/pages/notifiedfatalitiesmonthlyreport.

plant and equipment hazards as well as the handling and placement of hazardous materials, appropriate site trafficking and compulsory safety equipment and protective clothing not just for workers but also for site visitors.

Similarly, at handover, and after the defects liability period, a commissioning safety plan is required in which there is a nominated role to ensure safe handover and conformance with safety monitoring, certification and clearance of safety legislation.

Civil engineering activities must continue to strive towards better work practices in line with specific safety management framework(s) that provide an umbrella for hazard identification, a job safety analysis and safe work method statement, safety targets identified within key performance indicators for safety, a task-specific project risk register, adequate training and toolbox safety instruction beyond the provision of personal protective equipment, an occupational health and safety (OHS) committee for sites and activities with more than 20 workers; structured site inspections; full supply chain and subcontractor compliances; compensation arrangements and on-site medical/first aid obligations.

> To address risk and ensure safe working practices in specific work tasks on-site and throughout the supply chain, reference must be made to a wide number of acts and standards, namely, the National Standard for Construction Work and Occupational Health, as well as the Occupational Safety and Health Management Plan as part of the National Occupational Health and Safety Commission (NOHSC:1016 [2005])[16]; which might be suggested as being couched within Australia's overarching Occupational Safety and Health (OHS) performance strategy towards improvement as a result of reducing incidence, improving management of OHS, prevention of occupational injury, elimination of hazards at the design and specification stage and a strengthening of policing safety at a state and federal level.[17]

Australian National Standard Code of Practice guidance notes include[18]

- Asbestos Safe Removal of Asbestos, 2nd edition (NOHSC:2002 [2005])
- Code of Practice for the Management and Control of Asbestos in the Workplace (NOHSC:2018 [2005])
- Guidance Note on the Membrane Filter Method for Estimating Airborne Asbestos Fibres, 2nd edition (NOHSC:3003 [2005])
- Atmospheric Adopted National Exposure Standards for Guidance Note on the Interpretation of Exposure Contaminants Atmospheric Contaminants in the Occupational Environment (NOHSC:1003 [1995])
- Standards for Atmospheric Contaminants in the Occupational Environment, 3rd edition (NOHSC:3008 [1995])

- Guidance Note on the Interpretation of Exposure Standards for Atmospheric Contaminants in the Occupational Environment, 3rd edition (NOHSC:3008 [1995])
- Carcinogenic Substances National Model Regulation for the Control of Scheduled Carcinogenic Substances (NOHSC:1011 [1995])
- Code of Practice for the Control of Scheduled Carcinogenic Substances (NOHSC:2014 [1995])
- Competencies in Industry Guidelines for Integrating OHS into National Industry Training Packages
- Confined Spaces Safe Working in a Confined Space—Australian Standard AS2865-1995 (NOHSC:1009 [1994])
- Construction National Standard for Construction Work (NOHSC:1016 [2005]). Induction for Construction Work Precast, Tilt-up and Concrete Elements in Building Construction
- Noise National Standard for Occupational Noise (NOHSC:1007 [2000])
- National Code of Practice for Noise Management and Protection of Hearing at Work, 3rd edition (NOHSC:2009 [2004])
- National Standard for Plant (NOHSC:1010 [1994])

Educational guidelines towards occupational health and safety including

- Guidance Note for the Development of Tertiary Level Courses for Professional Education in Occupational Health and Safety (NOHSC:3020 [1994])
- National Guidelines for OHS Competency Standards for the Operation of Load Shifting Equipment and Other Types of Specified Equipment (NOHSC:7019 [1992])
- National Code of Practice for the Preparation of Material Safety Data Sheets 2nd edition (NOHSC:2011 [2003])

4.4.1 Industry initiatives in occupational health and safety in civil engineering

A number of ongoing initiatives seek to offer independent profession-based industrywide opportunities to address health and safety in civil engineering. Two such ventures are Standing Committee on Structural Safety (SCOSS) and Confidential Reporting on Structural Safety (CROSS).[19]

SCOSS is an independent body of some four decades' standing that seeks to review civil engineering matters affecting safety to identify in advance developments increasing risk to structural safety. CROSS is a somewhat younger scheme to improve structural safety by using confidential reports to highlight lessons learnt.[20] CROSS publishes a periodic newsletter highlighting case studies of note and has subscribers across Europe, Southeast

Asia, as well as Australasia. Reports and case study reviews seek to link organisations such as the Institution of Civil Engineers[21] and the British Constructional Steelwork Association towards providing updated widespread dissemination of safety initiatives.

Case studies made available by these initiatives review temporary works as well as false-work and formwork failure and seek to highlight the key issues that allow future projects to learn from the mistakes of the past.[22] In addition to these dissemination sources, the Office of the (Australian) Federal Safety Commissioner works with industry to develop case studies on various occupational health and safety issues that present 'anonymous' findings in lieu of the Privacy Act, which limits explicit specific coverage.

Given previous discussion generally regarding occupational health and safety legalisation compliance, and Section 4.3, which seeks to highlight a more specific recognition of case studies as a way to review lessons learnt from structural applications, it is perhaps applicable at this stage (and in Section 4.4.2) to discuss safety in more explicit terms as it relates to concrete and formwork erection on-site.

4.4.2 Safety: Concrete and formwork

This section discusses concrete installation and, in particular, investigates Australian formwork practice and the relevant documents that seek improved safety on-site.*

Accidents and incidents have occurred, at a more than acceptable rate, in formwork; although most formwork failures are found to concern differing inadequacies in practice, the occurrence of formwork failures may perhaps be prevented by targeting reform at a more fundamental starting point, namely, by examining the guidance and legislation governing formwork practice. The following discussion presents an analysis of documentation governing Australian formwork practice, alongside industry perceptions, and provides a review of current practice.

Accidents and work practice incidents can be costly, time-consuming and damaging for company reputations, and even more so they significantly affect the health, safety and welfare of site personnel and workers. To prevent incidents, safety in practice must be a primary concern for all civil engineers. The Engineers Australia Code of Ethics (EA, 2010 discussed in detail in Chapter 6) makes specific reference to the importance of health and safety in its charter, placing responsibility for this aspect above private interests, including profit. However, despite this emphasis, safety standards can lapse and incidents occur; formwork is one such area apt to lapses in procedure.

---

* Input from Paul Brandis,[23–40] received with thanks.

To ensure that a high standard of practice is pursued by all engineers, the current safety standards in Australian formwork practice must be addressed. Standards Australia (SA 2010a) defines formwork as the temporary works erected to mould and support cast *in situ* concrete. Even though, as this definition indicates, formwork is temporary, it should not be subject to fewer safety standards compared with any other permanent structure

A panel of senior representatives from industry as well as the professional and standards authorisations bodies were approached for their personal perceptions of current formwork safety legislation. The following discussion references both representative opinions and highlights case study lessons learnt from formwork (mal)practice.

Previously perceived as a means to an end, today's design preparation and installation of formwork is suitably robust. This may be a reason for principal contractors' now largely taking project management roles and subcontracting out formwork practice. Construction technology has also developed over time, and proprietary formwork systems are now frequently used in the Australian concrete construction industry. These systems are largely preengineered and prefabricated, leading to economies in the design and construction of formwork. However, studies have shown that assumptions made in the design of such systems are often inadequately communicated. Care must also be taken with regard to the continued reuse of such systems, which removes the opportunity to seek economies in design on the premise of its 'short life'. With new entities becoming involved in the construction process, whether it is formwork suppliers or subcontractors, issues with cooperation, coordination and communication have become apparent.

Project management is complicated by the fact that its key parameters—cost, schedule, safety and quality—are often competing objectives. In the context of formwork, its use is on the critical path and it can contribute 40% (and even up to 60%) of the total concrete construction costs. Formwork will therefore often determine project success in terms of both schedule and budget. With most businesses driven by their bottom line, it is important to remember project cost-saving measures, which may lead to inadequate formwork resulting in formwork failure that must be avoided.

Safety of formwork practice in Australia is governed by state-based occupational health and safety legislation, which varies from state to state. Common to all legislative frameworks is the condition to comply with the National Formwork Standard (AS3610) for structural integrity requirements. Codes of practice are sometimes provided as advice on acceptable ways of complying with the obligations under state legislation and should perhaps be adopted as a method of risk management. Under Western Australian legislation, AS3610 is referenced as a code of practice for formwork.

4.4.2.1 Formwork failure: Case study examples

There have been significant legislative developments in formwork practice over the years, although there are still worrying repetitions of poor practice. In 1966, formwork for a bridge in Welshpool collapsed due to poor standards of design, which were replicated some 40 years later when, in 2010, a bridge formwork collapse in Canberra was attributed to a similar level of design flaw(s).

Industry must continue to review and apply lessons learnt, firstly, to avoid putting any more lives at risk and, secondly, to address rework construction costs of millions of dollars. The lessons learnt from three recent Australian formwork failures cautions designers and contractors of the dangers in structural civil engineering activities include:

Robert Sergi bridge collapse
> In 2000, a formwork collapse occurred during the construction of a bridge in Corio, Victoria. This collapse demonstrated the full destructive potential of formwork failures with the bridge now posthumously named to commemorate the fatality of Robert Sergi. A report released by the State Coroner's Office found the collapse was largely attributed to false economies, with the principal contractor requesting a change in the formwork design to reuse material that was already in stock. Furthermore, it is believed that the principal contractor altered the headstock beams with little consultation, under pressure to get the job finished within existing timelines.
>
> The findings of this collapse draw attention to key considerations that when using subcontractors, the principal contractor cannot extricate itself from the works, and that education and training of engineers needs improvement.

Christchurch Grammar School slab collapse
> In 2009, a formwork collapse occurred during the construction of a suspended slab at Christchurch Grammar School in Claremont, Western Australia. No injuries were recorded. A report released by the Construction, Forestry, Mining and Energy Union of Western Australia detailed findings of preliminary investigations that the formwork failed due to insufficient bracing, which led to buckling of the formwork standards (supporting column elements). Compounding these design flaws, it seems that there was a lack of suitable management, where flaws went unnoticed.

Gungahlin Drive Extension collapse
> In 2010, a formwork collapse occurred during the construction of a bridge forming part of the Gungahlin Drive Extension works in Canberra, Australian Capital Territory. The collapse was catastrophic: 15 workers were injured and the project was delayed extensively.

A report released by Snowy Mountains Engineering Corporation (SMEC 2010), found that the collapse was attributed to the main girders (spanning 13 m) not being braced to prevent lateral movements induced by the 3% cross fall of the bridge deck.

A number of additional shortcomings in the formwork design and construction practice were also noted, with several differences between the 'as-constructed' details and the documented design. The collapse occurred after substantial completion of the concrete pour; potentially, other cases in the Australian construction industry exist where formwork could have been considerably close to failure and no one was even aware. The collapse was attributed to design flaw and once again a lack of suitable management in detection compounding issues.

Despite improvements in the industry over time, the above case studies demonstrate that formwork incidents are still occurring in modern construction. The Privacy Act of Australia limits construction failure information dissemination unless the incident is particularly dramatic or results in loss of life. The Office of the Federal Safety Commissioner, however, works with industry to develop case studies on various occupational health and safety issues, which present anonymous findings in lieu of the Privacy Act to allow lessons learnt to be made available to industry in general (this perhaps compliments the more informal dissemination opportunities provided by SCOSS and CROSS described in the Section 4.4.1).

Having looked at case studies, it is perhaps timeous to consider the Australia Standards code of practice for formwork.

The Australian Formwork Standard, AS3610,[41] was last fully revised in 1995. Two years after its release, the Standards Development Committee for Formwork (BD-043) started reviewing AS3610 in an attempt to keep up with developments and innovation in the industry. Since then, the Development Committee has produced two drafts for public comment, one interim amendment and partially republished the Standard with the introduction of AS3610.1-2010 covering 'Documentation and Surface Finish' in February 2010. Draft revisions of AS3610 focussed on enhancing design guidance; proposed amendments to the Standard in this respect thus far remain outstanding. AS itemises the various proposals linked with AS3610 as follows:

- AS 3610-1995, Formwork for Concrete; under revision: DR 02319 CP, DR 05029 CP, DR 08167; superseded by AS 3610.1-2010 (in part), supersedes AS 3610-1990, DR 93275; Amendments: AS 3610-1995/Amdt 1-2003; Supplements: AS 3610 Supplement 1-1995; AS 3610 Supplement 2-1996.

Guidance governing Australian formwork design practice, given its age, is perhaps due for review.

Although part two of the standard remains undeveloped, inconsistencies might be deemed to exist through the coexistence of AS3610.1-2010 and AS3610-1995. Releasing the standard in parts, where both standards review similar material, may risk the potential for misapplication in industry and potentially lead to inconsistency.

With the recent changes in the way Standards Australia (SA 2010b) have developed their respective standards, it becomes necessary for stakeholders to propose a 'net benefit case' demonstrating the positive effect that redevelopment will bring.

Industry professionals actively involved in Australian formwork practice have presented alternative views on the suitability of AS3610 in its current state. Some suggest ongoing change is needed or at least desirable, and that there is always room for improvement. Others perhaps suggest that all remains well and that the net benefit case excludes official rerevision, albeit that a somewhat independent initiative towards a 'Formwork Design Handbook', is ongoing. Generally speaking, although compliance with AS3610 is a primary expectation, it is not the only requirement. Engineers, as always, require due diligence.

In addition to its use as an Australian design standard, AS3610-1995 is also referenced under Western Australian Regulations as a code of practice. Given that Australia is governed by a 'State-based Regulatory Scheme', the legislative approach to governing formwork practice varies nationwide. New South Wales and Queensland have both developed independent codes of practice to govern formwork practice. Western Australian legislative framework, under the State Government, has not adopted the same approach. A comparison between AS3610 and the New South Wales and Queensland codes of practice indicate anomalies, perhaps in regulating safety for form workers:

- Given that AS3610 is a 'performance-based' standard, it tends to list the required outcomes rather than giving practical guidance on suitable processes to achieve those outcomes as the interstate codes do, perhaps given that Australian standards are not legal documents and must be applicable on a national level. Thus, the use of AS3610 to provide practical guidance on meeting the statutory obligations set by differing state legislation is worth further review.

Training may be considered as a means to complement and enhance compliances:

At the industry level, there is some degree of education and training available; however, it is perhaps somewhat rudimentary.

- Formwork engineers practising in large corporations are generally trained using in-house educational packages and on-the-job shadowing.

- Independent practitioners largely rely on their own initiatives in reviewing the available documentation.
- To support developments in guidance and legislation, formwork design education and training needs are perhaps required in a more structured format.

Having reviewed the dangers inherent in formwork provision to facilitate the installation of *in situ* concrete, Section 4.5 looks at the supply and installation of prefabricated components. Prefabrication and off-site work is discussed in Section 4.5 in terms of opportunities to enhance productivity.

4.5 PREFABRICATION AND MODULARISATION PRODUCTIVITY

This section reviews ongoing industry trends towards off-site building methods for engineering and construction projects and reviews productivity.*

In the past few decades, the construction industry in many countries has suffered from poor performance and low productivity. The labour-intensive nature of the industry and diminishing levels of specialist skills and craftsmanship are perhaps major factors hampering productivity growth in construction. A key suggestion for resolving the productivity constraints of traditional on-site (*in situ*) construction has long been design specifications that embrace off-site construction, which uses methods such as prefabrication and modularisation to try to help efficiency and standardise the management of quality. Off-site production is suggested as producing a positive way forward, although prefabrication versus *in situ* discussions are somewhat overly anecdotal and lack an empirical objective means to clearly define the parameters that lead to positive gains.

In most countries, construction is a key economic contributor; nevertheless, the industry has historically suffered from poor performance in comparison with other sectors. In Australia, the United Kingdom and the United States, skills shortages and gaps have been identified as a particular problem. This is significant given that labour skills are an important determinant of productivity in the industry. Indeed, in the past few decades, productivity growth in building has been lower than in other industries, with some developed countries, including Australia, facing declining productivity. It has long been suggested that on-site wet trade activities (*in situ* concrete and masonry tasks) are somewhat less predictable than prefabricated off-site approaches and that this affects productivity.

---

* Input from Faisal Alazzaz,[42–71] received with thanks.

Productivity generally is the amount of output per unit of input. Given that the construction sector is highly labour-intensive, a key determinant of sector productivity is labour productivity, which is measured as the output per unit of labour input. Despite the apparent simplicity of the definition of labour productivity, this indicator is difficult to track consistently, largely because of the complexity of quantifying and comparing diversified outputs in construction. For example, is output the formwork and false-work that contributes up to 60% of an *in situ* concrete component, or is it just the finished concrete component after the temporary works have been removed?

Broadly speaking, off-site construction is where subelements are built remotely then transported to and installed on site, as distinguished from *in situ* methods built directly on site. These off-site (prefabricated) construction approaches, in lieu of *in situ* methods, may be a way forward for addressing calls to improve productivity management issues in building. Off-site construction requires more explicit definition.

Off-site construction is a process of prefabricating elements and components for subsequent installation into a desired on-site location. Generally, there is an overall construction project requirement on-site, which can require the manufacture of component parts in a factory setting that subsequently requires the transportation, installation and ultimate integration of such elements at the building site.

Off-site construction can be divided into distinct parts:

- Basic component manufacture (doors-with-furniture/architraves, window units)
- Element preassembly (prefabricated panels, slab portions and the like)

As distinct from tilt-up panels, which are somewhat prepared on one part of a site and 'tilted up' in an adjacent part of that site

- Module component preassembly (toilet and WC [water-closet] drop-in modules and the like)
- Modular self-contained 'buildings' (usable buildings with minor installation on-site)

Off-site construction is gaining increasing attention in the fourth category; namely, modular building category has become synonymous with the term *modular construction*, which to an extent excludes the other three groups from the preassembly principle.

Off-site work, including processes such as prefabrication and modularisation (from a craft, into a standardised and controlled manufacturing process), has long been touted to lead to perhaps greater efficiency, reduced costs and increased quality. A key suggestion is that in off-site production methods, productivity gains can be achieved.

Indeed, off-site (fabrication of subelement components) production has been found to result in an up to 40% gain in efficiency per employee. Nevertheless, off-site production itself faces limitations. Similar to on-site construction, a significant cause of defects with prefabricated units in off-site construction is poor craftsmanship albeit that continuous improvement and employee motivation policies (touched on briefly in the discussions related to operational management) may, however, improve craftsmanship quality and in turn allow for a reduction in defects.

Prefabrication or off-site construction is certainly not a new phenomenon and the use of portable buildings has been reported since the nineteenth century. In the period following the Second World War, off-site construction or prefabricated buildings gained popularity largely due to a surging demand for public housing and economies of scale in standardised mass production. Driven by advances in technology and reflecting the needs of the society, off-site construction is now being increasingly used in a more diversified range of areas (hotels, schools, commercial/industrial buildings, retail hubs and petrol outlets) that use uniform repetitive designs and specifications.

In Western Australia, particularly in the mining and resources industry, the need to supply quick affordable facilities to service and provide infrastructure for mine production operations, and not least to provide accommodation units and 'villages' for miners and workers and their families (beyond a basic 'donga' provision) is somewhat driving local calls for 'modularised construction'. Largely, socioeconomic drivers seem to influence perceptions regarding the value of off-site construction.

Off-site construction does seem to be able to add value to construction and economies generally. In studies of the value of the off-site construction industry (in the United Kingdom), total gross output of the off-site construction sector has almost doubled in the last 10 years (to almost £6B); the total value added by the sector also doubled over the same period to almost £1.5B. Albeit that the overall share by off-site construction in the construction sector remains low, making up only 2.1% of the total value of (the United Kingdom's) construction sector.

The rise and fall of this sector seems to be significantly influenced by the status of the general economy. It is unknown whether the trends found in the United Kingdom can be applied to Australia, where construction, and in particular infrastructure and civil engineering ventures that service the mining and resources industry, seems to be able to buck international trends, not least to remain stable throughout the global financial crisis. Indeed, *in situ* brick and block still dominates domestic Australian market, with off-site prefabrication and hybrid 'tilt-up' construction only seen as emerging technologies. Generally, perhaps uptake of off-site construction is partly influenced by stakeholder perception towards the benefits of off-site construction.

4.5.1 Assessing benefits and disadvantages of prefabrication

Off-site construction is often suggested as being able to bring about several benefits to the construction process such as improving productivity performance and improving the quality of built facilities in a cost-effective manner; studies and reports from the United States, Australia and the United Kingdom indicate off-site construction gains; however, evaluations of off-site construction may be somewhat too grounded in anecdotal evidence rather than rigorous data. There is currently a lack of research that objectively measures the benefits of off-site construction. Instead, most research focuses on subjective studies of experiences with off-site construction. A number of factors assess prefabricated benefits:

Time
> The most significant benefit of off-site construction is argued to be time-savings. By transferring a significant proportion of the construction work to an off-site facility, time spent on-site is reduced. The more predictable conditions of the factory and the economies of scale generated may ensure that construction deadlines are met more effectively than in a traditional on-site environment. In studies, a significant majority of respondents regularly choose time/speed as the main reason for preferring off-site construction. Time benefits are regularly reported by almost half of any surveyed group of practitioners as the main reasons for using off-site construction. Reductions in on-site assembly time are perhaps one of the main benefits of off-site methods over traditional methods.

Quality improvement
> Another benefit, cited by many stakeholders, is quality improvement enabling tighter control than an on-site environment. A perception exists among clients that elements made off-site, in a factory, are more consistent and had gone through a greater degree of quality control and testing than elements made on-site. Less time spent on snagging (remedial works) was also mentioned as a benefit. Previous studies find that quality is the second most significant factor (after time) reported by clients in choosing off-site construction: indeed, a significant number of study participants (almost one-third of clients and designers responding to request for information) cite quality improvement as their first choice, whereas 15% of contractors do the same.

The potential for improvement over time is key and significant given industry's move towards continuous improvement globally (via Bechtel and other dominant players). There remain opportunities to apply operational management (total quality management) tools/techniques towards the measurement of current and subsequent improvement of future efficiency in the civil engineering industry.

Relieving skills shortages

A third major factor behind stakeholders favouring off-site construction addresses skill shortages. Off-site construction enables construction processes to be 'outsourced' to another environment, requiring less labour to be invested into traditional on-site processes, addressing skill shortages. Indeed, almost two-thirds of house-building firms (in the UK) cite this as a driving force. Compensating for craftsmen skill shortages, remains within the top six reasons for general contractors using off-site construction. Australia, like many other countries, currently recognises skill shortage by active recruitment and apprentice schemes; the extent to which this is more viable in the urban centres and in factory-based environments, rather in rural on-site locations, may be an added factor.

Cost

Of all perceived benefits of off-site construction, cost was the most controversial; some sources citing it as a major advantage but some listing it as a major disadvantage. Improved cost certainty of off-site construction may well be perceived to lessen the likelihood of cost blowouts; however, other studies find that key stakeholders did not rate cost as a benefit at all. One study finds that more than 90% of respondents believe higher initial off-site construction costs were an inhibitor to wider use and a significant limitation to off-site construction, with another study showing 50% of clients/designers and 40% of contractors and builders listing cost as a main barrier to off-site production uptake (inherently perhaps as a result of the huge knock-on transportation costs that can accrue). These results somewhat fly in the face of other studies that argue that almost half of all clients and designers questioned regularly cite reduced initial cost as an advantage of off-site construction; indeed, with 40% of respondents citing reduced whole life cost as an advantage to be gained from off-site prefabrication and modularisation.

Productivity

A broad view indicates that greater productivity can be viewed as the overriding benefit of off-site construction, with the reduced time, higher quality and lower cost of projects ultimately meaning that the process is more productive per unit of input than on-site construction.

Indeed, the future work proposed by this study has, as one of its aims, a structured means and method to identify and measure the variables of productivity objectively.

Disadvantages in off-site construction

There are significant perceived disadvantages to off-site construction among key stakeholders. A (UK) survey of house-building firms finds that only one-third of respondents were satisfied with the performance of off-site methods within their own organisations and only half of those who had tried prefabrication were satisfied with the performance of off-site construction generally (as compared with respondent satisfaction levels related to more traditional on-site *in situ* construction with a happiness level of more than two-thirds).

On balance, off-site construction is still perceived negatively and is seen as being less effective than on-site construction. The biggest disadvantages for stakeholders were

- The poor build quality of prefabricated components
- Incorrect designs
- Delayed delivery time
- The volatile nature of the supply chain
- An inability to make changes on-site
- Related long lead (lag) times
- 'Designers' favour customisation rather than standardisation
  - Albeit that perhaps the current Western Australian client need (for component supply/installation of uniform solutions), expressed by mining and resources industry clients, favours standardisation and does *not* seek architecturally significant solutions.

Regulatory guidance was also a major barrier preventing the full exploitation of off-site construction. There is a perceived lack of guidance codes and standards particularly related to off-shore supply chains. Generally, civil engineering and the construction industry are traditionally risk-averse and there is insufficient knowledge of objective benefits of off-site construction beyond perception.

There is somewhat of a lack of an Australian focus in much of the current available literature; insights developed overseas require contextualisation for Australia (and Western Australia). Although investment in off-site construction is currently nominal in Western Australia, resources and mining industry clients are in fact increasingly seeking prefabricated solutions.

Extending the theme that case studies help to better assess the relevancy of off-site construction, Section 4.6 explores design specification choices related to precast in lieu of *in situ* concrete.

4.6 PREFABRICATION AND DESIGN SPECIFICATION DECISIONS

The following discussion highlights design specification decision making in rural Australia and reviews precast and *in situ* concrete design alternatives.*

The process of design specification decision making must quickly go beyond an assessment of technical competency; alternatives (proven as structurally sound design solutions) must instead be assessed and chosen based on factors other than (an already) established fit-for-use functionality.

The following discussion reviews two such alternative design solutions, namely, precast or *in situ* concrete as building component options in civil infrastructure projects to service the mining and resources industry in the remote Pilbara region of Northwestern Australia. The work identifies and compares the merits and demerits of precasting civil concrete components that might traditionally be cast *in situ*, and examines factors such as increasing labour and accommodation constraints faced by clients and contractors in rural areas, which contributes to project durations and costs. Experienced stakeholders were consulted with regard to their perceptions of the decision-making processes involved in preparing civil engineering design solutions, with a view to summarising what might be felt as the risks, benefits and opportunities that arise from implementing either *in situ* or precast concrete methods.

Discussion is made of how the range of variables encountered on project sites influence the decision-making (value management) process from a design engineer's perspective; generally (somewhat unsurprisingly), factors of time and cost are regarded as influencing the decision-making process. These cost and time variables are explained next in the context of site works.

Fit-for-use materials
  Building and civil engineering turnover involving design services, contracting and materials manufacture represents approximately 10% of the gross national product for most nations. Stakeholders within the construction and engineering industry must, however, avoid becoming complacent, especially in light of the recent global financial crisis, which saw a stalling of many major/mega projects. More competitive times have meant that building design and contracting companies, on behalf of increasingly cash-conscious clients, have had to make changes in all aspects of their businesses, not least by becoming more cost-effective in their choice of fit-for-use materials and the resources required for their installation.

Rural location variables
  Notwithstanding the necessary environmental impact assessments (EIAs) and environmental protection agency compliances (discussed

---

* Input from Graeme Bikaun,[72-87] received with thanks.

in Chapter 5), Western Australia continues to benefit financially from an industrial boom that has seen a major increase in infrastructure projects in the north of the state (a state that is comparable in size to Western Europe, has a population of only 2.3 million and houses the world's most isolated mainland state capital of Perth). Since the discovery of resource deposits in this isolated region, much time and effort has been devoted to its extraction and subsequent supply to a wide array of buyers.

The construction of adequate facilities to fulfil the infrastructure requirements in the Pilbara and Kimberly regions of Western Australia places a lot of commercial and logistical pressure on the stakeholders involved. Barriers to efficient infrastructure installation include limited local labour resources, constraints on worker accommodation and fly in/fly out journeys into this isolated landscape, cost and time effects of daily travel to the site from new/neighbouring workers' encampments and the effects on productivity of these restraints.

The problems that clients and contractors face in projects influence total capital expenditure as well as the useable life cycle of the projects. Obligations and responsibilities remain high for, on the one hand, the provision of a fit-for-use infrastructure product and, on the other hand, payment for services rendered. Often, a lack of appreciation for the vast distances involved lead to distorted expectations amongst the stakeholders, resulting, in turn, in conflict and disagreement over appropriate service provision and remuneration.

By way of example, there is a huge labour cost difference between an urban construction worker compared with a similarly qualified tradesman working rurally when accommodation, meals and travel are factored in. These indirect costs are compounded somewhat with the increased need for rural environment occupational health and safety inductions/implementations, daily travel time to and from the site, and given the high turnover of personnel as a result of worker dissatisfaction with isolated environs, increased prestart meetings and lost production time during subsequent personnel handover.

In an effort to address these project production efficiency concerns and reduce capital costs, clients and contractors are becoming increasingly interested in an assessment of just how much civil work can be taken off-site (doing away with traditional on-site *in situ* installation), seeking prefabricated solutions in urban centres such as Perth and transporting readymade components intact for simplified site assembly. More work done in urban precast yards potentially addresses the resources industry's need for quick and inexpensive infrastructure, saving clients and contractors money and reducing the duration of projects. Issues raised include precast production, component installation and the need for comparison of alternatives.

Precast production

Precast production is the process of having concrete components made off-site in factory conditions. The use of precast techniques might be expected to affect time, overall costs and labour utilisation. Using the precast concrete method over other historical (*in situ*) techniques means that the product is constructed under controlled conditions, uncoupled from site processes and delays. Some comparison studies between *in situ* and precast concrete find that precast concrete removes operation dependencies involved in site-based pouring of *in situ* concrete, which has a major influence on the overall duration of the project; fast-tracking, concurrent (noncritical path) activities achieves efficiencies where precast manufacture in Location A, is concurrent with earthworks undertaken in Location B, with units transported and lifted into place in one operation when the schedule demands it.

Acknowledgment of design team specification decision making, however, contrasting precast with *in situ* methods, must address the full range of benefits/disbenefits to allow realistic comparisons.

Component installation

The more traditional site-based component installation method of *in situ* concrete is a method of placing concrete straight into its designated (formed/formworked) position on-site.

The typical advantages of using *in situ* concrete might be argued to include that it is amenable to almost any shape, connections are homogenous with the rest of the structure, the entire production activity is controllable on-site, alterations can be made at the last minute and the design can proceed as the structure is built.

Disadvantages of using *in situ* concrete may also be suggested as follows: subsequent alteration is difficult if change of use needed, errors (and corrections) to reset formwork/false work (once concrete is cast) are very difficult and expensive, all activities involve plenty of resources and plant on-site, and reinforcement and formwork/false work tend to be labour intensive.

Differences in *in situ* and precast methods can be summarised in terms of quality, cost, time and safety:

- Precast quality can be achieved more easily due to controlled conditions, better maintained equipment and highly skilled labour; albeit a need exists to balance quality between time and cost in product development.
- *In situ* methods reduce risk of off-site manufacturing and delivery issues transferring to site.
- Precast methods don't experience delays due to site-based constraints such as weather.

- Precast methods purportedly save cost via reduced labour utilisation.
- Safety aspects may be easier to maintain in a controlled factory floor environment.

It is somewhat difficult to place a weighting on the advantages/disadvantages between *in situ* and precast concrete production as each method needs to be considered on its terms in specific circumstances. Although precast concrete elements, including footings, foundations and retaining walls, have the potential to offer advantages over *in situ* concrete work, it is recognised that *in situ* concrete is currently the traditional/typical, common and familiar method adopted on project sites in the Pilbara/Kimberley regions of Western Australia, where familiarity remains a major antecedent to efficiency. Indeed, it might be stated that precast components, rather than being cheaper, have anecdotally always proved more expensive than *in situ* concrete work due to the high transport costs.

There remains further need for empirical comparison in terms of supply and the management of supply.

Supply factors

Concrete production in rural (Western Australian) regions is typically cast *in situ* and generally either supplied from the closest township or by a mobile concrete batch plant arrangement. These decisions are usually made based on factors such as distance from the local township, cost of local supply, amount of concrete required, weather conditions and accessibility of delivery.

Whatever overall advanced specification design decision is made, it is the construction manager/site project manager (and given the lack of personnel, seldom a suitably qualified resident engineer) of the civil infrastructure project who becomes charged with overseeing the deliverance of the whole project and bares the sole responsibility of all of the work under contract by defining the project plan, creating the project team, keeping track of all progress, managing the delivery process and directing resources and evaluating any changes that may occur during the duration of the project including weather, mistakes and variations.

Managing supply

Management of the (installation) process becomes important. Often, the site project manager is somewhat of a generalist who integrates contributions from a variety of specialist teams to achieve primary objectives whilst considering resources' availability; specialist knowledge of the supply and installation of *in situ* concrete may be less than detailed.

In the precast industry, however, quality management is a priority; high-quality success rates in the manufacturing industry stem from a good (technical) management structure. Controlling and communicating the technical and practical aspects to skilled workers by management cannot be overemphasised in any field of work and seems to lend itself to prefabricated factory processes.

One important role for the site manager is that of resource estimating and planning project human resource management; both are essential components of a comprehensive project management plan to execute and monitor a project successfully. All contractors, large and small, must devise management strategies and put together plans that enable their resources to be efficient and effective to work with the project plan. At the micromanagement level, increased productivity means decreased unit cost and indicates project performance. At the macro level, productivity serves as a vital tool in countering inflationary effects and determining wage policies.

Given the discussion of the factors deemed to affect the choice of either precast or *in situ* concrete specification above, a sample of experienced practitioners were approached for their expert opinion.

The respondent panel had experience of

- Working on a range of projects in isolated locations of the vast state
- Dealing with and extensive familiarity with both precast and *in situ* concrete construction techniques
- At least 10 years professional experience in their respective fields

Case study analyses of suggested component alternative specifications sought to compare choices in terms of cost and time.

Comparison data was subsequently requested from Company A, in line with an objective to assess alternative concrete pedestal components' specification cost variables for a project located in rural Western Australia approximately 1700 km north of the closest main city of Perth.

Two alternative specifications were considered, namely,

- *In situ* concrete pouring on site
- Mixed method installation of *in situ* concrete bases and precast concrete pedestals and bases

The mixed method installation option (as mentioned above for *in situ* concrete bases and precast concrete pedestals and bases) requires

- Traditionally, high man-hour items
  - Bolts, ligatures, pedestal reinforcement and pedestal formwork
- Its precast component manufactured 1700 km away in the city of Perth which is transported to site, after which the remaining concrete

connecting the pedestals is poured *in situ* once the precast units have been installed

Figure 4.2 gives a generic illustrative comparable footing plan, whereas Figure 4.3 gives a generic illustrative section view of a comparable footing using a mixed method precasting approach.

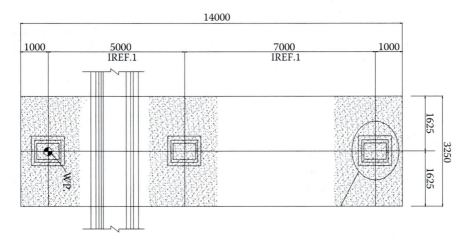

Figure 4.2 Generic illustrative comparable footing plan.

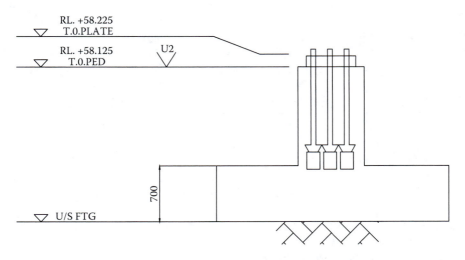

Figure 4.3 Generic illustrative section view of comparable footing (mixed method precasting).

In terms of cost

- The 'mixed' (precast and *in situ*) option was cheaper than the '*in situ* alone' option by 12%.
  - Unlikely on its own, to recommend its choice.
- Transportation costs for the prefabricated sections came to a 14% share of the total cost.

The significant factor in this comparison is deemed as the amount of on-site hours that the mixed option reduces.

The cast *in situ* option results in 437 h to construct the footing (which is equivalent to having seven employees working 10 h a day for just over 6 days).

On the other hand, the alternative mixed (prefabricated/precast) option means that a labour team needs spend only 90 h on-site (which equates to four employees on-site working 10 h a day for 2.25 days).

Table 4.2 presents a summary of the specification (value management) comparison.

A considerable time-saving in overall or total hours spent to complete the footing was found: (i) *in situ* option man-hours of 437 h compared with the (ii) precast/mixed method requiring just (90 + 255) 345 h, is a significant saving when extrapolated beyond the installation of one footing.

The time saved by using the mixed method outlines the much higher productivities that can be gained by opting to use the precast specification, especially when considering small, awkward, high-volume repetitious items. An example of this can be seen in the form of supplying concrete on-site, which is rated at 2.5 man-hours/m³ compared with the supply of concrete off-site, which is 1.2 man-hours/m³; implying a 52% increase in productivity factor.

A number of variables influence specification choices. Reflections, beyond the (value management) case study specification comparison above,

Table 4.2 Tabulated summary of method comparison

	Cast in situ	Precast pedestal/in situ base
Total cost ($)	83,726	74,409
On-site man-hours	437	90
Off-site man-hours	0	255
Site breakdown	7 men for 6 days	4 men for 2 1/3 days
On-site man-hours per cubic metre	11.8	2.4
Total man-hours per cubic metre	11.8	9.3
Cost per cubic metre ($)	2263	2077
Productivity rate for placing concrete in pedestals (man-hour/m³)	2.5	1.2

highlighted a number of key factors to assist design/choice decisions to best realise infrastructure needs. Variables to be factored into specification choices include

- Accommodation (in rural areas) is integral to construction method(s) used; worker encampment issues influence precast choice.
- The cost of labour (installation) is higher in remote (Pilbara) areas than in urban areas.
- Unavailability of skilled workers rurally adds cost.
- The cost of materials (supply) hikes cost, influencing anecdotally preference for off-site work.
- Use of precast elements reduces loading of critical path(s).
- Precast standard/safety compliance is deemed easier to achieve than *in situ* operations.
- Components most suited to 'precast and transport' are repetitious, small (less than 24 T) and installation-awkward items.
- Off-site productivity factor can improve on on-site works by up to 52%.
- The relationship between time and cost is not linear in rural locations: the deployment of more resources does not necessarily lead to improved production times.
- In both 'cost plus' and 'lump-sum' contractual situations, benefit mainly go to clients; there is limited incentive for contractors' innovative methodologies.
- Conventional *in situ* concrete approaches may skew estimates unless prospective risk (of potential labour unavailability on-site) is not factored in.
- At an estimating stage, unconventional methodologies similarly require risk review.

Cost, time, quality, safety and transport are considered vital checklist items when deciding on a construction method.

Quality and safety standards are more predictable in the precast/prefabrication yard. The cost of precast compared with *in situ* concrete has become a lot more competitive in recent years. Transport issues are arguably the biggest reason against using precast concrete in regions such as rural Pilbara/Kimberley.

The difference in off-site labour compared with on-site labour costs is quite apparent.

- Labour in (Western Australia) rural regions is as much as three times the cost of labour used in urban (prefabrication yard) areas, if all living expenses are taken into account.

Clients in the mining construction industry in rural Western Australia are more interested in being *undertime* than *undercost* because the faster the infrastructure is in use, the earlier the company can start profiting from the excavation and export of its product.

Time is directly related to cost and cannot be looked at separately when deciding on construction methods because it is the ultimate cost of the project that needs to be considered, which includes the potential cost of delays and project overruns (liquidated damages). The risks involved with each method need to be evaluated; however, it might be argued that because of the inconsistent productivities found with on-site concrete pouring, these risks tend to sit higher with *in situ* concrete projects.

One of the key motivators for using precast over *in situ* is the time-savings that can be gained. If a precast section can be brought onto the site and put into place in less time than *in situ* placement, outside a project's critical path, then this is a preferred methodology. In summary, it can be seen that precast methods have many advantages over *in situ* concrete techniques because it eliminates much of the risk associated with on-site work. Putting safety and quality aside as fixed variables of specification and quality management system compliance, then cost and time are key factors.

Because time is the (mining and resources sector) client's number one priority, a method that reduces project duration is desired because this allows an installation to be active and productive quicker. Implementing precast options (as discussed above) offers the required time-savings when compared with *in situ* construction.

Having emphasised time and cost factors and having somewhat regarded quality (thus far) as a compliance requirement, it is perhaps pertinent to now return to a discussion of the variables that constitute compliant quality. The discussion in Section 4.7 focuses on the occurrence of defects during the construction process and describes the extent to which rework may be minimised throughout the initial construction/installation stage.

4.7 PREDICTING DEFECTS IN CIVIL ENGINEERING ACTIVITIES

This section reviews the extent to which early defects prediction in on-site activities can play a part in achieving efficiency gains in construction projects.*

A significant factor contributing to cost/schedule blowout is construction defects. In line with defects is perhaps the construction industry's overlapping tasks within on-site work programmes as part of an integrated critical path, work breakdown–structured approach. The discussion reviews

* Input from Abdullah Almusharraf,[88–110] received with thanks.

the mechanisms for the identification and mitigation of installation defects resulting from concurrent work packages during the construction phase. A main challenge here is an accurate means and method to predict, at the early preconstruction planning stage, cumulative deficiencies and better manage overlapping tasks and, in turn, avoid negatively influencing project cost and duration.

The identification of actual defects during the construction phase might go towards a prediction of future defect susceptibility for specific activities. In addition, the identification of key overlapping task(s) variables and examination of the behaviour of each variable and its effect on the overall project quality in general are important for civil engineering projects. The following discussion highlights how defects appear principally as a result of design stage factors ($\approx 33\%$) and engineering installation stage factors ($\approx 66\%$); summarising the work discussed, it can be said that

On-site observations and analyses determine the stage at which defect occur, namely,

- The 'design stage' contributes to one-third of civil engineering defect occurrences.
- Whereas, on-site activities during the 'construction phase' contribute to a significant two-thirds of defect occurrences in civil engineering works.

In addition to a discussion of the stage at which defects appear, the following discussion also explains how both 'project process/methods', and 'people on-site' are predominately responsible for all on-site defects; both contribute to approximately 40% (respectively) of the material installation defects on a civil engineering work-site, whereas the remaining 20% or so of the defects noted stem from ambiguities such as design/documents.

Generally, it might be stated that an investigation of error and defect in the civil engineering and building design process can begin to make construction more efficient. On-site, there are opportunities to, on the one hand, examine multiple factors related to overlapping tasks during the construction phase and then, on the other hand, try to predict (resultant) defect occurrence and related efficiency levels.

A number of previous studies have sought to address and explore defect at the early stages in construction processes; the ideas can be summarised along the following lines:

Overlapping activities
   There is a suggestion that one way to seek a reduction of sequential defects is by examining overlapping processes (termed by some in the manufacturing sector as *concurrent engineering*, although this

technique is somewhat uncommon in civil engineering). Preceding task and succeeding task activity pair overlap is determined through the nature and range of information exchanges, with links termed as dependent, semidependent, independent and interdependent.

Activity pair sensitivity

The concept of overlapping activities evolves by presenting a number of frameworks that determine overlapping parameters in preceding (so-called upstream) and succeeding (so-called downstream) task activities. Evolution is a function of overlapping activities as an evaluator for information over upstream activity time, generating initial information. Sensitivity of information downstream is then 'rework' needed as a result of a change in upstream activities.

Concurrent and overlapping task activities strategies

Generally, overlapping activities are well acknowledged as a project tool and as a means to reduce project duration and cost; however, it has been argued that overlapping, although useful, can produce errors, defects and rework, increasing project risks. To overcome this, construction defects need identification and distribution into execution phase activities and tasks to measure susceptibility for specific defect exposure and give appropriate overlapping between activities.

Defects description

Terminology such as error, fault and failure are commonly used to describe technical problems and construction defects. Not all, however, necessarily lead to rectification or rework; this variation in defect description affects the understanding of the problem and may lead to inappropriate reactions. Design phase defects occur due to quality deficiencies/omission/change. Although construction defects at the execution phase often occur due to workmanship/supervision, where rework reinstallation to allow compliance and certificate of practical completion sign off.

Defect effect

Construction mistakes, errors or defects usually reduce project satisfaction levels and often create conflict and dispute between stakeholders; the search for monetary and nonmonetary compensation becomes common to address deviations from originally forecast costs, schedules and quality that can, in some cases, increase a project sum by up to 5%.

Defect classification

Quality problems might be assessed in terms of their respective deviation from compliance with design and construction drawings and specifications, as a result of variables such as supply chain noncompliance and worker installation noncompliance.

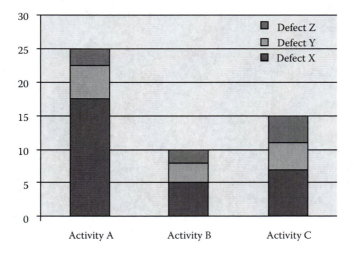

Figure 4.4 Hypothetical link: Activity susceptibility to defect exposure.

Case studies can help understand and identify which activity or task has low or high sensitivity in relation to defects and rework towards distributing effort, improving quality and simultaneously controlling cost and time. Activities and tasks are categorized based on susceptibility to exposure to defects.

Figure 4.4 gives a hypothetical classification. In this example, if the rate of exposure to defects of an activity 'A' is 25%, and defect is classified as exposure to a defect 'X' is 70%, defect 'Y' is 20% and defect 'Z' is 10%.

The degree of overlapping between activities A, B and C is identified based on susceptibility to defect exposure and rates of defects for each activity. Activity A has the highest susceptibility to defect exposure.

The following work develops this approach (links susceptibility and defect) towards an early indication of when progress reaches a critical point in overlapping tasks (a critical point representing a probability that defects will occur increasingly). It is suggested that perhaps understanding this behaviour begins to mitigate defect effect.

Civil engineering defect distribution

Building on work identifying the traditional design-bid-build stages—concept, preliminary design, detailed design, construction and start-up—defects may be distributed appropriately, where potential sources become applicable, namely, structure, people, process, internal uncertainty and external uncertainty.

After which the main steps to distribute the factors of construction defect are to

- Review causes of construction defect from secondary research and distribute construction defects into project phases and sources for each defect
- Classify susceptibility for specific activities and related exposure to defects, and measurement of a rate of each activity defect based on exposure times
- Identify all variables of overlapping activities (from secondary research) that will affect work positively or negatively, then develop a formula that measures overlap through least squares regression and multiple regression analysis
- Prepare an overarching estimation to predict and send an early alarm signal to a site manager and the respective foreman of the (increasing) probability of defects occurrence(s) in that range of specific on-site tasks
- Validate the defects prediction approach by further detailed case study

Groups of variables in two fields were assessed: (a) construction defects in the execution phase and (b) overlapping activities processes. For the first field variables, parameters and appropriate classification techniques for construction defects were based on existing previous work.

From the residential/construction industry, 342 construction defects cases and factors were assembled (causes and factors at design, construction, start-up and the like were generated from sources related to variables such as people and processes). This database was organised in terms of distributing construction defects across the project phases.

- Three hundred and forty-two defects were distributed in construction project phases (concept, preliminary/detailed design, build and start-up).
- Construction defects (for specific rankings) were divided into different sources causing the defects (structure, people, process and internal/external uncertainty).
- The rate of construction factors for each project phase and the source that caused the defects.

Generally, critical path methods allow a task B to start with A; resultantly, construction defects may increase, thus ongoing work will develop formal measurement (and validation) of the range of work progression(s) for both A and B, to predict an amount of cumulative construction defect(s) through a least squares regression and multiple regression analysis to measure a critical point of maximum overlap.

Table 4.3 Defects appearance (%) and subphase

Subphase	Defects appearance (%)
Project concepts	1
Design	30
Preliminary design	7
Detailed design	93
Construction	67
Late design	10
Site management	70
Production management	2
Materials and equipment	14
Fabrication	3
Transportation	1
Start-up	2

Table 4.3 presents the rate of the defect appearance in each project (sub) phase.

- The greatest rate of defect appearance occurs at the construction phase (67%), followed by the design phase.
- Interestingly perhaps, although previous studies stated that the 'design phase' causes most of the construction defects, this ongoing study is beginning to suggest that most of the defects occur in the construction phase.

Table 4.4 describes 'processes' and 'people' as being responsible for 43% and 38% of defects, respectively.

Overcoming construction defects and work undertaken to remedy defects during overlapping activities in the execution phase is a key problem for construction and civil engineering.

As mentioned previously, the 'construction phase' seems to be the stage at which most defects are attributed (contributing to 67% of defects) followed by the 'design phase' (to which 30% of the defects can be attributed). 'Process' causes the most defects (attributed as 43%) followed by 'people' (38%).

Mechanisms able to help mitigate defects during the construction phase will always be welcomed by civil engineers, site managers and foremen, especially if they allow a relatively easy way to identify (the specific factors of) a specific task or series of work tasks that are most at risk from cost and time blowouts.

In general terms then, despite quality management systems (as part of the project documentation required by standard forms of general conditions

Table 4.4 Percentages of defects and sources (contributions)

Subphases	Structure (%)	People (%)	Process (%)	Uncertain internal (%)	Uncertain external (%)
Project concepts	—	100	—	—	—
Design	5	37	40	19	—
Preliminary	—	72	29	0	—
Detailed	5	34	41	20	—
Construction	3	36	44	7	8
Late design	—	45	41	9	5
Site manager	2	38	43	9	8
Production	—	100	—	—	—
Material	10	21	62	—	6
Fabricate	—	43	29	—	29
Transport	—	25	50	—	25
Start-up	—	28	57	14	—
Total (%)	4	38	43	11	5

of contract, such as clause 29 of AS4000) to improve project processes, defects do still occur throughout both design and construction.

Having recognised the place of project and contract documentation in attempting to address quality, quality compliance, occupational health and safety, continuous improvements through modularisation and prefabrication, as well as defect occurrences (see previous discussions), it is now perhaps pertinent to examine the contract documentation explicitly. Chapter 5 discusses contract documentation in detail.

Chapter 5

Contract documentation for civil engineering projects

Quality is embedded within civil engineering projects by explicit contractual compliance requirements, set out and described in the complete design drawings and specifications that detail all elemental material supply, installation and compliance tests. Cost and time values are similarly stated (and broken down) in unambiguous explicit terms in the project's contractual documentation. Parties have a responsibility and an obligation to conform to all stated quality, cost and time parameters as set out in four key contract documents that bring civil engineering projects to fruition.

The four principal documents needed to realise a civil engineering contract are

i. *General conditions of contract* (GCC) charting responsibilities and obligations and incorporating contract period and a timeline/programme schedule.
   A. Timelines are defined in specific GCC clauses (such as AS4000 clause 32 and AS2124 clause 33).
ii. *Specification* documents dictating material makeup and installation processes.
   A. Specifications are directed by specific standard contract form clauses (such as AS4000 clause 8).
iii. *Costs/rate schedules* within bills of quantities (BQ) cost breakdown(s)
   A. Cost breakdown and prices are set out in a BQ that forms part of the standard forms of contract (as AS4000 clause 2).
iv. *Drawings*, ranging from general arrangement and location, to substructure and superstructure drawings, detailed sections and finishes and services drawings.
   A. Drawing supply is also guided by specific clauses in standardised GCC (such as AS4000 clause 8).

The four typical sets of documents for project realisation (contract, specification, BQs and drawings) are discussed with reference to their respective components.

This chapter presents discussions of the four key documentation requirements in the following subsections:

- Contractual and legal practices and procedures   (Section 5.1)
  - Contract constituents
  - Procurement paths and contract types
  - Project progression
  - Scope variations and payment claims
  - Clause interpretation examples
  - Standard forms selection criteria
  - Nonstandard bespoke forms of contract
  - Dispute resolution
  - Environmental requirements
- Specifications   (Section 5.2)
  - Specification systems suppliers
  - Exemplars and types
  - Specification decisions practice examples
  - Rough terrain infrastructure requirements
  - Environmental factors in specification
  - Sustainable options for demolition materials
  - Waste reuse and recycling
  - Building information modelling opportunities
- Rates schedules and bills-of-quantities   (Section 5.3)
  - Measurement of work under the contract
  - Structured itemisation of project elements
  - BQ preparation
- Design drawings   (Section 5.4)
  - Design criteria
  - Selected guidance for design (drawings)
  - Drawings examples cross-reference

5.1 CONTRACTUAL ARRANGEMENTS

Clients traditionally engage a design consultant to prepare a civil engineering design solution that is able to meet their specific needs; this detailed design is then made available to prospective builders. The client's representative design team then review constructors' tender-bid submissions towards recommendation and choice of the most suitable firm to build the client's required facility. The parties subsequently sign a contract that clarifies all obligations and responsibilities for project realisation.[1,2] This traditional sum/value agreement between parties to carry out the work is industry's most common contractual category type, in which a detailed design is known, bid for and awarded.

Table 5.1 Arrangement

Civil engineering contractual arrangement examples	
Category type[1]	Form of contract
Traditional lump-sum	General standard conditions of contract (AS4000 or AS2124)
Cost reimbursement	GCC (AS4000/2124)
Design and build	Design and construct conditions of contract (AS4300 or AS4902)
Management	Management of trade conditions of contract (AS4917)

As well as the traditional lump-sum contractual arrangement, a number of other options exist; these include

Cost-reimbursement, where the full extent of the works is as-yet unknown
Design and build, where one specialist firm can offer a complete package
Construction management method, where the whole of the works are let in a series of separate specialist contracts by a management firm (Table 5.1)

National standard forms of contract (and their related standard GCC) compliment category types (Australian Standards examples; Table 5.1).

Category types of civil engineering contracts by Australian Standards, but with categories somewhat applicable generally throughout the commonwealth of 54 intergovernmental independent member nations is represented by the following listing showing conditions of contract between parties for construction work in Australian:

- AS2124 (1992) GCC
- AS4000 (1997) GCC
- AS2545 (1993) Subcontract conditions
- AS4300 (1995) GCC for design and construct
- AS4902 (2000) GCC for design and construct
- AS4917 (2003) Construction management trade contract—General conditions
- AS4905 (2002) Minor works contract conditions
- AS4919 (2003) General conditions for provision of asset maintenance and services
- AS4122 (2010) General condition of contract for consultants
- AS2987 (1987) General conditions for supply of equipment with/(out) installation
  - AS2125 Applicable general conditions of tendering and form of tender

- AS2127 Form of formal instrument of agreement
- AS4120 Tendering compliance regulations and codes

Clients make themselves aware of potential builders (or designer/builders) initially (prior to signing contracts to build) by soliciting expressions of interest.

Clients (and client representatives) may seek expressions of interest through a competitive process open to all and widely publicised, or by a competitive process between a small number of suitable contractors selected for their expert knowledge, or by direct negotiation with one specialist leader in the field. The main goal of this process of (open or selective) competition or direct negotiation (or indeed other procurement path options[3]) is to identify parties willing to be contractually bound to enact the works.

5.1.1 Elements of a contract

Signing a contract to build a civil engineering project replicates contract law generally, where a contract is

> a legally binding agreement between two or more parties; with the intention to address the (building) scope of a project via an offer by a party with the capacity to carry out the work and an acceptance by a client who needs the facility (built); alongside, a reasonable period of consideration of the obligations involved.

This is formalised somewhat by (national) standard forms of tender-offer and is typified by Australian Standards (AS) standard forms such as AS2125 (1992) Tender form (OFFER).[4] The terms of a contract are usually contained within three or four dozen clauses (within standard forms of contract) that use explicit vocabulary to describe roles. Definitions are given where the ordinary meaning of the words need greater clarification. The builder is obliged to build the facility (defined within the scope) and the client takes on a responsibility for payment of construction within the scope.

Contracts, once signed, continue until practical completion and rectification of any outstanding identifiable defects in build quality. If a job does go astray, the parties can, in very extreme situations, bring the contract to an end (discharge the contract) before the facility has been built. This usually involves one or the other party seeking compensatory damages for poor performance or parties not fulfilling obligations.

Standard forms of contract are widely used in civil engineering and construction. Utilisation of available standard forms is encouraged by various learned bodies and the professional institutions charged with their preparation. Standard forms can be argued to reduce conflict (because all obligations and responsibilities are clearly set out), balance risk (between the client who needs the facility built and the builder who seeks payment to construct the works) and generally smooth working relationships (through well-established and long-familiar clauses that define role and duty).

Off-the-shelf standard forms of contract reduce the cost of preparing new, project-specific contract conditions for every job. The use of standard forms of contract, although voluntary, is widespread. The following shows the typical constituents of the general conditions of a construction contract.

GENERAL CONDITIONS OF CONSTRUCTION CONTRACT:
TYPICAL CONSTITUENTS[22]

- Definitions, documents that confirm scope
- Assignment of work, intellectual property, protection of site-works, insurances, bonds
- Roles, responsibilities, representatives
- Site conditions and set-up, cleaning-up
- Quality system requirements, testing
- Programming sequences, working hours
- Suspension when site becomes unsafe
- Time and progress, delay expectations
- Defects liability and rectification
- Variations to the original scope of works
- Payment for acceptable performance
- Termination if default occurs
- Claims for work beyond the original scope
- Dispute resolution procedures
- Appendices that clarify parties, roles and contract parameters

From the 1920s to the 1950s, institutions of engineering helped formulate standard forms and conditions of contract. Various national joint contracts tribunal–type organisations began to take over ownership of this process (typified historically by joint contracts tribunal in the UK[5]). In Australia, the main standards authority, Australian Standards (AS), has been increasingly empowered to prepare, regulate, review and update standard forms of contract.

A standard form of contract, such as the clause coverage typified by AS4000, in which numbered clauses and their respective intentions are (highly) summarised as follows:

1. Interpretation: Definitions of standard terms (terms written later)
2. Nature: Obligations, contractor does the work, client (principal) pays

   3. Provisional sums: Amount inserted if not possible to determine cost (M&E)
   4. Portions: Separate 'phased' completions if needed
   5. Security: Money paid by contractor held by a third party as a performance bond
   6. Evidence of contract: Documents constitute the contract
   7. Service of notices: Correct address implies receipt
   8. Contract documents: Responsibility for correct info, dimensions not scales
   9. Assignment: Subcontractors may be used within reason
  10. Intellectual property: Patents and rights respected
  11. Legislative requirements: Contractor to adhere to all other laws
  12. Protection of people and property: Contractor does what is necessary or payment is reduced
  13. Urgent protection: Contractor does what is necessary or payment is reduced
  14. Care of work/reinstatement of damage: Contractor does what is necessary, except nonresponsibility for war, invasion, radiation and the like
  15. Damage to persons and property other than work-under-the-contract (WUC): Contractor does what is necessary
16–19. Insurance: Contractor to insure for loss or damage
20, 21. Superintendent: Fulfils all aspects of the role, in good faith
22, 23. Contractor's representative, contractor or representative supervise WUC competently
  24. Site: Date of site handover, 'artefacts' go to client
  25. Latent conditions: Contractors expected to inspect site, irregularities to be 'expected'
  26. Setting out the works: Client to indicate survey, contract to adhere
  27. Cleaning up: Contractor to remove surplus or payment reduced
  28. Material (labour, plant): Contractor uses enough resources for WUC until practical completion
  29. Quality: Contractor to establish and adhere to a conforming quality system
  30. Exam and test: Before defects liability, superintendent tests work at client's expense
  31. Working hours: Working hours and days specified
  32. Programming: Sequence by contractor, except if superintendent specifies order and time
  33. Suspension: Unsafe work stops on-site operations with payment reduced where appropriate
  34. Time and progress: Notify all parties if delay expected
      Extension of time—Requirements and procedures

Assessment—Superintendent assesses contractor's effort to prevent/ mitigate loss

Practical completion (PC)—Main contractor (MC) requests 'Certificate of PC' given (or not) in 14 days

Liquidated damages—Payment by contractor if PC not reached

Bonus for practical completion—An agreed amount

35. Defects liability: Contractor rectifies defects as soon as possible after PC
36. Variations: Contractor receives variation order, certified and paid for by client
37. Payment:

Progress payments—Interim staged payments made to the contractor

Certificates—Contractor makes progress claims, superintendent processes for payment certificate

Time—Client pays 7 days after superintendent's certificate, or 21 days after progress claim

Approval—Payment does not signify that work is acceptable

Unfixed plant and materials—On-site materials payment agreed beforehand

Final claim—Contractor claims all money 28 days after defects liability

38. Payment of workers and subcontractors: Contractor provides evidence of payment
39. Default or insolvency: How parties may terminate the contract

Contractors default: Examples of breach by contractor

Taking out—Work deemed unsuitable, resources retained/used by client

Adjustment—Superintendent assesses costs to complete work taken out

Principals default—Payments not made specified by contractor

Insolvency—Bankruptcy terminates contract

40. Termination by frustration: Events outside of the parties' control terminates obligations
41. Notification of claims: Superintendent assesses claim and decides validity in 28 days
42. Dispute resolution:

Notice—Differences or dispute acknowledged for attention

Conference—Meeting to resolve 14 days after receipt of notice

Arbitration—Meeting with third party if unresolved seek default (default is usually via the UN Commission on International Trade Law)

43. Waiver of conditions—Waivers and variations to be agreed in writing

5.1.2 Contractual category type selection: Practice example

Choosing an appropriate contractual category type requires both an under-standing of the options available as well as an ability to align a category option with the circumstances and requirements of a particular project. Civil engineers charged to advise clients on the choice of best option of contract category type should take five key steps:

Step 1: Identify and summarise, via appropriate in-house past-experience and available texts, all contractual category type options[6]

Step 2: Review the characteristics and requirements of the project at hand

Step 3: Align project requirements with a suitable contractual category type

Texts[7] that may assist in an alignment of 'contractual and procurement arrangements' with 'project characteristic specif-ics' include Deciding on the Appropriate Contract by the Joint Contracts Tribunal (JCT) , UK, and Relationship Contracting and Optimising Project Outcomes by the Australian Constructors Association (ACA). Noteworthy is that, although somewhat pre-disposed perhaps towards partnering-arrangements in lieu of more traditional approaches, the ACA's 'Project Delivery Method—Suitability Matrix' does suggests a workable 10-point scale based on a weighting of project variables such as early delivery, green-field versus brownfield site, technology and complexity, risk, guar-anteed maximum price expectations, environmental aspects and the like.

Step 4: Suggest means to select/appoint an appropriate party to carry out the job

Step 5: Identify an appropriate standard form of GCC

A scenario might involve the need for a civil engineer to advise clients on an appropriate contractual arrangement for the following 'range' of pend-ing projects:

a. A large site has been acquired on the city outskirts but the client is unsure if finance can be secured for *all* of the planned development parts such as leisure complex, then/plus retail units, then/plus accom-modation apartments.

b. Various city hotels/serviced-apartments are owned by a client, who dislikes the current haphazard, disjointed nature of repairs, mainte-nance and renovation.

c. A head office building has had to close due to fire damage, the extent of which is unknown; normal working conditions are required as soon as possible.
d. A client seeks advice regarding projects with different design complexities.

Step 1
A summary of the potential range of contractual category types might include a review of options including: traditional full-design BQ lump-sum, bill of approximate quantities contracts, cost reimbursement contracts, target cost, continuity contracts, serial, term contracts, two-stage tenders, design and build contracts, drawings and specifications, management contracting, partnership/alliance contracting and build operate (own) transfer BOT/BOOT procurement approaches. These procurement options are summarised in Table 5.2.

Steps 2 through 5
Table 5.3 summarises project characteristics and alignment of

- Project particulars (step 2 column)
- An appropriate contractual category type (step 3 column)
- Method of appointment (step 4 column)
- An appropriate standard form of (general conditions of) contract (step 5 column)

Having identified suitable contractual arrangements, the client (and the client's representatives and consultants) must then compare the available constructors and justify selection of one preferred builder.

To review tenders (builders' offers to construct/design-and-construct a project), a client needs an objective means of comparison. Selection criteria items are often given in advance to firms bidding for a job. The organisation most likely to win will be the firm that (in the eyes of the client) best addresses selection criteria items such as requirement conformity, value for money, quality assurance, technical resources availability, good track record of previous performance and an innovative approach.

A number of guidance texts, from public sector government (and international) departments and learned bodies and associations, assist in procurement and builder selection criteria matrices, these include

- AS4120 (1994) Code of Tendering from Australian Standards[8]
- Australian Procurement and Construction Council (APCC)3[9]
- Guidelines for Tendering by the Australian Constructors Association[10]
- Public Private Partnership Guide by the Australian Constructors Association[11]

Table 5.2 Categories

Contract category	Summary of particulars
Traditional full-design BQ lump-sum	Suitable for projects if all planning has been completed at the tender (invitation to price) stage; does not provide for builder participation at the design stage
Bill of approximate quantities contracts	Common in civil engineering works; detailed design work incomplete at tender stage; enables early start; prices/rates used for re-measured/unforeseen work
Cost reimbursement contracts	Scope of work unknown at early stages; contractor reimbursed (given-back) actual cost of materials and labour plus a fee for overheads and profit
Target cost	As cost reimbursement but with incentive targets for work to minimise costs
Continuity contracts	Uses contractor already on-site to save time/money on potential next job (phase); original BQ/rate schedule (materials/work prices) used in new work
Serial	As continuity but future contract approximate size known; uses original BQ
Term contracts	Project cost unknown; competition via schedule of rates; successful firm works over agreed extended period; used for large-scale maintenance/repair jobs
Two-stage tenders	Contractor involvement in planning and programming work tasks; successful firm agrees schedule of rates to value the work when final design is complete
Design and build contracts	Uses contractor's skill/expertise in specialised fields for all design work then all building work; efficiency gains from building own design
Drawings and specifications	Simplistic/standard projects where BQ preparation unjustified. Contract terms for cost/payment/time/variation need clearly stated to avoid future conflict
Management contracting	Main contractor uses managerial expertise in design/planning/ administration of contract to facilitate all work carried out by specialist subcontractors
Partnership/alliance	Risk-sharing with builder reimbursed directly for all direct costs in all circumstances; motivated by profit-sharing resulting from good performance
Build operate (own) transfer BOT/ BOOT	Design, build, operate and maintain then transfer to client after a set recovery period; see also engineer, procure, install, commission (EPIC)

- Achieving Excellence in Construction (UK)[12]
- Procurement and Contract Strategies (UK)[13]

A client's acceptance of a builder's tender-offer is but the beginning of a journey towards project completion; the following discuss project progression relative to conditions of contract[1,14–16] that structure requirements and milestones to be met.

Table 5.3 Step selection

Step 2	Step 3	Step 4	Step 5
Project particulars	Contractual arrangement	Appointment approach	Standard contract form
a. Development of potential/ subsequent parts/ phases	Continuity contractual arrangement	Selective competitive tendering then further negotiation based on existing phase-one BQ's	GCC AS2124/4000
b. Structured maintenance requirement	Term contractual arrangement	Selected competitive 'schedule of rates' comparison between known firms	Asset maintenance and services AS4919
c. Fire damage needing early work and timeous reopening	Cost-reimbursement contract	Negotiated appointment with one specialist/expert firm	GCC AS2124/4000
d. Projects of differing design complexities and variable degrees of technical innovation	• Complex, unfamiliar, innovative, high-specification designs and projects might seek to 'dictate' quality via a traditional approach (or perhaps in an alliance), where a specialist (local or overseas) design engineer first prepares all designs in advance and then invites qualified (local) builders to bid for (or align together) to construct exactly what has been given to them (power stations and the like) with obligations made explicit in GCC (traditionally using AS2124/4000) • Noncomplex, commonplace and standardised functional designs and projects might benefit from a design-and-build arrangement, in which experienced (local) firms, familiar with standard requirements, are invited to submit their own design(s) to construct (roads, schools, etc.), formalised by GCC (AS4300/4902)		

5.1.3 Project progression via GCC: International/national standard forms

Standard forms of contract are published by learned bodies and related professional industry organisations and are commonly used with respect to civil works. Standard contracts provide continuity and familiarity, and seek an equitable balance of risk between the client and their builder/design builder.

Nonstandard, stand-alone, ad hoc, bespoke contracts used in mining/ resources industries are somewhat less common in civil engineering in that

• *Bespoke contracts* (new, from-scratch, unique, project-specific contracts) with one-of project-by-project clause developments are somewhat expensive to generate, albeit they offer a client an opportunity to establish their own form of contract in a structure that addresses that company's own risk profile (unilaterally, it might be argued) for execution of contract work, alongside their own (unilateral) project organisational/management structure adaptations.

Globalisation in civil engineering workload alongside cross-national sources of funding increasingly requires more internationally applicable GCC. Indeed, many countries find that their construction industry makes use of not only local standards of GCC's (such AS2124 or AS4000 in Australia) but also of international (neutral jurisdiction) standard forms of contract.

The New Engineering Contract, principally the NEC3 Engineering and Construction Contract (ECC), is one such standard form drafted specifically for an international application;[17] indeed, NEC is being increasingly used throughout the Australasian region, particularly perhaps in New Zealand as an alternative to NZS3910.

Seven NEC contract forms exist, all of neutral jurisdiction, to complement the various procurement options for civil engineering projects. Formats include: priced and target contracts, cost reimbursement and management contracts, short (minor/simple projects) contracts, term service contract and framework contracts.

Other widespread, globally recognised contracts include the suite of standard forms offered by the International Federation of Consulting Engineers (more commonly known as FIDIC—Fédération Internationale des Ingénieurs-Conseils),[18] used between employers and contractors on international construction projects.

FIDIC publishes a number of forms of contract for different procurement category types, including

- CONS: Conditions of contract for construction, facilitating progression of works designed by an employer's engineer, then built by a main contractor (traditional)
- P&DB: Conditions of contract for plant design build, for mechanical and electrical (M&E) asset installation via one contractor's design then construction
- EPCT: In which all engineering, procurement, construction and subsequent 'turnkey' operational status at handover, is by one design-and-build contractor
- Short form of GCC used in small capital value projects

This FIDIC suite of standard forms of contract is endorsed by, amongst others, the World Bank,[19] which provides a source of financial and technical assistance to developing countries and newly developing regions around the world.[20]

Progression of works in all World Bank projects is by (stipulation of the use of and) application of FIDIC standard forms of contract, which clarify the responsibilities and obligations of employers and contractors on a wide range of international construction and engineering projects.

Standard forms of contract, both international and national, share a common goal; namely, to facilitate the smooth administration of civil engineering work on-site. Standard forms of contract (and the clauses therein) clarify expectations related to

- On-site civil engineering project commencement
- Appropriate mechanisms for the posting of notices related to claims and disputes
- Progress and more particularly progress payments
- The means to address any suggested variations to the scope of the works
- Appropriate certification of the completion of all work on-site

Australian Standards (AS) standard form of GCC, AS2124 (1992) GCC is somewhat of a precursor for the 43 clauses described within the newer AS4000, AS standard form of GCC. It might be suggested that the older AS2124 is perhaps more prescriptive and, therefore, some might suggest, somewhat more explicit in its description of obligations and responsibilities compared with the newer AS4000, which provides more opportunities for stakeholders to review and agree upon project-specific timelines for compliance and the like. Both AS2124 and AS4000 have their supporters and detractors.

Whether AS4000 or AS2124 is selected, both forms share a number of key stages to bring a civil engineering project to fruition including commencement, project notices, progress payments, variations and completion certificates. These project stages are tabulated below alongside clauses that direct the contract parties.

Commencement of the civil engineering project and on-site start-up expectations are clarified in standard forms of (general conditions of) contract clauses.

The following table uses a GCC standard form of contract (AS2124[21,22]) to show linkages between commencement activity, party responsibilities and applicable GCC clauses (in AS2124).

Commencement	
Activity	*Requirement and AS2124 clause*
Access/possession	Principal gives Builder authorisation to take possession as a notice in writing (clause 27.1)
Commence work notice	Builder gives client representative 7 days notice of a start date; within 14 days of gaining access (clause 35.1)
Security/bond	Builder gives third party a security bond 28 days after tender acceptance (clause 5.4)
Insurance	Contractor seeks insurance before starting the work (clause 21.1)

Notices applicable to the civil engineering project and appropriate mechanisms for posting such notices for contract claims, disputes and the like are tabulated next.

Notices

Notice	AS2124 clause
Disputes	Clause 47.1
Default	Clause 44
Extra work claims	Clause 21.4
Time extension claims	Clause 35.5
Suspension of work notices	Clause 33 and 34

Progress, *timeline* and *progress payments* for civil engineering projects are referenced in contractual clauses that state the programme of the work to be done and staged payment procedures such as those illustrated in the following.

Progress payment

Activity	Requirement and AS2124 clause
Progress	Programme of work (clause 33)
Payment claim	Builder gives client representative claim as a percentage of completion of works (clause 42)
Payment certificate	Client representative authorises payment in 14 days or remeasures (clause 42)

Variations applicable to the scope changes and the like of intended civil engineering works are open to consideration by the parties based on contract clauses such as those shown next.

Variations

Activity	Requirement and AS2124 clause
Variations	Clauses may require a Builder to vary work and correct defects (clauses 30 and 40); pricing variation items are made explicit (clauses 40, 40.5 and 41)

Completion (both practical and final completion) of a civil engineering project requires appropriate certification, clarified in contractual clauses illustrated next alongside instructions and timing of the release of retention money withheld by the client.

Completion certification	
Activity	*Requirement and AS2124 clause*
Practical completion	Builder requests practical completion certification from client representative (clause 42.5)
Final completion	Certification occurs approximately 28 days after the defects-liability period (clause 42.7 and 42.8)

Although structural integrity in design and build quality is the raison d'être of the civil engineer, cost/payment, time and design change concerns remain ever on the table.

5.1.4 Project progression: Variations, quality, extensions of time and claims

Contract clauses acknowledge ever-present anxieties over the dollars and days to achieve quality/fit-for-purpose construction. Variations, extensions of time and claims often dominate civil engineering project administration.

Clients have a right to instruct *variations*, and contractors are obliged to carry out work requests for suitable remuneration.

The client (principal) gives the project administrator (superintendent) the power to order variations, which change the WUC and alter or modify the design, quality or quantity. Variations may also apply to site access, working space work-hours or work-order. Variation orders in turn lead to an adjustment in the contract sum and the completion date (see Variations [AS4000 clause 36]).

VARIATIONS (AS4000[22] CLAUSE 36)

- Builder varies WUC upon receipt of written direction, *if* variation is capable under provisions of contract.
- Contractor receives variation request and notifies *Superintendent* of cost/effect on programme/date for practical completion.
- Costs valued by Superintendent following prior agreement, reasonable rates (priced BQ with profit/overhead mark-up).
- Superintendent alone cannot remove work from work-under-contract.

Quality workmanship (couched within national common law) is a builder obligation. Indeed, fit-for-purpose working is implicit and a builder's skill and judgement is expected. Similarly, fit-for-purpose specified materials must be used (as national supply-of-goods laws). If discrepancy or divergence from contract documents occurs, then the contractor tells the superintendent who instructs action to deal with differences.

QUALITY (AS4000 CLAUSE 29)

- Implicit that materials are new/as-specified and installed appropriately.
- Builder to use conforming quality system (ISO 9000/1) towards contract compliance.
- Work/materials (via superintendent review as nonconforming, need removal/replacement/demolition/or reconstruction.
- Contractor to comply in 8 days or done by others at cost to the Main Contractor.
- Superintendent may *accept* defective work (clause 29.4) but must issue a variation.

Defects and nonconforming materials or installations identified (by the superintendent) during construction require removal from site. Indeed, the contractor remains liable for defective work and latent (hidden) defective work until the end of the defects/liability period if there is a link between the contractor's breach of duty and a loss (injury or damage) to the client.

Until practical completion certification, contractors are obliged to make good on a (superintendant's prepared) schedule of defects within a reasonable time.

DEFECTIVE WORK/DAMAGE (AS4000 CLAUSE 14)

- Contractors are responsible for defects from commencement date until the date of practical completion (PC).

Extensions of time requests often follow variation instructions and, if accepted (by the superintendant), formally adjust the completion date.

EXTENSION OF TIME (AS4000 CLAUSE 34)

- Contractor entitled to extension of time (beyond clause 32) if delayed in reaching PC by due cause (Appendix 23 of AS4000).

Claims for additional costs or extensions of time to complete a project arise when a Contractor perceives that they're entitled to payment for loss or expense incurred due to the actions of others. Contractor's claims arise due to

- Perceived delays in the project timeline and a need for an extensions of time
- Changes to the scope of the project and the work-under-the-contract

- Disruptions caused by the client or the client's representatives
- Variations made to the contract by the client or the client's representatives

Claims arise from a need for the main contractor to perform work that was not anticipated when the contract was originally agreed. Payments resulting from claims made for a contractor's loss and expense are described as damages and a client is often willing to calculate and pay damages due without fear of further litigation.

Types of claims might include

- Contractual claims, in which a contractor experiences loss or increased expense resulting from the client's representatives error
- Common-law claims resulting from breaches of implied terms in the contract
- Quantum merit claims, in which no price has been set originally and a client/employer and a contractor agree reasonable and fair payment for (additional, varied) works
- Ex-gratia claims, in which a long-term benefit might result between the parties

It is worth noting that vague mentions of 'delay' and 'loss and expense' by parties in site-meetings or similar venues are unlikely to constitute grounds for a claim. Written applications are required to list

CLAIMS NOTIFICATION (**AS4000** CLAUSE **41**)

- Written notice to the parties setting out the general basis and amount of claim
- Superintendent's assessment of the contractor's claim and notification of a decision (within a reasonable time frame)
- Opportunity for contractor, if unhappy, to then seek a further dispute resolution

loss and expense, the determination of liability (and ultimately, superintendant acceptance or rejection of liability) and determination of a quantum or a price based on a superintendant's acceptance or proportional acceptance of a contractor's value calculation.

5.1.5 Contractual administration practice examples

Civil engineers who administer projects and seek to determine solutions to interparty disagreements[23] need look no further than a one-step reflection upon the GCC clauses.

Table 5.4 Contract administration

Issue	Reflection upon GCC	AS2124
a. Labour dispute	Suspension of works by contractor to be only by superintendant's express written permission	Clauses 34 and 34.2
	Liability for cost of any such suspension due to labour dispute attributed to the contractor	Clause 34.4
	Date for practical completion not to be affected; although unlikely, an extension of time *may* apply *if* the superintendent gives permission under clause 34.3	Clause 34.5 Clause 35.5
b. Damages due	Contractor required to pay the client a daily liquidated damages entitlement for extending beyond completion date	Clause 35.6
c. Inclement weather	Unseasonably bad weather, assessed via statistical norms supplied by the Bureau of Meteorology, will result in additional time and cost payment to the contractor	Clause 36

The key reference material related to contractual problems guidance might be found in '150 Contractual Problems and their Solutions' by Roger Knowles.

A scenario might involve a main contractor, already engaged on a large road project for many months, experiencing a range of difficulties including

a. Labour disputes causing disruption; a query arises if the contractor can suspend work and, if so, who is liable for cost, and does action affect date of completion?
b. If contractor fails to reach practical completion, are damages applicable?
c. Does the acknowledgement of inclement weather[24] result in time/cost payments?

Following consultation with a standard form of contract, such as AS2124, decisions for the way forward are drawn from GCC clauses, as summarised in Table 5.4.

It is in the best interests of all parties if disagreement regarding administration of the work can be resolved through an agreed interpretation of progression with reference to the contract conditions. If agreement is not reached, dispute can fester and escalate. Dispute resolution is often a necessary requirement to bring projects to fruition.

5.1.6 Contractual dispute and resolution

Disputes can arise in civil engineering projects where differences of opinion regarding contract progression between the parties go beyond disagreement to become altercation and argument about a perceived unfair advantage gain at the expense of others. Disputes stem largely from differing interpretations of a projects' contractual frameworks, documentation and on-site events. Managing problems within the form of a contract can avoid escalation into so-called summary relief sought through the law courts.

Simple conflict escalates to serious dispute stepwise: (i) unexpected event, (ii) suffering of additional cost/delay, (iii) compensation request, (iv) denial of damages due, (v) rejection of decision, (vi) escalation of dispute towards third-party (arbitrator) involvement.

Dispute resolution/arbitration (GCC AS4000)	
• Dispute notice to 'conference' which, if ineffective after 28 days, leads to arbitration	Clause 42.1–4
• Arbitration procedures and parties described/nominated	Annex item 32

Most standard forms of contract direct the parties towards an informal means of dispute resolution. An initial 'conference' (as AS4000 clause 42[22]) is suggested where parties first seek agreement through mediation, which although not binding provides an opportunity to take stock and confer towards resolution for a period of 28 days or so. Beyond unresolved mediation, parties then seek adjudication through the formal adversarial dispute resolution process of 'arbitration', in which an expert arbitrator is instructed upon the technical and administrative issues of the conflict at hand.

Arbitrators are usually nominated by the professional body and preselected by default in form of contract (AS4000 annexure 32) as the president or member of the Institution of Arbitrators and Mediators Australia.

Arbitration and its adjudication mechanism seek dispute resolution through the principle of 'natural justice', where all hear, see and reflect without bias. Often, overarching Construction Contract Acts[25] dictate standard adjudication procedures, even if standard forms of contract do not expressly incorporate such Acts.

In Western Australia (WA), for example, the Construction Contracts Act 2004 seeks to provide a rapid adjudication process to resolve disputes that arise over payments when individuals and organisations enter into contracts for building and construction work. The Act allows for the appointment of an adjudicator and establishes time limits on the adjudication process. The Act does not cover salary or wage disputes; similarly and interestingly perhaps, the Act does not cover (perhaps due to intensive high-level lobbying?) mining and resources-extraction activities.

An arbitrator's decision is deemed binding for the parties in dispute. It is hoped that having received a verdict by an independent arbitrator, the parties would then be able to live with the resolution and move towards project completion. Challenge to the decision is only to be based on irregularity such as failure to comply with natural justice or uncertainty as to the effect of the award.

Litigation (legal court proceedings) is the last resort for resolving disputes when all other attempts at conference-mediation and arbitration-adjudication have failed. It is expensive, time-consuming and very public and is instigated by the issuing of a writ (a court command) and relies on legal professionals to represent the parties and prepare the materials to be heard by a judge and the court; proceedings (if available) go to specialised technological courts to review the points and determine if a higher court is required. Litigation is adversarial, requiring the party raising the claim to prove that they have been disadvantaged, rather than expect the other party to disprove this claim.

Extending discussion of the law and its impact on tendering and project administration, Section 5.1.6.1 examines in more detail the range of criteria and the processes that influence civil engineering project procurement routes and the selection of an appropriate standard form of contract to realise a design proposal.

5.1.6.1 Standard forms of contract selection guideline

The qualitative selection processes for the existing range of standard forms of contract available to the Australian civil engineering and construction sector is discussed.*

The following discussion considers opportunities for a structured decision-making criteria guide towards the selection of the most appropriate standard form of contract, in a given environment, for the Australian (WA) civil engineering/construction industry; this section describes decisions required to choose an effective standard form of contract (from a wide array of options) and the factors that influence these decisions.

Structured guidance available for a *local context* is somewhat sparse, paving the way perhaps for the review below, in conjunction with a range of expert stakeholders in the Australian engineering construction industry, for a more (West) Australia-centric guidance tool to help improve the speed and accuracy in the choice of *which* standard form of contract is most suitable from the wide range of (locally) available options. Such structured guidance may perhaps also serve as a contract training aid for civil engineers (at their formative stages) who find themselves involved in procurement activities in the construction industry.

This section then, explores contract alternatives in the WA civil engineering/construction industry, with a key focus on the decisions leading to the choice of which standard form of contract is most suitable for a given Australian development project, where it might be remembered that the

---

* Input from Erin Macpherson,[26-37] received with thanks.

purpose of a contract is essentially to ensure that what is agreed will in fact be carried out by the relative parties.

> In essence, a civil engineering contract acts as a mechanism to mitigate the risks involved in bringing a design solution to fruition, by outlining the requirements, obligations and responsibilities of all of the respective stakeholders.

However, the very existence of a contract cannot guarantee the omission of risk. An effective contract is one that outlines the responsibilities of the parties in a clear and precise manner. Although this remains the constant goal for industry, in reality, this is very difficult to achieve due to respective stakeholders' varying interpretations of clauses.

Industry strives for continued improvement, that is, the minimisation of interpretations and the protection of the interests of the respective parties. A large number of alternative Australian standard forms of contract have been developed to address this concern; however, this in turn adds to the complexity of the contract choices and the decision process due to the sheer number of standard forms of contract options available.

Indeed, as mentioned previously (in the discussion concerning critical chain project management in Section 3.7), West Australian professionals identify a number of contractual concerns, particularly significant liquidated damages associated with projects conducted in their unique home state.

To make a decision as to which standard form of contract to use, experienced contract engineers often follow a somewhat innate, and largely subjective, decision process. First, a decision must be made as to which procurement strategy is likely to prove most suitable. Second, an appropriate tender process to engage contractors and consultants needs to be selected. A decision also needs to be made as to which type of contract will meet the appropriate needs of the chosen procurement strategy and finally which standard form of contract will best serve the needs of the project.

Figure 5.1 outlines decision relationships, in which column four represents the choice/range of potential standard forms of contract available for (West) Australian construction and civil engineering activities.

This section looks to formalise the somewhat innate, subjective processes related to local procurement routes by reviewing opportunities for a structured, objective decision-making guide towards the choice of the 'best' alternative form of contract in a given environment. The discussion draws on a variety of sources to provide an overview of the factors to be considered to make an appropriate decision.

There are a markedly large number of standard forms of contracts available to contract decision makers; this range of options is confusing to the

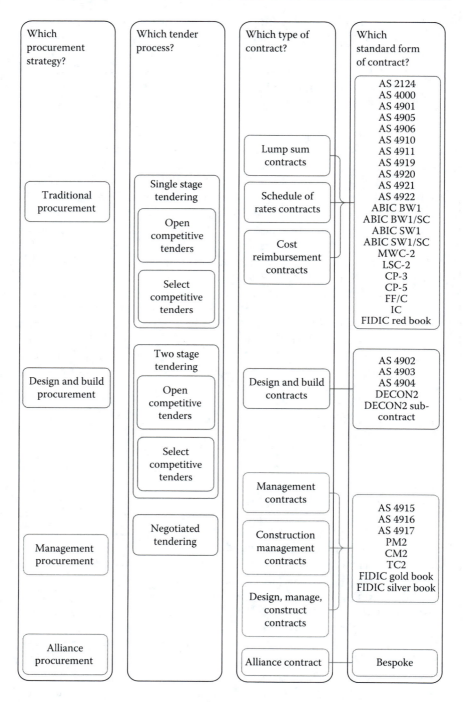

Figure 5.1 Contract decision relationships for (Western) Australia.

uninitiated and is somewhat aggravated by a lack of independent guidance on which standard form is most suitable.

The large number of choices available and the number of decisions that are needed to be made makes the decision process a difficult task requiring considerable time and therefore monetary investment to select an effective standard form of contract. Without a clear contract selection guide, it is easy to neglect the full range of factors given the number of considerations that must be taken into account. Also, as a result of the complexity of this decision, the task is generally reserved for experienced civil engineers who must draw on their years of individual experience to ensure a comprehensive consideration (which remains a personalised and subjective consideration) of the relevant factors.

This problem can be seen to offer three opportunities for improvement:

- First, to improve the efficiency of the contract decision-making process
- Second, to improve the effectiveness and objectivity of the contract decision-making process
- Third, to pass on contract knowledge in such a way as to allow less experienced engineers to take on more responsibility in making contract decisions

These opportunities would be best met by a structured approach to contract form selection. A need therefore exists within the WA civil engineering and construction industry for a structured approach to making contract decisions.

Civil engineering consultants are often asked to advise unknowledgeable clients on which standard form of contract to use for construction projects. It is important to choose an appropriate contract that balances the project parameters and allocates risk appropriately to avoid poor quality of construction, time and cost overruns, and reduce the likelihood of claims and litigation.

A compromise is also often required between conflicting project parameters such as cost, time and quality. For example, high product quality is likely to be achievable at a higher cost or longer timeline. Similarly, to achieve shorter project programmes than thought to be ideal, product quality may be reduced or increased costs may be experienced. The challenge of successful consultants is to find a balance between these priorities through suitable contracts.

The contract decision is one which is very complex, requiring careful consideration of a number of different variables to reach a final decision. It is widely accepted that the success of a contract, and therefore a project, is dependent on the appropriateness of the contract, being objectively assessed against the characteristics of the project and the project priorities, on an individual basis. With no structured approach to contract decision making, consultants run the risk of investing vast amounts of time into making correct contract decisions or, alternatively, they may rush the decision or neglect to consider important variables resulting in a less than optimal contract being implemented.

A number of studies (largely UK based) have sought to develop a structured decision procedure for particular aspects of the contract decision-making process. However, it might be suggested that these do not seek to attempt to offer a complete solution to the contract decision; concentrating on either procurement strategy or the type of contract. Extending such work, the following discussion describes the search to identify the variables deemed most important to the decision-making process and the extent to which 53 variables were deemed to influence the contract decision. Of these, just over a dozen variables (as presented in the following) were deemed as the key influencers of the choice of form of contract. A measure of the significance (or weighting) of these 'project priority' variables against the 'project characteristic' variables for various situations (looking at how the project priorities in WA influence the contract decision for different types of projects) is discussed. The discussion is based on a pilot study. Pilot study data collection was undertaken in the form of semistructured interviews with procurement and contract professionals within the civil engineering industry in (Western) Australia, where all participants were very senior personnel with extensive knowledge of contract and procurement practices.

The panel raised the issues of client–contractor relationships and the difficulties in changing the industry ethos. The relationship between the client and the contractor affects every factor taken into consideration in the decision process and is, therefore, fundamental to the success of the project. Although a contract defines the relationship between the parties in terms of liabilities and responsibilities, it cannot define the attitudes of the parties on which these will be based. Therefore, to achieve positive project outcomes in terms of low costs, short schedules and high quality, it is important that both clients and contractors conduct themselves in a manner that encourages respect and cooperation between the parties. The benefits of such are expected to extend beyond the client in terms of improved value for money for the project, and to the consultants and contractors as a result of an increased workload (and therefore profits) brought about by a positive industry reputation and repeat work with the client.

The tendency within the (West) Australian construction industry to oversimplify construction administration without due consideration to the increased legal environment in which they operate was also raised among participants. The increased emphasis on reducing the project schedule (time constraints) was cited as the most likely reason for this industry culture. Similarly, time constraints and lack of familiarity with all alternative options was indicated by participants as the most likely reason for implementing a contract out of familiarity (because it had been used on a similar project) without due consideration of all the relevant factors. On this basis, the use of a 'selection guide', such as the one subsequently developed and discussed in the following, is likely to eliminate the trade-off between speed and suitability by facilitating an accelerated contract decision (and therefore project start) whilst ensuring a suitable contract is selected.

A total of 28 factors were identified (through consultation with the panel) as influential in the contract decision process. Of these 28 factors, a number of factors influence more than one decision, as summarised in Table 5.5.

It can be seen that a total of 18 factors affect the choice of procurement strategy, 16 factors influence the choice of the type of contract, 11 factors influence the decision of which tender process will be used and 12 factors affect the decision of which standard form of contract is most suitable.

These factors, to some extent, cannot be assigned a specific weighting given their interdependence on one another and the priorities of the individual project. The factors identified as influential to the standard form of contract decision, outlined in Table 5.1, were used to develop a series of related decision flowcharts (represented by the illustrative Figure 5.2) to aid in the selection of a suitable standard form of contract for Australian civil engineering construction projects.

The flowchart (Figure 5.2) directs the user to complementary pages of a final research report[38] (go to page 92, and the like), where a final decision suggestion arises from a yes/no series of project status questions within the model; a number of definitions are also made available within the final report to assist the user in identifying/objectifying respective yes/no onward responses. This model flowchart specifically applies to the factors influencing the choice of procurement strategy, as the first step in the four-stage process (illustrated in Figure 5.1, related to which procurement strategy, which tender process, which type of contract and which standard form of contract).

The collective decision flowchart(s) developed in conjunction with representative industry personnel might be of assistance in a number of ways:

- Save time by reducing the number of hours required by contract engineers to make a standard form of contract decision.
- Reduce the requirement for legal consultation, therefore, reducing the time required for negotiations between clients and contractors or consultants.
- Enable contract knowledge to be passed on to new contract staff more efficiently and comprehensively.
- Improve corporate memory by reducing the risk of contract and procurement knowledge being lost when experienced employees leave the company.

Table 5.5 Key decision factors

No.	Factor	Procurement strategy	Type of contract	Tender process	Form of contract	Total
1	Schedule/timing	•	•	•		3
2	Minimum cost	•	•	•		3
3	Cost certainty	•	•			2
4	Complexity	•	•	•	•	4
5	Nature	•	•		•	3
6	Expected cost/value	•	•	•	•	4
7	Scope	•	•			2
8	Size	•	•			2
9	In-house capability	•	•			2
10	Quality	•	•	•		3
11	Allocation of risk/responsibilities	•	•		•	3
12	Market conditions		•	•		2
13	Flexibility	•	•			2
14	Level of competition	•	•	•		3
15	Time certainty	•				1
16	Control over subcontractors	•	•		•	3
17	How well-defined scope is	•	•		•	3
18	Evidence of transparency			•		1
19	Evidence of competition			•		1
20	Number of vendors able to supply goods/services	•		•		2
21	Dispute avoidance	•				1
22	Project administrator				•	1
23	Location				•	1
24	Funding				•	1
25	Procurement strategy leads to difficulty in evaluating tenders			•		1
26	Domestic/commercial				•	1
27	Separable portions required				•	1
28	Client profile				•	1
	Total	18	16	11	12	

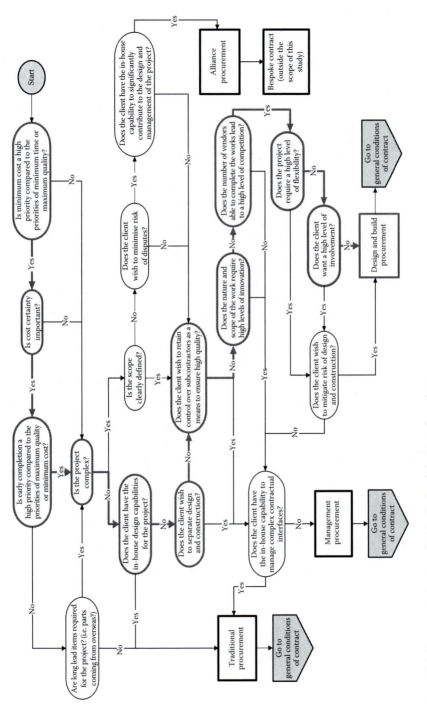

Figure 5.2 Representative sample of a finalised decision flow chart.

- Allow less experienced contract engineers to take a more active role in the contract selection process, in turn reducing the involvement of more experienced engineers and therefore allowing these experienced engineers to be available to perform more advanced tasks
- Ensure all important variables are considered for every project leading to the most suitable contract selection
- Implementation of effective contracts leads to improved client–contractor relationships and reduced litigation
- Address the cost of legal consultation in developing a bespoke contract
- As a result of improved corporate knowledge, reduce costs by maximising the efficiency of experienced, more costly, engineers
- Reduce litigation and litigation fees as a result of suitable contract selection

The discussion above focussed on the decisions that need to be made to select an effective standard form of contract and the factors that influence these decisions. Four key components to the form of contract decision process were identified. These components required decisions to be made between four identified procurement strategy options, five tender process options (including single stage and two-stage tendering), eight types of contract and more than 48 standard forms of contract that are available within Australia. A total of 28 factors were identified as influential in the form of contract decision processes, which could not be assigned a specific weighting given their interdependence on one another and the priorities of the individual project.

The standard form of the contract decision flowchart(s) developed as part of this pilot study might be expected to improve the speed and accuracy by deciding which standard form of contract to use, and to serve as a contract training aid for civil engineers in their formative years. Having discussed at length the standard forms of contract, it is perhaps pertinent to now review bespoke (from-scratch) forms of construction contract as they relate to civil engineers. Section 5.1.6.2 presents a brief discussion of bespoke forms of construction contract.

5.1.6.2 Standard and bespoke forms of contract

This section presents a brief examination of hybrid forms of contract for civil engineering and seeks to compare standard forms with 'bespoke' contract options.*

The following discussion focuses on comparing the performance of Australian Standards, as the most commonly used standard form(s) of contract, with their hybrid counterparts, which have been developed (in the public sector) by

---

* Input from Tachella Atmodjo,[39–45] received with thanks.

Water Corporation[46]
> A business enterprise, owned by the West Australian Government and accountable to the Minister for Water

Main Roads[47]
> Responsible for Western Australia's highways and main roads representing almost 30% of the state's total assets and one of the largest geographically spread road agencies in the world, covering 2.5 million/km².

Although, as discussed in Section 5.1.6.1, there are many commercially available standard forms of contract, there remains in industry a preference by some for bespoke (from-scratch) forms of construction contract or, more commonly in the public sector, *hybrid* forms of contract that alter and add to the already available and widely used standard forms of contract. This is somewhat a result of a shift in contracting trends towards consultants (and in-house legal departments) keen to ensure value for money for their clients and redistribute risk appropriately. Indeed, there is a perception by the mining and resources industry that current AS standard forms of contract are insufficient to coordinate big/mega complex projects.

Generally stated, reasons for the use of bespoke contracts or hybrid contracts (that vary the standard forms) include

- A wish for a perceived bettering of a balance of risk by the principal (stakeholder)
- State and territory's own government and administrative organs' local provisions requiring standard forms with amendments towards increased equity

Perhaps a key reason for some to move away from off-the-shelf standard forms of contract include the growth in frequency of contractual conflicts, namely, dispute over scope of work and (perceived) variations to the scope of work which in turn leads to disputes over extension of time, and claims for entitlement of payment in recompense.

Indeed, most reported cases of dispute (in databases such as 'WestLaw AU', one of the many available full-text legal databases with information covering case law, commentary and judgements) centre on damages/claims for extensions of time and payments related to perceived variation of the scope of contract in construction and civil engineering projects.

Both public sector organisations Water Corporation and Main Roads amend, alter and add to the two principal standard local forms of contract in Western Australia (AS4000 and AS2124). Illustrative amendments are shown in Table 5.6.

Table 5.6 Amendments

Clause	Contents	Amendment	Rationale
Water Corporation amendments to AS4000			
11	Legislative requirements	Clause 11.1 Added paragraph on documentary evidence	Added process
		Added clause	Viewed as missing
		• Clause 11.3—Other requirements	
		Superintendents authorised to remove persons failing to comply with Operating License and OH&S requirements	Company's policy
		• Clause 11.4—Policy requirements	
		Compliance audits cooperation	Company's policy
Main Roads amendments to AS2124			
8	Contract documents	Clause 8.6 Confidential Information: Added paragraph stating principal's rights to terminate whole or part of the contract upon contractor's failure to present Formal Instrument of Agreement	Added process
		Added clause 8.8, Public Disclosure of Contract Details:	Viewed as missing
		Addresses issues regarding public disclosure including which project expected to be disclosed, liabilities/ expense caused and authority of Auditor General for WA	
18	Insurance of the works	Clause 18 Principal Controlled Insurance Programme—Contract works insurance and general liability insurance	Company's policy

The two organisations seem to lack records of evidence showing any improvement in performance as a result of amended or special clauses. Similarly, however, such amended clauses have not resulted in any perceived additional conflicts/disagreements, in comparison with having the clauses unaltered. Largely then, it is unclear whether amendments have helped greatly with the clarification of role, responsibility and obligation. The amendments made to the documents certainly did not change the lump sum nature of respective contracts, perhaps suggesting that Australian Standard Project Delivery methods are still very much current.

It might be argued perhaps that the use of standard forms *with amendments* suit local provisions; document comparison notes that this government requirement may be identified as a reason for amending standard forms of contract, as shown in Table 5.7.

An arbitrator consulted in the process of section discussions suggested that amending parts of a standard contract will alter the meaning of other clauses within the document and create ambiguity. The case between *Multiplex v. Queensland* was made as a reference, in which amendments were made by the principal to clause 36 of AS2124, and the Supreme Court judge held it to have altered the meaning of the word 'occur' in the unaltered clause 35.5. The decision was later overturned on appeal. However, this does highlight the danger and resultant confusion of amending standard clauses without careful review and evaluation of the effect on other clauses within the standard.

That being said, parties do seem to try to remain vigilant. Main Roads implement a prequalifying system for contractors who manage to go through this process to remain well informed of specific conditions of contract. In addition, Main Roads executives are involved with West Australia Road Construction and Maintenance Industry Advisory Group Representatives, where changes made to current version(s) of Main Roads' conditions of contract are communicated and fed back to stakeholders.

Generally, it might be suggested that Australian Standard Forms of Contract remain sufficient to provide for current contracting practice based

Table 5.7 Requirement amendments

Clause	Contents	Amendments
Main Roads amendments to AS2124		
20	Insurance of employees	Clause 20.2 Insurance of Employees: Contractor obligation to insure against contractor's statutory and common law regarding claims from persons employed by contractor
Special	Enforcement of 'buy local policy'	Ensure contractor complies with the buy local policy of the government of Western Australia and the 'buying wisely policy' published by the State Supply Commission of Western Australia
Water Corporation amendments to AS4000		
45	Civil Liability Act 2002	To address suitably Part 1F of Civil Liability Act (WA) 2002
Special	Regional employment support	Drafted to ensure sustainable workforce. Water Corporation accomplishes this by reducing a percentage from tender-bid prices, if the bidder is from the region, as a comparative tool when evaluating tender proposals

on a process of drafting by a committee of industry representatives, catering to the needs of all industry sectors. Perhaps it might be suggested that the reasons for amending standard forms of contract seem to be where roles, responsibilities and obligations are viewed as missing or require added procedure or process particularly towards clarification of the company/government's requirements for, for example, buy local and the like.

Having addressed Australian contract conditions that seek to clarify the roles, responsibilities and obligations of the parties, Section 5.1.6.3 reviews overseas contractual arrangements' applications, using, as an example, the fellow commonwealth nation of Malaysia.

The discussion in Section 5.1.6.3 seeks to provide civil engineers with a wider perspective and recognition of Australia's position within Southeast Asia. Section 5.1.6.3 continues the theme of civil engineering standard form of contract payment clauses, but seeks to view them from a wider (Malaysian) perspective.

5.1.6.3 Overseas contractual arrangements: Malaysian payment clauses

This section considers overseas contractual arrangements; particularly analysis of payment effected by standard forms of construction contract in Sarawak, Malaysia.*

In this fellow commonwealth nation of Malaysia, there are several standard forms of GCC used as formal tools for contractual relationships between the parties in the local construction industry. This section seeks to create a better understanding of the explicit clauses in alternative standard forms relating to payment in Malaysia and how, in their application, construction contract administration affects efficient project realisation.

> The most commonly used and dominant standard form in Sarawak is that of the Malaysian Public Works Department 75 (PWD 75); the CIDB Standard Form of Contract for Building Works 2000 Edition (CIDB 2000) presents a somewhat newer alternative, whereas a third popular contract document option can be stated as the Malaysian Standard Form of Building Contract 1998 Edition (PAM 1998 form).

In this section, (Malaysian) contract clauses are examined. An analysis is made of three standard forms options in East Malaysia (PWD 75, CIDB 2000 and PAM 1998). Generally, this section finds that the prevalent standard forms in Malaysia's largest state of Sarawak require further consideration in terms of penultimate claims, account preparation procedures, time frames for settlement and submission of final claim.

---

* Input from Ting Sim Nee,[48–63] received with thanks.

An engineering construction contract refers to an agreement, which is enforceable by law, executed between the employer and the contractor together with the documents forming the proposed scope of construction work. In Malaysia, such a contract is a promise or a set of promises that are enforced under the Contract Act 1950 (Act 136).

In Malaysia, the usage of standard forms of contract in construction is extensive. These fundamental standard forms are largely categorised according to their specific main purpose. For example, the Public Works Department (PWD) forms are used for works in the public sector, the Pertubuhan Akitek Malaysia (PAM) forms are essentially used for private sector building projects and the Construction Industry Development Board (CIDB) forms are intended for use in both public and private sector projects.

PWD 75 is the standard form of general conditions of civil engineering contract used extensively in the public sector construction projects in Sarawak, Malaysia's largest state. The PWD 75 under study was revised in May 1961. It was originally based and closely modelled on the Royal Institute of British Architects Standard Form of Building Contract, which was a common standard form of conditions of contract used in United Kingdom. It is important to note that the PWD 75 standard form of contract is, as it has remained for the past 40 years, the principal public sector standard form in use in the state of Sarawak.

Due to the independency of the body politic and as agreed in constitution when the state of Sarawak joined the Malayan Union to form Malaysia in 1963, the state government of Sarawak is independent and has full authority in certain fields, with public works being one of them. Hence, PWD 75 is still used extensively in all the public sector projects in Sarawak, including all federally initiated projects, state-initiated projects and other projects in relation to the state government. PWD 75 is also widely applied in private sector projects in Sarawak. The private companies, clients or semigovernment authorities who use the PWD 75 as the conditions of contract would maintain most of the clauses whilst making minor amendments to suit their projects in a somewhat piecemeal invalidated fashion.

The Malaysian Public Works Department (nowadays referred to as Jabatan Kerja Raya) has taken the initiative of revising this contract document since 2000 and rolled out PWD 75 (Version 2006) in 2007. Despite the 2006 revision, it has yet to be fully rolled out in Sarawak or embraced by a state industry conditioned by four decades of PWD 75 dominance. With information gathered from Sarawak's Public Works Department, many private and governmental organisations are currently still using the older version due to reasons such as unfinished projects, familiarity with old forms and, in some cases, no awareness of the existence of the new form despite efforts to promote its usage.

In a typical Malaysian (or international) engineering and construction standard form of contract, the effecting of payment to the contractor in return for performance is one of the primary obligations of the employer and the right of the contractor. One of the most common methods of effecting payment to the contractor is through the process known as 'interim' periodic payment, to relieve the contractor of the burden of a totally negative cash-flow, the cost of which may otherwise be an overhead reflected in the tender sum.

If there is any overcertification or undercertification, there can be a commensurate revision in the subsequent certificate. Failure of the contract administrator to issue the relevant interim certificates, in line with the stipulations of the contract, can expose the employer to a possible claim of breach of contract by the contractor. Owing to the significance of payment and interim certificates, all alternative standard forms of GCC in Malaysia have incorporated payment provisions. These are shown in Table 5.8.

The main standard forms in Malaysia experience discrepancies in the interpretation of the clauses of payment. Issues related to interim certificates and payments have always been problematic for contract administrators; claims and disputes over payment are the common sources of problems. It may be worth noting that in relation to payment disputes alone, findings from questionnaire surveys conducted in connection with this section show that more than 60% of Malaysian contractors have experienced late payment problems, whether in government-funded projects or in private projects.

Given the occurrence of potential discrepancies between parties regarding payment, the most readily available standard forms in Sarawak, Malaysia, (PWD 75, PAM 1998 and CIDB 2000) require further review; particularly those clauses relating to payment and how they achieve efficient, appropriate and timeous remuneration for construction work. The discussion for the remainder of this section identifies and reviews explicit clauses of certification and payment in each of the three standard forms of contract under study, and details application issues experienced 'on-site' in Malaysia.

PWD 75

PWD 75 is the dominant standard form of contract used in Sarawak, Malaysia. Clause 35 in PWD 75 deals with all interim certificate

Table 5.8 Payment clauses in Malaysian standard forms of contract

	Malaysian standard forms of contract		
	PWD 75	CIDB 2000	PAM 1998
Principal clauses related to payments	35	42, 43	30.0

matters and regulates payment under this form. It governs all aspects of the subject of payment including interim certificates, retention monies, final accounts and the effect of certificates.

The panel of experts consulted reveal that, in common practice, payment to the contractor is seldom made by the employer within 14 days; the period stated in this clause is seldom adhered to locally. Another 'more reasonable' time frame is argued as needed to update PWD 75—30 days for payment pursuant to the issue of interim certificates, is suggested to make payment to the contractor, given that this represents current practice in industry.

Clauses 35(c) and 35(d) are related to retention monies to be held by the employer. Analysis of this clause found that the employer does not necessarily inform the contractor in writing of a reason, if he exercises any right of deduction from monies due under the contract, as the contractor should know the reasons behind such deductions and what amount is to be deducted.

Clause 35(e) covers matters concerning the settlement of accounts, whereas clauses 35(f) and 35(g) contain provisions with regard to final certificate, respectively: issuance and nature and effect. Although there is no specified time frame for the employer to make final payment to the contractor, the panel of respondents comment that normal practice is that the employer makes final payment to the contractor within 12 months after the defects liability period. Again, it is argued that to avoid disputes, explicit expression might enhance administration under PWD 75.

CIDB 2000

Like the other standard forms of conditions of contract, CIDB 2000 contains contractual provisions dealing with certificates and payments. Clause 42 stipulates procedures for payment claims to be made and assessed.

Clause 42.1 requires the contractor to provide the Superintending Officer (SO) with a statement of work done for amounts up to the last day of relevant interval. In contrast, clause 35(b) of the PWD 75 provides that the amount so stated in the contractor's application is up to a date not more than 7 days before the date of such application. The respondent's panel suggests that the CIDB clause reflects better a more timeous payment of work done.

The respondent panel warns, as a general point of caution that, in practice, the determination, definition and qualification of a SO, and consequently, their power for certification, is open to debate in CIDB 2000.

Clause 42.2 is related to the issuance of interim certificates. This sub-clause authorises the SO to issue an Interim Certificate within 21 days of receiving the Statement of Work Done, as compared to within 14 days in clause 35(a) of the PWD 75, arguably a more practicable time frame.

The amount stated in an Interim Certificate would be the total amount stated in the Statement of Work Done, less retention monies, any amount previously certified and any sum certified to be deducted under the contract. In contrast, clause 35(b) of the PWD 75 does not include the last item, which is identified as an area of contention by the panel of respondents.

A significant feature in the CIDB form is the requirement under subclause 42.3(c) (ii) for the employer to deposit the amount of retention monies deducted in a separate banking account held in trust by the employer, so as to protect it in the event of the employer's insolvency or to deflect a contractor's claim over possession of its interest accrual. Case law (MLJ 444) highlights previous discrepancies between contractor and client on this issue and perhaps justifies the inclusion of this CIDB clause.

The provisions contained in subclause 42.8, Preparation of Final Account and the issue of Final Certificate, are quite different from and, in the respondents view, considerably clearer than those stated in clause 35(f) of the PWD 75, particularly with respect to the explicitly stated time frame. Overall, the respondents to this study comment that general comparisons between PWD 75 and CIDB 2000 highlight that the later document attempts to identify, review and clarify a perceived imbalance of risk conceded by contractors.

PAM 1998

PAM Standard Forms of Contract, used essentially for private sector building projects, detail clause 30 as covering matters related to certificates and payment under the contract, with clause 30.1 describing issuance of certificates and the provision for the (superintendant) architect to correct previous intermediate certificated discrepancies.

Clause 30.5 ([i] and [ii]) set out the nature and purpose of retention and the rules on its treatment and, in essence, duplicate CIDB 2000 (Subclauses 42.3[c] [i] and [iii]). It is argued here that the stated need to justify and make transparent the employer's right to deduct described by PAM and CIDB, is found to be lacking in PWD 75 clause 35.

PAM 98 clause 30.6 bears a similarity to clause 42 of the CIDB 2000 and sets out the procedures and the various steps needed before the Final Certificate can be issued; namely, measurement/valuation of works 6 months from practical completion with the contractor to assist (under 30.6[i]) with supporting documentation; and issuance of a penultimate certificate of payment with summary copy of valuation to the contractor. In comparison with clause 35 of the PWD 75, the procedures and time frames outlined in PAM 1998 are argued, by the study's respondent group, to clarify requirements greatly.

In line with the risk apportionment philosophy of the PAM 1998 form, conditions remove the conclusiveness of the final certificate and address the potential for dispute illustrated in case law both locally and overseas (MLJ 16, All ER 121). Carrying on from the discussion above, clause 30.8 indeed clarifies that no certificate issued by the architect is conclusive evidence that any work, materials or goods are in accordance with the contract.

Overall, the respondents to this study comment that general comparisons between PWD 75 and PAM 1998 highlight that the latter document clarifies procedures, whilst maintaining a positive employer's position.

Notwithstanding the general points stated above, Table 5.9 summarises the specific differences in payment procedure that exist between the three Malaysian standard forms of contract. PWD 75, CIDB 2000 and PAM 1998 differ in subtle ways in their efforts to clarify payment responsibilities and obligations (Table 5.9).

Based on the above discussion, it is generally suggested that civil engineering stakeholders in the Sarawakian construction industry might benefit from

- Acknowledgment of the alternative contract forms and project suitability review
- Explicit standard form clauses that address penultimate claims, account preparation and time frames
- Inclusions of scenario-based clauses for matters such as payment and interim certification
- Procedural formats with simple language to enhance understanding amongst participants

The preceding suggests that PWD 75 (the dominant and most widely used standard form of construction contract in Sarawak, despite a recent revision) is lacking in terms of completeness because it does not cover issues such as penultimate claims, account preparation procedures and time frames for settlement and submission of final claim, whereas other standard forms that are less widely used in civil engineering and the construction industry in Sarawak, such as PAM 1998 and CIDB 2000, do present viable, more complete alternatives.

Given that many items (detailed above) are not specified in PWD 75, it might be suggested here that this carries an inherent difficulty for contract administrators to manage the various procedural chain-of-events, which may in turn lead to imbalances of risk, differing interpretations concerning

Table 5.9 Malaysian comparisons

	Malaysian standard form of contract		
	PWD 75	CIDB 2000	PAM 1998
Relevant clauses	35	42, 43	30.0
Interim certificate			
Documentation submitted for issuing certificates	Detailed written application	Statement of work done	Details and particulars
Amount due in interim certificates	≤7 days	Last day	≤7 days
Period of interim certificates	Fill-in by agreement	1 month if none stated	1 month if none stated
Time frame for issuance after submission of documentation	14 days	21 days	Not specified
Minimum amount stated in Appendix	Yes	Yes	No
Notification if no certificate issued	No	Yes	No
Period of honouring certificates	14 days	21 days if none stated	14 days if none stated
Suspension for nonpayment	No	Yes	No
Retention monies			
Retention percentage	10%	10% if none stated	≤10% if none stated
Limit of retention fund	Fill-in by agreement	5% of CS if none stated	≤5% of CS if none stated
Employer's interest is fiduciary as trustee	No	Yes	Yes
Separate bank account	No	Yes	No
Contractor's beneficial interest	No	Yes	Yes
Contractor to be informed in writing of deduction	No	Yes	Yes
Release of one moiety	Yes	Yes	Yes
Release of second moiety	Final certificate	Interim certificate or final certificate	Relevant certificate
Measurement and valuation			
Right to correct certificates	No	Yes	Yes
Time frame for measurement and valuation at practical completion	Not specified	3 months	6 months if none stated
Penultimate certificate	No	Yes	Yes
Submission of documentation for final claim	Not specified	30 days of making good defects certificate	6 months after practical completion
Computation of contract sum	Yes	No	Yes

(continued)

Table 5.9 Malaysian comparisons (Continued)

	Malaysian standard form of contract		
	PWD 75	CIDB 2000	PAM 1998
Relevant clauses	35	42, 43	30.0
Final settlement			
Final account preparation procedures	No	Yes	No
Issue of final certificate (occurrence of specified events)	No time frame specified	30 days	3 months
Effect of final certificate	Yes	Yes	No
No certificate as conclusive evidence	Yes	Yes	Yes

Note: Malaysian standard forms comparison: summary of clauses for certificates and payments.

payment and certification and, ultimately, dispute occurring from a lack of comprehensiveness. Alternative forms of PAM 1998 and CIDB 2000 cover procedural matters in detail, adding to their respective comprehensiveness; however, those current usage rates are significantly less than PWD 75.

Given the high proportion of public sector development in Malaysia's largest state, it is perhaps logical that the PAM conditions, aimed principally at the private sector, are less widely used than the PWD document. Uptake of CIDB 2000 remains particularly low, with industry stakeholders somewhat intractable in their avoidance of the CIDB document.

> Although a somewhat academic analysis of Malaysia's CIDB 2000 suggests that this form provides a more appropriate balance of risk between the parties, industry practitioners disagree and comment that it is *perceived* to be too contractor-friendly, offering less protection to the client than the more traditional PWD 75 and certainly PAM 1998.

As is most often the case, employers and their representatives decide on which form to use, and it is perhaps little wonder that after four decades of continuous and mostly unchallenged use, PWD 75 remains the standard form of choice in Sarawak, despite perceived shortcomings.

Overall, this section on standard forms of contract in Malaysia suggests that stakeholders might be directed to recognise that there are a range of alternative forms available that may be better suited to available procurement processes. Although familiarity with one particular standard form may well ensure (rote fashion) efficiency in contract administration, it may also perhaps breed complacency and might limit attempts towards best-practice betterment of project realisation.

This section has extensively reviewed contractual dispute and resolution, opportunities for a standard form of contract selection guideline, nonstandard so-called bespoke forms of contract and overseas contractual arrangements using Malaysian payment clauses by way of example.

Going beyond such standard contractual discussions, Section 5.1.7 shall now discuss the extent to which today's civil engineer must embrace and work within environmental law. Section 5.1.7 discusses environmental laws and legislation compliance requirements for civil engineers.

5.1.7 Environmental law

Civil engineering development is subject to environmental law couched within common law and legislation in which the environment is protected and regulated by both Federal/Commonwealth law and State law and national standards.[64]

Many projects are likely to have a significant effect on the environment and subsequently require an environmental impact assessment (EIA) as part of a planning and development application. Local communities, acting through elected members of a local authority, are ultimately responsible for (engineering) development plans that provide a framework for dealing with development applications for planning permission approvals and appeals.

A number of texts set out environmental law and environmental assessment in Australia;[65–67] these include

- Environmental Protection Act 1986, Environmental Impact Assessment (Part 4 Division 1): Administrative Procedures 2012
- ISO 14000—Environmental Management (where the ISO 14000 suite provides tools for companies and organisations looking to identify and control environmental impact and performance)
- ISO 14001:2004 Environmental Management Systems
- ISO 14004:2004 Environmental Management Systems
- AS IEC 60300.3.3-2005 Dependability Management Part 3.3 Application Guide—Life Cycle Costing

A development plan[68] is a statement of regularly reviewed planned policies for the community, providing a basis on which proposals are considered. Permitted categories are set out, as are areas/buildings deemed too sensitive to be disturbed.

Guidelines relating to the control of future development cover a wide range of environmental issues. Major industrial developments are subject to planning where planning applications are assessed by a planning authority.

At a precontract stage, a client/developer submits proposals to planning authorities who seek public comment and also consult EIA directives before making a decision.

EIA CRITERIA

Characteristics of development

- Use of natural resources, production of waste, pollution and nuisance, risk of accident from substances and technology

Location of development

- Existing land use, quality/abundance regenerative capacity of natural resources, wetlands, coastal/mountain/forest areas, nature reserves, areas classified or protected under legislation, conservation of wild birds and natural habitats of wild fauna and flora

Characteristics of potential impact

- Extent of environmental impact (geographical area and size of affected population); duration, frequency and reversibility of the potential environmental impact

An EIA is a systematic assessment of a civil engineering project's likely significant effect and sets out to measure impact and identify subsequent technical mitigation measures that a civil engineer might employ to reduce impact. It might be worth mentioning that the term *significant* environmental impact is somewhat ill-defined by law. Significance remains somewhat subjective and open to interpretation:

- There is no legal definition of significance in the Environmental Protection and Biodiversity Conservation Act (EPBC Act) for Administrative Guidelines of Significance, in which 'national environmental significance' includes (Part 3 Division 1):
  - World Heritage properties
  - Ramsar wetlands related to the Ramsar (Iran'71) Framework for International Cooperation for Conservation and Wise Use of Wetlands[69]
  - Threatened species (such as the quokka and the loggerhead turtle) and listed migratory species
  - Commonwealth marine environments
  - Nuclear energy activities (e.g. uranium mining)

An EIA should ultimately communicate to both public and relevant authorities all decisions made. Planning authorities often welcome meetings between developers and the public to discuss the issues before planning decisions are made.

A number of legislative guidance flowcharts[64,70] as well as Federal and State Environmental Protection Authorities archives[71] are available to assist in the interpretation and application of environmental impact mitigation measures. Many State and Federal Environmental Protection Authorities (EPAs) provide access to archived EPA Reports for previous road/civil engineering projects. By entering search words such as 'road', websites allow access to hundreds of project report lessons learnt.

Key Western Australian environmental protection references include

- Environmental Protection Act 1986, Environmental Impact Assessment (Part 4 Division 1) Administrative Procedures 2012.
  - The 2012 Procedures supersede Administrative Procedures 2002. The superseded 2002 procedures may be viewed alongside superseded 2002 EIA process flowcharts on pages 576 to 580.
- Revised Environmental Assessment Guidelines timelines for environmental impact proposals;[72] review and application of relevant EIA process flowcharts on pages 5957 to 5959 can assist with civil engineering project application/determination.
  - In 2013, the Environmental Protection Authority Western Australia revised its assessment timelines.

The several hundred historical case studies archived by the EPA can assist civil engineers in determining lessons learnt from previous jobs towards an assessment of the key environmental issues of their current project and ultimately help in the development of technical solutions appropriate to mitigate potential environmental impact.

EIAs are a feature of all environmental plans, requiring development of an environmental impact (protection) statement to inform decision makers and the general public of the potential impact of activities on the environment.

Often, the legal status of an EIA depends on the extent to which a civil engineering project is predominately a Federal or a State development. In Australia, there is no legal hierarchy between State and Federal/Commonwealth; both may be involved, either together or independently. Similarly, State and Federal methods of assessment differ somewhat.

EIA TYPICAL FORMAL ASSESSMENT LEVELS[64]

- Assessment on proponent information (API) and/or environmental protection statement
- Environmental scoping document (ESD)
- Public environmental review (PER)
- Recommendation, summary acceptance or proposal unlikely to be environmentally acceptable and/or further public inquiry

Federal environmental protection agencies enact Australian Commonwealth Acts:

- The National Commonwealth Act sits within the EPBC Act of 1999, where Commonwealth Minister for SEWPC (MP) presides for the Federal Australian Government.

- EPBC Act 1999 Commonwealth of Australia (C'th) sets out national involvement in EIA and determines 'actions' with significant effect on matters of national importance in which 'action' includes a project, development, activity or an alteration

State authorities in Australia determine environmental impact using state-based legislation. For example, in Western Australia

- The state of Western Australia applies the Environmental Protection Act (WA) 1996, where the Environmental Protection Agency (EPA-WA) polices (independent from the legislative assembly of WA) the environment on behalf of the State Minister for EW (MLA)

Australia has no legal hierarchy for EIA between State and Federal/Commonwealth:

- State and Federal/Commonwealth may become involved in EIA, either together or independently, although usually the Commonwealth secede to State activities.
- State EIA seems to have greater public input/participation and the like.
- Effectively, EIA sits under the two umbrellas of State and Commonwealth, or indeed EIA can shelter under only one of these.
- This dual coverage is a quirk of the six States of the Federation retaining extensive (but not complete) powers over education, police, the judiciary, transport, the environment and the like.

5.1.7.1 Methods of environmental assessment: Federal and state

Australia Federal/Commonwealth National Method of Environmental Assessment:

- The matter is referred to the Commonwealth Environment Minister (Minister for the Sustainability Environment, Water Population and Communities).
- Proposals may be referred by a proponent (those in support of the development) or Commonwealth or State government agency; the Minister may also 'call in' referral from the person proposing the activity.
- The Minister has 20 business days from the date of the referral to decide whether the action requires approval or not; the Minister decides to approve action in 30 business days of assessment report or 40 days following a public enquiry report (Section 130).

- The Minister is also empowered to vary, suspend or revoke approvals
- Commonwealth Federal National Method of Assessment allows five options (or an updated rolled-out variation of the five options) under Part 8 of the EPBC Act, traditionally described as
  - Accredited assessment process, assessment on preliminary documentation, public environment report (PER), environmental impact statement (EIS), public enquiry
- In deciding upon the method of assessment, the Minister takes into account:
  - Economic and social matters, the principle of ecologically sustainable development, the precautionary principle, responsibility to intervene and protect the public from exposure to harm
- Public EIA/reports prepared by proponent
- Matters prescribed by legislation

In addition to the Federal progression above, civil engineering projects are subject to the state authority's dictates as below.

Australian State method of environmental assessment in, for example Western Australia, involves

- Environmental Protection Agency (for WA) is authorised to enact legislation under Part 4 Division 1 of Environmental Protection Act 1986 (WA); thus, the EPA (WA) has the power to assess proposals that are likely to have a 'significant effect on the environment', where any person can refer a proposal to the EPA (Section 38).
- EPA WA Authority's objectives for EIA:
  - Ensure proponents take primary responsibility for protecting the environment influenced by their proposals; ensure best practicable measures taken to minimise adverse effects on environment, and that proposals meet environmental objectives and standards; to protect the environment and implement principles of sustainability; provide opportunities for local community and public participation, as appropriate, during the assessment of proposals; encourage proponents to implement continuous improvement in environmental performance and apply best practice environmental management; ensure that independent, reliable advice is provided to Government before decisions are made (under EPA [Part 4 Division 1]: 2002).[73]
- State (Western Australia) environmental assessment proposals require EPA (WA) to consider a number of factors when deciding whether a proposal must be assessed (merits an EIA) including
  - Environmental values; impact on health, comfort, welfare and amenity of people; impact on social surroundings; principles of sustainable development and a balance of environmental,

economic and social factors); the likely level of public interest in compliance with clause 4.1.2 Environmental Impact Assessment Administrative Procedures (WA) 2002.

Decisions made by the relevant environmental protection agency are communicated in reports, which are then distributed and archived for future reference. Report content complies with relevant legislation and must be an (understandable) assessment of the specific proposal and report the final outcome of the assessment of the proposal, setting out what the authority considers to be key environmental factors, alongside recommendations for proposal implementation and any conditions and technical (civil engineering) procedures to be met by the design and its construction methodology.

The section above addresses civil engineering practices and procedures related to

- The need to comply with environmental law   (Section 5.1.7)
- What constitutes a contract   (Section 5.1.1)
- Contractual category type and procurement paths   (Section 5.1.2)
- Project progression using standard GCC   (Section 5.1.3)
- A project's potential for variations and payment claims (Section 5.1.4)
- Examples of a civil engineering job's contract admin-
  istration with reference to GCC clause interpretations (Section 5.1.5)
- Methods to address contractual dispute and processes
  for resolution, alongside review of contract option selec-
  tion and (local and overseas) clauses interpretation(s) (Section 5.1.6)

Alongside legal procedures and environmental legislation compliance, civil engineers must determine appropriate specifications. Next, Section 5.2 describes specifications for materials supply and installation for design solutions.

5.2 SPECIFICATIONS FOR DESIGN SOLUTIONS

Section 5.1 clarifies contractual responsibilities and obligations; one of the key responsibilities for the civil engineer is the development of a design's specifications, in conjunction with an obligation for structured compliance with these preprepared, project-specific instructions during building work on-site.

Specification writing has long been part of the design process for civil engineers. Historical texts that provide an insight into the importance of this essential activity include the still widely available 1913 book by Kirby, entitled *The Elements of Specification Writing: A Text-Book for Students in Civil Engineering.*[74]

A specification is a one-of document that gives detailed materials and work-installation information for the project at hand; it is a detailed description prepared by a consulting design engineer that explains to the builder everything about the materials and workmanship that cannot be interpreted from the available design drawings. Specification writing is somewhat more succinctly defined as 'any information not in drawings'.[75]

Specifications communicate design decisions, link explicitly to contractual obligations, complement drawings and help define quality. The specification is a legal, written record of design decisions and is used as a document on which estimations and tender action are based. Specifications are used as an on-site work activity document and a dispute resolution tool. Compliance with a given design specification must link explicitly with the contract's quality system, which checklists material and installation appropriateness and clarifies tests to be conducted throughout on-site activities.

5.2.1 Specifications systems suppliers

A number of organisations (such as NATSPEC[76] and the Building Code of Australia [BCA][77] in Australia, alongside Australia–New Zealand proprietary materials certifier Code-Mark,[78] as well as Government infrastructure bodies and States' roads authorities,[79] and long-standing UK organisations such as the National Building Specifications [NBS][80]) provide guidance and often stock reference items in the form of off-the-shelf general specification descriptors towards assisting civil engineers in their development, adaptation and application of project-specific specifications; classifications are usually structured along elemental and subelemental workgroup items. It must be noted that specification systems suppliers themselves readily acknowledge that they provide guides and tools and stock examples for adaptation, and do not seek to give an explicitly specific specification. Project-by-project, one-of specification preparation remains the responsibility of the design engineer.

Specification providers include NATSPEC:

- NATSPEC's (the trading name of Construction Information Systems Limited) National Classification System

Specifications might refer to specific parts of the BCA:

- The BCA exists in two volumes of the National Construction Code. The BCA is produced and maintained by the Australian Building Codes Board on behalf of the Australian Government and State and Territory Governments.

Propriety products might be included in specifications policed by schemes such as Code-Mark:

- The 'Code-Mark Scheme' seeks to provide a mark of conformity that proprietary building products are based on technical documentation alongside regular review of manufacturing and quality control processes that monitor compliance with the BCA.

National Building Specifications:

- In the UK, this is made available by the Royal Institute of British Architects and provides a detailed international source of information.

Government Agencies:

- Specifications are provided by many government infrastructure bodies and states' roads authorities. In Western Australia, for example, Main Roads WA gives specification.

It perhaps goes without saying that 'specifications' should seek to be 'specific' and certainly avoid ambiguity and vague instructions and avoid excessively glib instructions that 'all work and materials comply with building codes authorities and all relevant (Australian) Codes and Standards'.

- Similarly, phrases to avoid in the preparation and dissemination of a design specification document are excessively glib suggestions that 'components be supplied and installed to industry standard/as appropriate/as required'.

On the other hand, a good specification clarifies a design's materials and installation requirements in ways that can be verified explicitly (in conjunction with a quality system checklist of tests and controls) towards provision of a solution that best satisfies the client's original brief.

5.2.2 Specifications exemplars

Specification exemplars are very widely available from respected public, private and professional organisations such as Main Roads WA,[81] NATSPEC,[82] NBS,[83] and the International Construction Information Society.[84,85]

Main Roads WA Specifications exemplars:

- Specification Development Guidelines for Custodians by Main Roads Western Australia 2007+

NATSPEC Specification exemplars:

- NATSPEC Learning Work-section 2009

National Building Specifications:

- NBS exemplars

International Specification examples:

- The International Construction Information Society provides a range of Specification Exemplars, alongside an extensive discussion of specifications internationally (across several countries including Australia, Canada, Czech Republic, Finland, Germany, Japan, the Netherlands, New Zealand, Norway, Sweden, the United Kingdom and the United States).

5.2.3 Specification types: General, technical, prescriptive and performance

The civil engineering project usually requires specifications that provide instruction relevant to both general items as well as technical requirements.

General specification sections in the specifications document aim to clarify preliminary information for the project such as confirmation of the form of contract to be used to administrate the project, and perhaps restate and cross-reference inclusions and specific clauses and appendices in the signed (standard) form of contract and bills of (approximate) quantities that indicate site access provisions, the staging of work, hoardings, existing services and the provision of temporary services.

The general specification might also allude to how the work has been described in trades (and again perhaps cross-reference other documents such as the project's BQ or provisional BQs that similarly give work classifications). General specifications might confirm provisional sums (for the amount set aside for unforeseen work) and prime-cost sums (money set aside for nominated specialist mechanical and electrical work), as well as perhaps provide a schedule of rates (and day-works rates) if not documented in the contract anywhere else and also clarify the requirements and conditions for any cost adjustments to allow for the potential of price fluctuations in long programmes or in environments of economic or labour access instability.

Technical specification sections provide information related to the design's desired quality of materials and expected installation practices, and give detailed instructions (in conjunction with the contract's quality system) for operations, inspections and testing during construction.

Technical specifications (with design drawings) provide tenderers with the full range of information required to prepare a project estimate and submit a tender-bid and similarly give client's representatives the confidence that the selection of a builder is made with regard to consistent qualifications and expectations.

There are four main approaches in preparing a (technical) specification; namely, as a proprietary specification, as a reference specification, as a performance specification or as a prescriptive/descriptive specification (see Specification approaches).

SPECIFICATION APPROACHES

Proprietary specifying:
- Specification of an item in a project with reference to a supplier, trade/brand name or catalogue reference

Reference specifying:
- An identifiable printed/published document (by a national/international standards authority) incorporated and referenced

Performance specifying:
- Prescribes a desired end-result and criteria by which a result is judged for its acceptability related to tolerance compliance and the like

Prescriptive/descriptive specifying:
- Describes in detail the materials, workmanship and installation/construction procedures required to be used

Concrete beams might be prescriptively specified with reference to component parts

- *Concrete*: Aggregates, cement, water, chemical admixtures, slump/strength testing, sampling/certification, hand/mechanical vibration installation, curing and protection
- *Reinforcement*: Bar or mesh strength/grade/size/finish, chairs/tie wire support, cutting length, hooks/cranks, stirrups and tie fixing and cover and lateral spacing
- *Formwork*: Ply sheet/grade/form/size, support joists/bearers/floor/centres, variations from plumb/level/line, props, cambers, stripping/striking, cleaning and reuse

Performance specifications, on the other hand, might dictate

- General compliance with drawings x, y and z
- Loading of x live loads and additional y and z dead loads

- Deflection limits without damage to fixture w and x, no physical impact and not exceeding comfort level y and z
- Bearing where x shall not overload the specified bearing capacity of soil z
- Fire-rating in which, in the event of a fire, the structure shall retain integrity x for the safe evacuation of y at a rate of z

Materials specifications might discuss fit-for-purpose expectations such as

- Concrete specified and based on physical properties such as strength, modulus of elasticity and mix design
- Steel specified by yield strength, modulus of elasticity and ductility
- Materials testing, which requires evidence that specified properties are obtained
- Either composition or physical properties, not both due to possible incompatibility or restriction of supply

Workmanship specifications might discuss

- The usage and handling of the materials, fabrication into the structure, the method and order of installation, the quality and qualifications of the personnel required, the standard of workmanship, tolerances permitted and safety precautions

The Konya disaster in Turkey[86] provides an important and tragic case study in which inappropriate policing of standards and specifications lead to devastating results. In this case, it is perhaps most worrying that although building codes are in place and do exist, they are proving ineffective. Review of the disaster includes evidence that

- Forty people died when an 11-storey block of flats collapsed in Konya, Turkey.
- Families were not comforted by claims that construction faults caused the collapse.
- Some pulled alive from concrete rubble described the concrete as just 'powdery earth'.
- Turkey is prone to earthquakes, but whereas the Standards and Specifications and Building Codes appear fit-for-purpose, it seems that they are enforced very casually and sporadically
  - Resulting in regular damage as a result of earthquake action that perhaps might otherwise be avoided.
- In Konya, however, there was no tremor—the building simply collapsed.
- Perhaps the policing of both specifications and a quality management system might be used to diminish future similar tragedies.

5.2.4 Specification decisions: Practice example

An approach similar to the previous 'quality system checklist practice example', described in Section 5.2.3, might assist the civil engineer to determine fit-for-purpose specifications (in, for example, locations of less assured quality compliance[86]) towards ultimately choosing from the different methods to specify materials, alongside related installation practices and supply chain variables for a particular project.

Step 1: Review characteristics/requirements of the project at hand and identify, summarise and adapt 'spec.' from available specification supplier exemplars.
Step 2: Write the project-specific design specification.

A scenario might involve working in a location where material quality is less than assured with poor adherence to available standards and specifications; specifically, a new apartment block is required in a location known for multistorey residency collapse resulting from construction faults and poor quality concrete.

Step 1
Concrete specifiers must first seek to identify and choose from the alternative reference specifications available nationally or internationally, such as Australian Standards, where concrete is generally specified in accordance with reference standards provisions from, for example:

- Methods of Testing Concrete, AS1012
- Specification and Supply of Concrete, AS1379
- Aggregate and Rock for Engineering Purposes; Part 1, Concrete Aggregate, AS2758.1
- Supplementary Cementitious Materials with Portland Cement: Fly Ash, AS3582.1
- Supplementary Cementitious Materials with Portland Cement: Slag— Ground Granulated Iron Blast Furnace, AS3582.2
- Supplementary Cementitious Materials with Portland Cement; Part 3, Amorphous Silica, AS3582.3
- Concrete Structures, AS3600
- Tilt-up Concrete and Precast Elements for Use in Buildings, AS3850
- Portland and Blended Cements, AS3972
- Quality Management and Quality Assurance Standards, AS/NZS ISO 9000

Equivalent international standards[87] might similarly be used in overseas locations, for example, EN BS 206-1, Concrete: Specification, Performance,

Production and Conformity in conjunction with complementary British Standard BS 8500, Concrete.

- BS EN 206-1 replaces British Standard for Concrete BS 5328 2003

For designed concrete, specify that the concrete shall be produced in accordance with relevant clauses of BS EN 206-1/BS8500 and also specify the following:

- Compressive strength class and exposure class or limiting values for concrete composition related to durability and abrasion resistance where in some cases it may not be necessary to specify a maximum water/cement ratio
- Nominal upper aggregate size and requirements for aggregates including physical and mechanical characteristics
- Type and quantity of fibres (if used), chloride content class and consistence class
- Permitted cement types and permitted admixtures

Beyond review, adaptation and (partial) incorporation of any relevant (inter)national standards to form a foundation of reference specifications, supply chain variables might be taken into account such that different specifying options might review specifications that are (i) designed, (ii) designated, (iii) standardised, (iv) prescribed or (v) proprietary where these options are summarised as follows:

- *Designed concrete*: Purchaser responsible for specifying strength and the producer responsible for selecting mix proportions to produce suitable performance.
- *Prescribed concrete*: Purchaser specifies proportions of constituents and for ensuring proportions will produce a concrete with the performance required.
- *Standardised prescribed concrete*: Selected from (inter)national standards above and made with a restricted range of materials.
- *Designated concrete*: Produced in accordance with a referenced standards specification given as above and which requires the concrete producer to hold current product conformity certification based on product testing and surveillance coupled with approval of quality system to ISO 9000/1.
- *Proprietary concrete*: Concrete for which the producer assures the performance, subject to good practice in placing, compacting and curing, and for which the producer has no requirement to declare the composition beyond compliances.

Recognise that, in this scenario/location:

- Designated mix design is a preferred option, to sit alongside;
- Quality systems, to checklist;
- 'Measurement and analyses' of concrete mix and target specifically;
- 'Monitoring, measurement and control of nonconforming products', where;
- Nonconforming concrete is identified in tests, measured against;
- Stated requirements/procedures to;
- Detect discrepancies between actual installed versus intended fit-for-purpose, such that;
- Interim stage-by-stage tests condemn work if concrete samples sent to an independent testing laboratory are lacking and fail to reach required strengths after set-periods of curing, thus;
- Nonconforming elements are condemned and removed;
- Necessitating reinstallation/rework;
- Prior to the issuing of any certificate of practical completion.

Step 2

Extending decisions made in step 1 above, the engineer then specifies the concrete (and all other subelement materials/installation requirements) in a structured format.

An illustrative example for concrete might include statements that define

- Scope
- Standards
- Reference documents
- Inspection
- Shop drawings
- Performance testing
- Sampling expectations
- Records retention expectations

SPECIFICATION PRO FORMA

Concrete: General

- Scope:
  This specification to be read in conjunction with drawings w, x and y, and update z, to cover forming, placing and finishing of work in location a, b and c.
- Standards:
  Australian Standards and the BCA specifically with reference to d, e and f

- Referenced documents:
  Specification and Supply of Concrete, AS1379
  Concrete Structures, AS3600
  Tilt-up Concrete and Precast Elements for Use in Buildings, AS3850
  BCA items x, y and z
- Inspection:
  All inspection arrangements as detailed in contract document quality system y, confirmed by GCC (AS4000 clause 29) must be followed as contract document g.
- Shop drawings:
  Related to x, y and z require copies to be forwarded for review to w, following approval procedure v of quality system y.

Concrete: Testing and sampling

- Performance tests:
  Test methods of testing concrete and x AS1012, and strength grade a, to follow method/compliance/recording as quality system y
- Sampling:
  Standard to be j
  Sampling location to be k
  Records to be retained in l, by m

Having described specification expectations for known entities, using accepted references and established pro forms, it is perhaps worthwhile to examine locations and situations in which the civil engineer might be less certain of the relevant reference material. Next, Section 5.2.4.1 explores design and specification expectations in situations where environments are less predictable (e.g. for infrastructure in rough terrain environments).

5.2.4.1 Specification requirements for infrastructure in mountainous regions

This section describes design and specification challenges that might be expected to exist for infrastructure construction in rough terrain and inclement environments.*

Infrastructural development is time-consuming and costly, especially for highways in rugged mountainous terrains. Factors known to cause road blockage and damage in mountain regions include flooding, mountainous debris, alluvial fan, road and drainage structure washout, landslides and roadside slope failures.

The situation is often exacerbated in developing (and newly developing) countries where existing engineering hydrological, geological and

---

* Input from Shariful Malik,[88–97] received with thanks.

geotechnical information is frequently limited or unavailable, aggravated further in some extreme cases by the need to address ongoing or past conflict zone mitigation measures.

Bangladesh, Pakistan, Afghanistan and northern India might be suggested as highlighting the importance and need for engineering assessments applied to both new-build design specification as well as the rehabilitation of existing roads in mountain areas where roads are often severely impassable.

In many cases, only limited engineering assessments have been carried out prior to road construction; only a few of the potential hazards have been identified during project feasibility study and design phases. Resultantly, postconstruction (postoccupancy) failure occurs more frequently as a result of this lack of problem analysis.

Civil engineering challenges such as those mentioned above perhaps identify the value in carrying out timeous and appropriately scaled engineering assessments because key contributions to the initial life cycle design and early specification by investigation, analysis and mitigation of potential hazards in a structured and objective manner.

The following discussion is made of the need for structures to categorise variables deemed essential, desirable and useful for the development of specific civil engineering guidelines related to, for example, rough terrain road design.

General development structures: Planning
The planning and locating of highway facilities are the first steps in a challenging process of providing a safe and efficient transportation system. All possible effects that highway construction may have on existing drainage patterns, river characteristics, potential flood hazards and the environment in general, as well as the effects of the river and other water (hydrologic and hydraulic) features may have on the highway, should be considered at an early design phase within project development.

General development structures: Coordination
Coordinating links between the various divisions of transportation agencies (likely to be involved with the project) must be established. Notifications of proposed projects must be made to other development stakeholders; permits and regulations applicable to the project should be identified as soon as possible. Often, project delays are due as much to legal process and local authority procedures as to the resolution of technical issues. Problems that may arise during design, construction or maintenance should be considered, as well as the need to determine environmental data. There is a need for a structured approach both to address and fully checklist technical problems, as well as for knowledge management of the process.

Infrastructural development in rough terrain mountainous regions is seldom easy: landslides and roadside slope failures are apt to result in road blockage, damage and recurrent (economically straining) refurbishment in mountainous regions around the world. Factors that limit road access and use in mountainous areas include: flood, rockfall debris, alluvial fan, extreme washout occurrences, adjacent slope failure and landslides. These environmental problems are often compounded in newly developing regions by a lack of existing engineering and geotechnical, hydrological and geological surveys.

Failure both during the construction phase as well as in the maintenance and operational 'postoccupancy' phase, engineering geological investigations, as well as field investigation and topographic surveys need to be carried out to determine both cause and effect using mechanisms such as field mapping, supplemented in places by trial pit investigation, road inventory, hydrological and hydraulic analysis, roadway planning, environmental assessment and stability analysis.

> In many world regions, addressing structural failure in rough terrain roads becomes more of a costly and time-intensive catch-up treatment of failure *after the event*, rather than appropriate design phase problem analysis at the early stages of development.

There is much need to carry out early appropriately scaled engineering assessments, as key contributions to the initial life cycle design and early specification by investigation, analysis and mitigation of potential hazards; unfortunately, in many regions, these engineering techniques are seldom applied in a structured and objective manner. There remains a need to address such initial design stage (value management/value engineering) procedures, towards a means to categorise variables deemed essential, desirable and useful for development of an optimum specific civil engineering guideline for road design, construction and maintenance. Practitioners in the design and construction of rough terrain infrastructure developments (and their respective future assessments) might be expected to gain much from the development of such a guidance procedure. A main aim might be to investigate the interactive components determining the hazards affecting highway performance towards the development of a structured objective planning guide for appropriate life cycle performance in rough terrain/inclement environments.

Towards improved guidance for infrastructure in rough terrains, a number of variables might be suggested, such as

- Examination of practicality and suitability of terrain characteristics coordination in relation to infrastructure problems
- Address factors often overlooked by low-level/conventional studies for rough terrain design and construction
- Explore sustainable road construction and geometry addressing (life cycle) cost in mountainous terrains towards integration and implementation into a whole-life approach

- Adaptation and recommendations for current highway design guidelines that provide compliance with existing standards and also address rough terrain variables
- Acknowledgement of risk mitigation required to address not only technical considerations but also knowledge gaps in environmental/geographical/conflict zone aspects

The road network generally plays a vital role in the economy of any country. Road design, construction and maintenance in steep and unsuitable landscapes, especially in rugged mountainous regions, requires a specialist's approach. There are many engineering problems contributing to road specification failure: land sliding, erosion, river flooding, slope failure and earthquake. Unfortunately, these problems are often compounded in many developing and newly industrialised countries by expansive tracts of remote isolated interiors with limited supply chains.

Systematic approaches to project assessment and design adopted on all roads irrespective of their intended function or standard require reconnaissance surveys during feasibility study, to summarise terrain conditions and to identify and compare design concepts and costs, and review environmental considerations.

Also essential in initial design considerations are

- An identification of preferred route corridor
- Hydrological investigations and hydraulic design for both feasibility stage and detailed design stage
- Geotechnical investigations at selected locations along the chosen route to define geotechnical parameters
- Design of alignment plans and profiles, cross-sections and associated drainage and retaining structures considering most common events and failures and risks

Typically, the more challenging the environmental conditions, the higher the design standards, and the greater the capital construction cost to fully satisfy rough terrain solutions. In the world's newly developing countries, there is an increasing requirement to build roads into the least accessible areas to address economic development demands. Often, geology plays a substantial part in (economic) underdevelopment, with life cycle maintenance and operational requirements particularly important for rough terrain infrastructure.

Full investigation of the parameters identified by a structured approach to rough terrain design guidelines is unlikely to be cheap. Indeed, it is hoped that the finances required to implement the best-practice guidelines will not prove restrictive to implementation. Capital cost then, must be addressed.

Although available material and labour are cheap in newly developing countries, infrastructure projects in these locations can be expensive. Often, a major reason for the expense stems from the need to mitigate the problems of constructing essential infrastructure in a current or recent conflict zone.

Additional costs beyond the usual technical application of labour, plant and machinery can relate to 'project demining' (clearance of mines and explosive devices), security costs, community development costs, specialist demining vehicle costs, as well as add-on expenses related to conflict zone 'specialist' transport costs, fuel cost, accommodation, overloading, rock-cut, blasting and hauling.

Anecdotal evidence from a number of rough terrain infrastructure projects in Afghanistan, for example, suggests that the measures to mitigate issues related to conflict can add up to 40% to infrastructure development costs; costs which unfortunately (but essentially) are typically redirected away from 'normal' construction technology applications and subsequently detract from the overall design, construction, operation and maintenance of the finished asset.

Beyond such conflict zone variables (argued here to eat into the time, cost and quality of set-aside schedules and budgets), guidance procedures for rough terrain infrastructure design, must be able to acknowledge, coordinate and manage a multitude of interrelated issues concerning the need to construct appropriate life cycle maintenance and operation solutions that address

- Embankment washout due to the potential for high-velocity flooding
- Deep mountain narrow gorge areas
- The locating of routes within gullies, narrow valleys and clear zones with sight distance difficulties
- Waste material management, where improper deposing of excess dirt and other materials can harm the river ecosystems and can adversely affect river courses

Indeed, guidelines should also seek ultimately to checklist and address

- Durability issues for constructed asphaltic flexible pavements that sit on (natural) rocky embankments and
- Improper slope cut/fill for mountain and embankment as a result of inconsistent and nonuniformly addressed geotechnical surveys, as well as a need for suitably detailed topographically surveys in deep mountainous areas

Design guides must attempt to recognise that currently there has been no/limited consideration of talus (unstable and unidentifiable) natural course material during design and construction, as roads pass through rockfall areas and resultant debris that block drainage routes. There is a need for the following:

- Widened carriageways in the deep mountainous areas (in compliance with normal road templates) able to take into consideration narrow gorges adjacent to water courses, as well as
- Adapting existing general drainage (water table analyses) design practices to suit deep mountainous areas
- The effect of rock-cut/blast during construction and resultant natural knock-on effects such as earthquakes and slips, as well as
- Appropriate snow-loading and freeze–thaw issues related to loading and potential damage/failure of pavement structures

Design and specification considerations must perhaps increasingly recognise user variables on asphalt surfaces such as large haulage transportation vehicles that expel oil/liquid in steep gradient locations to maintain vehicular passage; as well as improper management of alluvial fans in design.

Paved roads in extreme climates deteriorate in different ways compared with those in the more temperate regions of the world because of the harsh climatic and environmental conditions and a lack of suitable road pavement materials and road drainage. Location-specific (regional) guidance for civil engineering infrastructure, road design specifications and construction methodologies for rough terrain (in newly developing regions) is needed particularly for bituminous-surfaced roads in diverse and extreme conditions that suffer from accelerated failures caused by variable quality control during construction, high axle loads and inadequate funding for maintenance and refurbishment.

Although this section looks at the extent to which extreme natural environments affect the structural integrity of design solutions and specifications of materials and their installation, the following discussion in Section 5.2.5 reviews environmental considerations in a sustainability context.

In Section 5.2.5, the environment is considered more from the point of view of being one of the key triple bottom-line variables in developments. Section 5.2.5 also discusses the extent to which civil engineering designers are charged to ensure that material specification choices are sustainable.

5.2.5 Design specifications to facilitate environmental savings

Civil engineers and construction professionals and practitioners are increasingly charged to balance the capital cost of their design specification decisions with the potential for environmental saving, where future constructed assets, infrastructure and amenities are required to be flexible to accommodate changes in usage, seek to use less energy during construction, consume less energy during operation, but still be constructed economically and remain affordable to a client.

As with all materials and materials specifications, choice from a range of fit-for-purpose alternatives is a key responsibility of the engineering designer. Comparison and choice of a preferred specification alternative is made with regard to structural integrity, buildability, initial capital cost and life cycle cost in operation, as well as an additional selection criteria factor—that of environmental impact. There are four environmental impact factors to be taken into consideration by designers and materials specifiers when considering the range of potential alternative specifications, namely

- Effluent, waste and contaminated water production
- Embodied energy (CO_2) emission during manufacture, installation and operation
- Airborne emission in manufacturing, construction and in-use processes
- Solid waste and sludge by-product output

A number of international initiatives have produced a range of assessment methods that seek to encourage and assist specifiers to be more environmentally sensitive, including

- Green-Star Australia,[98] offering sustainable design and construction specifications via rating systems for all types of building projects and industrial facilities.
  - Green-Star Australia seeks to encourage sustainable design and construction.
- The UK's Building Research Establishment Environmental Assessment Method (BREEAM)[99] methodology, which is widely utilised and endorsed;[100]
  - BREEAM appeared in 1990 and remains one of the first such initiatives and currently offers a 350-page technical guide for specifiers and designers; the Netherlands Green Building Council advocates the use of BREEAM.
- The Green Building Council of the United States promoting the Leadership in Energy and Environmental Design (LEED)[101] environmental classification system.
  - The Green Building Council from the US launched its LEED system in 1998.
- Japan's CASBEE[102] system seeks to address problems peculiar to Asia
  - The CASBEE system was developed in Japan to enhance design choice environmental credentials.

BREEAM and LEED are the main environmental assessment methods currently in use. The Building Services Research and Information Association (BSRIA)[103] suggests that the main difference between the two methods is the process of certification:

- BREEAM has trained assessors who assess the evidence against the credit criteria and report it to the BRE, which validates the assessment and issues the certificate.
- Although LEED does not require training, there is a credit available if an accredited professional is used. The role of the accredited professional is to help gather evidence and advise the client. The evidence is then submitted to the United States of America Certification Body (US-GBC), which does the assessment and issues the certificate.

An oft-quoted encouragement towards environmental improvements in construction suggests that addressing standard dwelling specification by a 7% increase in capital costs can reduce annual running costs by more than 12% and reduce CO_2 emission by more than 17%;[99] similarly, careful element-by-element specification choices contribute much to the potential for environmental savings, with examples including

- Addressing humidity by including steam-based humidification systems to reduce, for an average office of 10,000 m², 40 tonnes of CO_2 emissions annually
- Recovering manufacture energy using 100-mm-thick mineral fibre insulation, recovering the energy used in manufacture in 3 years, in lieu of 300-mm-thick insulation, which takes 25 years to recover alongside its extra CO_2 generation at manufacture
- Volume and perimeter considerations in which reducing the building volume and perimeter curtain wall reduces initial capital cost, operating and maintenance cost and electricity/gas use for heating/air conditioning/ventilation

Alongside careful specification of new materials, specifications that encourage the use of recycled materials are increasingly important for sustainable civil engineering works, as discussed next in Section 5.2.5.1.

5.2.5.1 Specifications and the use of recycled materials

Increases in recycling over recent years have largely come from selective separation of materials prior to disposal, designer motivation to specific responsibly as well as landfill directives that encourage greater economic considerations in managing waste. Review of construction and demolition waste recycling across all elements and subelements towards uptake into new-build civil engineering products is increasingly important.[104]

A range of measures related to command and control regulation, and fiscal measures, can best encourage construction and demolition (C&D) waste reuse. If recycling is to achieve significant levels of sustained and economic viability in civil engineering activities, four conditions must be addressed (see Recycling viability: Four key factors).

Specification clauses are beginning to address environmentally conscious design. The need for easy-to-use specifications for recycled aggregate is a prime concern of a number of specifications websites[105] that provide designers with a source of guidance on construction materials made from recycled materials and advise contract-clause inclusions into tender documents to dictate the uptake of construction and demolition waste.[106]

RECYCLING VIABILITY: FOUR KEY FACTORS

Specifiers and clients

- Recognise C&D waste (aggregates, etc.) can replace fit-for-use primary sources
- Regional and local administrators encourage recycling by 'green' procurement/tendering plus design that encourages responsible specifications

Landfills

- Require regulation in accordance with legislation to avoid fly-tipping

Landfilling costs

- Must be reasonable (evidence from Europe suggests fiscal measures/fees encourage uptake of C&D waste in new construction in lieu of landfill)

Disassembly opportunities

- Sorting and processing must be increased
- Listings of recyclers needs dissemination

In the United Kingdom, 'AggRegain' is an online guide to sustainable aggregates for use in civil engineering and is offered in conjunction with 'WRAP' and 'National Green Specification'—UK-based waste resources and sustainable specification organisations.[107]

Support for the use of recycled construction and demolition waste in roadworks is evident in a range of specifications for highway works documents that already allow specification of recycled materials under a wide variety of applications.[108] Western Australia's infrastructure body, Main Roads, assists civil engineers in specifying recyclates via documentation such as the WA Specification 501 for Pavements, Main-Roads Western Australia,[109] whereas overseas, the Specification for Highway Works in the UK encourages the use of recycled materials.[110]

Standards such as BS/EN-12620 'Aggregates for Concrete' coupled with BS 8500 as an addendum to EN206 (EN206-1) are examples of an encouragement of greater certification and specification of recycled coarse aggregates (nonfines) for use in concrete.[111]

Alongside new standards for materials specification, revised codes of practice reexamine demolition to take account of advances in new technology, as well as construction health and safety regulations and environmental protection.[112]

Demolition's general guidance related to environmental legislation helping construction and demolition activities comply with environmental law includes

- Australian Standard AS2601-2001, The Demolition of Structures.
- Occupational Safety and Health Regulations 1996 defines demolition and includes regulations that relate specifically to demolition.
  - Regulations 3.114 to 3.128 in Subdivision 7, Division 9 of Part 3 of the regulations.
- Internet Regulations NetRegs website.
- The demolition protocol implementation document from the ICE.
- The Environmental Handbook for Building and Civil Engineering Projects, 2000; Part 3: Demolition and Site Clearance Publication C529 from CIRIA.
- The Control of Dust for Construction/Demolition Activities, 2003, report 456 from BRE.
- The Recycling Building Services, 2003, guidance note 16/2003 from BSRIA.
- The Code of Practice for Demolition BS 6187:2000 from the BSI.
- Site Waste Management Plans, Guidance for Construction Contractors and Clients, Voluntary Code of Practice, 2004, from the DTI.
- Construction (Health, Safety and Welfare) Regulations 1996; Environmental Protection Act 1990 (clause 43).
- Work-safe Australia.

The application of demolition protocols alongside specifications sympathetic to the uptake of recycled construction and demolition waste, if applied to current activities, can result in both environmental and economic benefits, which (in Western Australia, for example) can be translated into tangible and significantly measureable CO_2 emission reductions of 65,734.2 kg of CO_2 per small construction site, and an Aus\$14/tonne saving in the processing of waste materials arising from construction and demolition (these 'savings' are described in more detail in Section 5.2.5.2).

Having introduced the topic of demolition protocols, it is perhaps pertinent to now develop this discussion in more detail. Section 5.2.5.2 deals more explicitly with demolition protocol and recyclate specifications.

5.2.5.2 Demolition protocol and specifications for recycling

The following discussion reviews demolition protocols and opportunities to specify recycled materials in engineering and construction projects.*

The discussion reviews demolition practice and demolition protocol(s), both locally and overseas, towards analyses of potential environmental and economic benefits that might be gained from structured procedures for reusing waste and waste arisings.

Waste production is generally designated into three different streams (classifications used to describe waste materials produced from a particular source); namely, 'municipal solid waste', 'commercial and industrial waste', and 'construction and demolition' (C&D) waste, where C&D waste typically represents one of the largest waste streams generated.

Construction and demolition activities in WA currently produce most of the total waste being sent to landfill locally; the C&D waste stream contributes more than 2 million tonnes to landfills every year, accounting (it is thought) for 50% to 60% of the total landfill. Records of C&D material being sent to landfill in WA began in 1995. C&D material being sent to landfill in the Perth Metropolitan Region has increased periodically despite two main outliers: a decrease in 2001 might be attributed to the introduction of the new Goods and Service Tax (GST), whereas the decrease in 2009 may perhaps be attributed to knock-on competition effects related to the global financial crisis.

> It might be suggested that occasional drops in quantities of waste dumped, indicate that periodic financial concerns impact noticeably upon on C&D waste landfill disposal; it seems that measures and events that 'hit-the-pocket' result in landfill volume decreases.

The quantity of C&D waste recycled annually around Perth is thought to be 0.7 million tonnes, representing gradual increases in C&D material being recycled over the last 5 years. Despite these trends, the actual extent to which recycling is increasing is of a less significant extent when compared with the increasing amounts being landfilled as well as the targets being achieved both in the Eastern Australian states and overseas.

The potential to utilise demolition waste as recycled feedstock for building products and for supply to the construction industry is well recorded;

---

* Input from Nicholas Marshall,[37,113–118] received with thanks.

as a result, a range of stakeholders promote demolition protocol towards improved (waste) resource uptake. The application of these protocols has seen significant improvements in resource efficiency in demolition activity in the UK and Europe. The Waste and Resources Action Programme (WRAP), backed by government funding from the United Kingdom, has published a number of case studies highlighting the economic and environmental benefits achieved by applying these protocols. Despite such international advancements, the Australian Industry and in particular the Western Australian Industry is at the early stages in the implementation of structured processes to guide local demolition practice.

Three relatively well-established inert material recycling facilities are available to Perth (WA) Metro. Given that demolition material is segregated at the source, these recycling facilities provide a relatively inexpensive and resource-efficient alternative to landfill disposal in WA. These recycling facilities achieve recovery rates of between 96.6% and 99.5% for material entering their respective facilities. These are fed by industry whose willingness to recycle varies. In specific test case projects such as those initiated by the Canning District of Perth, best construction and demolition recovery rates resulting from efficient on-site segregation achieve recovery of upwards of 80%.

WA does recognise the need for action and a second draft Waste Strategy for Western Australia was produced by the Western Australian Waste Authority in consultation with the Western Australian community and was published recently. The document highlights areas of importance relating to the generation of waste in Western Australia. In addition, it provides direction towards dealing with this waste including government strategies, future regulations and areas of public opinion.

Waste authorities give two priority actions directly relating to the demolition industry: they recommend to governments that its agencies take 50% of the current C&D waste stream for use as raw material, and propose regulations to empower local government to require and implement waste disposal plans before authorising demolition. Although these strategies suggest the need for a greater diversion of demolition material away from landfill, a means to achieve this end is somewhat unspecified locally; perhaps it might be suggested that (more formalised development and) enforcement of a demolition protocol may help promote resource efficiency.

> Demolition protocols are somewhat assisted by initiatives such as explicit specification clauses allowing crushed recycled concrete within (the local authority's infrastructure body), Main Roads' Specification 501 for pavements.

As with a growing number of public sector organisations, Main Roads WA outlines the use of crushed recycled concrete in both sub-base and base-course applications. This specification has been successfully applied to (test) pavement sections on a range of new highway works.

To allow this section to further assess the opportunities for designers and specifiers to be encouraged in their ongoing uptake of recycled materials, opinions from a panel of experts in conjunction with on-site case study demolition measurements were conducted. The resulting analysis identified explicit local industry requirements such as suggestions that

- Significant recovery rates can be achieved by the demolition industry in Western Australia.
- High recovery rates by recycling facilities can only be achieved through the supply of clean, segregated demolition material.
- The existing specifications and proven performance of recycled products within Western Australia should result in increased procurement of recycled products.
- The current landfill tax levy within Western Australia should be increased from current levels towards a figure of more than $28/ tonne, to align with other Australian states.
- The landfill tax levy cannot be viewed as being the only factor that will influence the disposal practices within Western Australia.
- Legislation to make the recycling of demolition materials a requirement will work strongly towards creating a greater diversion of demolition material away from landfill.
- Tender specification of recycled products and legislative requirements to use recycled products are needed to achieve increased recycled product procurement.
- Significant economic savings can be made through the application of on-site segregation.
- Predemolition audits could potentially result in significant increases in the amount of demolition material being sent to recycling facilities.

The general opinion from the respondent panel of experts was that for greater resource recovery from demolition to occur, perceived barriers restricting recycling need to be addressed, and that any increases in resource recovery are unlikely without market forces that 'hit-the-pocket'. A senior respondent expressed the view that the industry will see a greater amount of demolition material being recycled in about 5 years' time, but added that more structured demolition protocol(s) could potentially see such an increase in the amount of demolition material recycled much sooner. Generally speaking, the continued applications of demolition protocols across the civil engineering/construction industry are expected to significantly assist in targets for more desirable increased demolition recovery rates. Pro forma demolition protocols, such as those promoted by the ICE, exist and include key processes, such as on-site segregation, predemolition audits, cost–benefit analysis and environmental benefit analysis.

On-site segregation
On-site segregation involves the separation of different structural components into similar waste streams. On-site segregation requires the assessment of a proposed demolition on a site-by-site basis, the definition of different waste streams and the production of a site design guide to allow for different waste streams. The potential benefit from waste material will be maximised if a structure or built-asset is taken apart systematically, and divided into waste streams such as concrete, masonry, steel, nonferrous metals, timber, plastic, glass and mixed waste.

Predemolition audit
A predemolition audit is implemented to accurately assess the resource recovery potential of a structure. To do this effectively, the materials/components of the structure must be outlined individually. These components must then be recorded with their corresponding tonnages, volumes, percentages of recycling/reuse opportunities and potential applications. These components are then used to create a BQ as a specific resource production guide for the proposed demolition.

There are three main steps that should be undertaken to improve the efficiency of the predemolition audit. These steps include desk study examining structural drawings, on-site measurement and visual assessment and quality assessment.

Cost–benefit analysis
The most applicable demolition cost analysis to (WA) industry is the comparison between demolition resulting in landfill disposal and demolition resulting in recycling facility disposal. All factors influencing the total cost–benefit of demolition options should be assessed and represented clearly for the client. For demolition resulting in landfill, the costs should include labour/plant, transport and disposal fee inclusive of the landfill tax levy. Similarly, demolitions resulting in reuse/recycling the costs should include labour/plant and disposal fee.

Environmental benefit analysis
Resource efficiency is an important step towards waste minimisation and greenhouse gas reduction. The application of a demolition protocol will significantly reduce both waste to landfill and CO_2 emissions associated with demolition projects. Recent studies find that recycling activities produce net environmental benefits. The results identify the amount of CO_2 per tonne prevented by recycling as opposed to landfill disposal. These individual material component indicators approaches can be applied specifically.

Toward further assessment of opportunities for C&D waste recycling, review of a site-based case study is now discussed, whereby (an internationally adapted and locally adopted) demolition protocol was used to address the demolition component involved in a prominent city centre redevelopment project. Although the city centre location chosen is deemed relevant, it might perhaps need to be placed into the context of urban transport costs. Indeed, given that the state of Western Australia equates to the size of Western Europe, but contains 2.3 million people, rural applicability for the applied measurements cannot be deemed appropriate. Nevertheless, from an urban point of view, the case study remains noteworthy. Notwithstanding the above caveat, the selection of the City Centre Demolition case study analysis was based on the site's high recycling potential, site constraint variables, extensive labour/plant requirements and a central location to provide reasonably indicative urban transportation distances. Results are discussed with terms of audit, B/C ratio and environmental analyses as follows:

Case study: Predemolition audit

A full predemolition audit was undertaken, involving a desktop study examining the structural plans of the existing structure proposed for demolition. The predemolition audit carried out here resulted in the creation of a full BQ for the arising volumes of demolition material. These volumes were then converted into the potential recovered tonnages and total tonnages to calculate the demolition recovery index (Table 5.10).

Case study: Cost–benefit analysis

The cost–benefit analysis of demolition resulting in landfill disposal as opposed to demolition resulting in recycling facility disposal was

Table 5.10 Demolition recovery index

Case study: City centre demolition site		
Material	Total tonnage	Recovered tonnage
Large/heavy concrete	834.92	834.92
Small/rubble concrete	221.04	210
Masonry	92.52	92.52
Metals	52.81	52.81
Timber	1.32	0.65
Glass	4.504	4.234
Miscellaneous	54.0	33.75
Total	1261.11	1228.88
Demolition recovery index	97.4%	

undertaken for the City Centre Demolition case study. The labour/plant costs were higher for recycling facility disposal due to the need for on-site segregation. However, this increased labour/plant cost was heavily outweighed by the economic saving made on lower disposal and transport costs associated with recycling facility disposal. The potential financial saving to be made on the City Centre Demolition case study, through the application of on-site segregation and recycling facility disposal, is shown in Table 5.11.

Case study: Environmental benefit analysis

The total reduction in CO_2 emissions, as well as the landfill capacity savings achieved by sending most of the demolition materials from the City Centre Demolition case study to recycling facility disposal, can be seen in Table 5.12. The reductions in CO_2 emissions are calculated and tabulated as follows.

The CO_2 emission benefit of 65K kg of CO_2 per small site is perhaps deemed to equate to the equivalent of taking 10 standard vehicles off the road for 1 year (based on a 'standard vehicle' producing 6600 kg CO_2/year).

Case study: Benefit

The City Centre Demolition case study presents a representation of both the economic and environmental benefits that can potentially be achieved by applying a resource-efficient demolition protocol, which addresses the relocation of demolition materials to recycling facilities as an alternative to landfill disposal, related to demolition activity (in

Table 5.11 Benefits

Potential economic benefit	
Savings per cubic metre of demolition material	$17.70/m³
Savings per tonne of demolition material	$13.91/tonne
Savings per semitruck load	$280.10/load
Total cost saving	$14,539.20/small site

Note: Potential economic benefit achieved by recycling demolition materials arising from the City Centre Demolition site case study.

Table 5.12 Environmental benefit

Potential environmental benefit	
CO_2 emission benefit	65,734.2 kg CO_2/small site
Landfill capacity saving	801.6 m³/small site

Note: Potential environmental benefit by recycling demolition materials (City Centre study).

WA). Participants in a follow-up verification exercise suggested that the potential economic benefits calculated were deemed reasonable and, if anything, somewhat conservative.

To encourage further resource recovery from demolition activities (in WA), structured demolition protocols might be applied to all demolition projects. Case study analysis provides evidence that the application of the demolition protocol can provide an economic benefit of $14/tonne savings on the processing of arising demolition materials as well as significant reductions in CO_2 emissions in small city centre demolition projects. The lessons learnt from demolition protocols, created and applied to projects nationally and internationally, are that demolition protocols can result in significant economic and environmental benefits.

Having reviewed the demolition protocols in this section, Section 5.2.5.3 extends review of general stream waste material reuse, to include analysis of best practicable environmental options (BPEOs) across the full range of materials that make up the constituents of the construction and demolition waste stream. Section 5.2.5.3 discusses sustainable options for recycling demolition waste materials.

5.2.5.3　Sustainable specification options for recycling demolition materials

This section discusses specification options that stem from the use of recycled materials and reviews BPEOs for demolition waste.*

Disposal routes for demolition materials include product salvage, low-level recycling as bulk fill, higher level recycling, landfill and, where appropriate, incineration for energy recovery. An assessment of the waste management options for concrete and masonry demolition arisings is discussed below towards BPEOs for construction and demolition waste materials. This section reviews opportunities to cross-reference waste management alternatives with environmental impact analyses and cost estimations.

National Waste Plans either explicitly or implicitly make use of BPEO methodologies to direct municipal waste management solutions that balance environmental, financial and social variables. Currently, plans seldom recognise BPEO approaches applicable to the arisings generated by construction and demolition.

Establishing a BPEO for any waste stream requires an understanding of the material generated, decision-making criteria for disposal and then a means to define, appraise, short list and, consequently, choose an optimum waste management option. The next discussion represents a methodology to establish BPEOs for the materials and products that comprise demolition waste.

---

* Input from Tom Dyer and Ravi Dhir,[119–131] noted with thanks.

To allow stakeholders to recognise the full potential of construction demolition arisings as a building resource, improved guidance is needed. Legislation-push from the aggregates levy and landfill restrictions, together with technology-pull from modern recycling techniques able to process more than 95% of demolished material, will continue to exert pressure on reuse and recycling rates for demolition materials beyond current levels. To address this need, guidance in determining BPEOs for demolition waste helps in assessing disposal routes and objectively addresses the environmental credentials of available options, identifies savings in disposal and transportation, generates returns from the arisings and reduces the demand for virgin raw materials.

Using concrete and masonry demolition arisings to illustrate the application of BPEO guidance techniques, a range of alternative disposal approaches are identified and compared. Stepwise plans to determine BPEOs allow analysis of each waste management option, together with a summary of the 'most' sustainable option for each subcategory of demolition waste.

The construction and demolition waste stream is said to constitute a majority of the controlled waste arisings (as described previously in Section 5.2.5.2, which discusses demolition protocols) and can be divided into coding subcategories. These divisions indicate the level of sorting, reuse and recycling possible for demolition waste across the construction industry and can be noted as being composed chiefly of concrete rubble and masonry products, and to a lesser extent, bitumen and soil, with the remainder a mix of products composed of gypsum, timber, glass, plastic, ferrous metal, non-ferrous metal, cabling and insulation.

Contract conditions and design specifications that are inclusive towards recyclates, such as using recycled aggregates, as permitted constituents for applications in the specification for highway works (again as mentioned previously in Section 5.2.5.2, dealing with demolition protocols), should continue to encourage the uptake of recyclable materials. Similarly, fiscal controls and regulations, as well as so-called 'green' procurement and specification processes can encourage demolition waste reuse and recycling, particularly perhaps in projects in the public sector.

Although technical expertise exists to utilise demolition waste arisings as new building products, it must be recognised that issues such as consumer confidence in secondary materials and transportation issues require addressing. Costs also represent a major determinant in the uptake and utilisation of demolition arisings, as well as industry's propensity towards disposal alternatives such as salvage and reuse, low-level processing as fill, high-level processing as feedstock for reconstituted building products, landfill and, where appropriate, incineration for energy recovery. Stakeholders must be able to make an objective comparison of the range of waste management alternatives available and identify the BPEO after consideration of respective environmental impacts, costs and relevant social safeguards. Environmental impact requires clarification.

Environmental impact and life cycle analysis

The choice of the BPEO to manage arisings requires EIA in terms of the constituent resources and energy associated with product inputs, as well as the pollution and waste from product outputs. Life cycle assessment (LCA) provides a basis on which to compare environmental impacts of building materials, by identification and quantification of the energy and raw materials used and emissions and wastes consequently released, the assessment and evaluation of potential impacts and an assessment of opportunities to bring about environmental improvements.

Judgements about different production processes and their effect on environmental impact comparison categories are seldom straightforward. Ranking and a simplified system of scoring or weighting environmental impacts, although somewhat subjective, allows data to be much more accessible. Weightings (representing a snapshot consensus of stakeholders) can be prepared for environmental impact categories such as climate change, ozone depletion, low-level ozone creation, ecotoxicity and the like.

Environmental impact expressed in terms of a single unit can assist comprehension. One such unit is the Ecopoint, in which all environmental impact values are estimated and a single collective impact is obtained. Low Ecopoint values indicate a lower environmental impact, so that a simple comparison of two alternatives identifies the lower Ecopoint value as the preferred option.

Waste management alternatives for each of the subcategories of the demolition waste stream can be described in a series of disposal schematics. Each disposal schematic depicts the disposal options and the range of processes that contribute to the overall environmental impact of each option. By way of example, Figure 5.3 illustrates, in general terms, concrete and masonry waste management options and identifies manual salvage, low-level bulk-fill sub-base-fill option, as well as high-level recycling options that involve the preparation of recycled aggregates as well as recycled concrete aggregates.

An approach to predict and subsequently compare the environmental impact of each waste disposal option builds on environmental life cycle analysis techniques for material production, plant use and emissions.

Environmental assessment data (environmental impact) can be presented in the form of Ecopoints, cost (economic) data is estimated in unit-rate costs-per-tonne, and a checklist of (everyone's) social safeguard legislation is identified. A case study that uses concrete and masonry waste material is

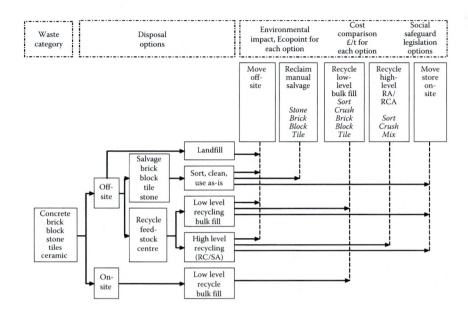

Figure 5.3 Disposal schematic for concrete/masonry waste.

discussed below to illustrate how the triple bottom-line three E's (environment, economics and everyone) criteria allow comparison of the range of available disposal options. Case study details for concrete follow:

Case study: Concrete and masonry waste materials

Primary, naturally occurring aggregates of crushed rock, sand and gravel are used in a wide variety of construction projects. Although the annual consumptions are met chiefly by quarrying or marine dredging, secondary recycled aggregates present an increasingly attractive alternative to virgin materials. The disposal schematic above describes the waste management options for demolition arisings made-up of concrete, brick/block, stone, tile and ceramic. Options include on-site and off-site reuse and reprocessing, off-site salvage, higher level use and landfill, in which

- Utilisation on-site of the processed inert rubble as low level bulk fill is common
- Reclamation and retail of cleaned (cement/mortar-free) stone, brick, block, tile, slate, beam, column and mouldings are possible
- Higher level recycling (sorting, crushing and combining of the arisings) as recycled concrete aggregate (95%–100% crushed concrete) and recycled and secondary aggregates (0%–94% crushed concrete) presents opportunities for resource efficiency within the construction industry

Environmental impact analyses of primary and secondary aggregate products require consideration in terms of input energy, output fuel emission, environmental impact, cost analysis and social safeguard legislation, as follows:

Input energy

The production of primary aggregates chiefly concerns the extraction and excavation of sands and gravels from dry and wet pits, as well as the crushing of igneous and other types of rock. Housekeeping of product stockpiles is carried out by mobile loading shovels and transport trucks. The processing of secondary aggregates and demolition rubble waste arisings requires similar levels of processing, crushing and stockpiling. The energy requirement for the production of crushed rock for construction is 15.4 kWh/t, with sand and gravel production requiring 10 kWh/t. Transportation and stockpile housekeeping, using a standard 15-tonne lorry to move 1 tonne of aggregate material over 1 km requires 0.014 kg of diesel.

Output fuel emissions

Secondary research provides the emission factors from diesel fuel and electricity generation required to crush igneous rock/stone, the emission factors for static diesel engines used in the production of fine aggregate and the diesel fuel emission factors for moving and housekeeping activities for aggregate over 1 km. Energy rates vary somewhat, with 1 tonne of diesel used in moving-plant equivalent to 11,600 kWh of energy, 1 tonne of diesel is equivalent to 11,889 kWh of energy in the production of fine aggregate and 1 tonne of diesel is equivalent to 12,667 kWh of energy in the production of crushed rock.

Environmental impact

Environmental impact values, measured in Ecopoints, associated with the production of 1 tonne of primary sand for construction purposes, the processing of 1 tonne of igneous rock and the environmental impact associated with the movement of 1 tonne of each material are tabulated in Table 5.13. The table also includes impact values for the production of secondary aggregates from demolition arisings.

Impact values calculated as Ecopoints (in which the lower the Ecopoint is, the lower the impact) can be tabulated as given in Table 5.13.

Cost analyses and social safeguard legislation

To assist the comparison of cost for each disposal option, a number of assumptions are required. Cost assumptions for the processing of concrete and masonry waste arisings might be deemed to include cost to dispose of inert demolition material to landfill of more than $9/t, cost to supply and deliver new virgin aggregates of more than $19,

Table 5.13 Impact

Impact category	Units	Impact/t		
		Sand production	Crushed rock production	Transport materials
Climate change	kg CO_2 eq (100 years)	1.83×10^{-4}	1.07×10^{-2}	0.1400
Acid deposition	kg SO_2 eq	2.79×10^{-4}	2.36×10^{-3}	0.0390
Human toxicity air pollution	kg tox	1.73×10^{-4}	1.90×10^{-3}	0.0040
Low-level ozone creation	kg ethene eq	4.33×10^{-7}	2.49×10^{-6}	0.0040
Eutrophication	kg PO_4 eq	8.01×10^{-5}	5.40×10^{-4}	0.0410
Waste disposal	tonnes	0.000	0.000	0.0000
Fossil fuel depletion	toe	1.92×10^{-4}	3.33×10^{-3}	0.0006
Minerals extraction	tonnes	0.200	0.6000	0.0000
Primary aggregate Ecopoint		0.611	0.6189	0.2283
Secondary aggregate Ecopoint		0.411	0.0189	0.2283

Note: Impact values for sand and rock production and transportation.

all-in cost to prepare recycled aggregate for feedstock of more than $13/t, minimum extra-over transportation fees of more than $4/t 24 km, costs to supply and deliver new bricks of more than $120/t and salvaged bricks of more than $90/t and general costs to salvage and prepare used masonry products of more than $7/t.

The location and nature of the demolition works and recycling facilities determine social safeguard legislative considerations. BPEO guidelines require stakeholders to have a checklist of all relevant legislation within a number of overarching fields including air pollution, contractual arrangement, contaminated land, dust nuisance, duty of care, environmental assessment, health and safety, landfill, noise nuisance, planning, public input, transportation and site plant. Although several/several dozen pieces of legislation are identified, a user checklist (such as that described in Section 4.4 on Occupational health and safety in construction) should assist compliance in safeguarding social values.

Transportation

Transportation is estimated (and tabulated in Table 5.13) to have an impact value of 0.2283 Ecopoints per t/km. Transportation therefore contributes greatly to environmental impact where a radius of 24 km for salvage and recycling is not uncommon. Again, this relates to an urban environment and is deemed not to apply to WA expansive rural climes. Nevertheless, for city-based projects, the scenario remains reasonable. Avoidance of double-handling during stockpile housekeeping, and limiting the amount of recycled product haulage will improve environmental impact considerably. Opportunities to reuse salvaged concrete and masonry products on or near the site of arising, as part of subsequent developments, should be explored.

Ecopoints

Ecopoints presented above for transportation and the production of primary and secondary sand and aggregate allow the assessment of the environmental impact of waste management options for concrete and masonry materials. Table 5.14 describes disposal options alongside their respective Ecopoint values and costs per tonne. The rankings tabulated for both environment and cost variables can be used to assist in the determination of the BPEO for the waste management of concrete and masonry, in which a rating of 1 signifies the most desirable BPEO and a rating of 5 signifies the least desirable option.

After product salvage and reuse locally, on-site crushing and material recycling as low-level fill gives the next most desirable environmental impact compared with other options.

Off-site product salvage provides a desirable BPEO depending on the nature of the product involved, although material recycling off-site is found to be comparable to the combined environmental impact(s) of landfill coupled with the supply of replacement primary material.

Costs associated with the salvage of individual quantities of bricks, blocks, ceramics and tiling depend on the type and quality of the available product. Cost analysis finds that product salvage and reuse near the site of origin, followed by on-site material recycling and utilisation, to be the most cost-effective waste management technique. Off-site material recycling is less expensive than landfill and virgin material supply.

Table 5.14 Waste ranking

Options	Variable		Ranking	
	Ecopoint	$/t	Environment	Cost
Landfill + primary material replacement	11.806	29	4	5
On-site material recycling	0.868	13	2	3
Off-site material recycling	11.826	16	5	4
Product salvage: on-site	a	7	1	1
Product salvage: off-site	11.187	12	3	2

Note: BPEO rankings for concrete and masonry material arisings.

a   Manual salvage: nominal impact.

5.2.5.3.1 Stepwise guide for demolition waste recycling

The following stepwise plan describes an approach to determine the BPEOs for recycling demolition waste.

1. Identify and categorise demolition arising.
2. Identify waste management options for demolition arising under examination.
   a. Describe disposal options and present disposal schematics.
3. Prepare Ecopoint spreadsheet.
   a. Replicate the Guide's environmental profile/Ecopoint spreadsheet.
   b. Where necessary, review the spreadsheet's environmental category weightings:
      i. Request new stakeholders to (re)rank environmental categories.
      ii. Adjust spreadsheet weightings via stakeholder consensus.
4. Identify environmental profile data references for processing waste arisings and primary production.
   a. Use search mechanisms to source existing databases for environmental profiles relevant to waste management options and product preparation.
      i. Establish input energy and resources in kg/t.
      ii. Establish output pollutants in kg/t.
5. Input environmental profile data into Ecopoint spreadsheet.
   a. Record spreadsheet's Ecopoint value for each disposal option.
6. Compare Ecopoints for each disposal option to find lowest value.
   a. Compare options against supply of primary resource (plus landfill).
7. Compare total costs for each disposal option.
   a. Identify unit-rate cost references relevant to product processing.
   b. Input costs to economic summary table(s).
8. Identify relevant social safeguard mechanisms for each disposal option.
   a. Source relevant planning and legislative documents.
   b. Consult documents to ensure necessary compliance for disposal options.
   c. Prepare and compare disposal option noise levels where appropriate.
9. Rank disposal options where lowest values receive a (preferred) ranking of 1.
10. Determine BPEO for demolition arisings.
    a. Compare rankings for environmental impact and cost to guide choice.

The stepwise method above has been used to analyse the disposal options (and associated haulage issues) of the demolition waste subcategories of concrete (as previous), glass, insulation, gypsum, bitumen, timber, plastic, ferrous metal, nonferrous metal, cable and soil.

The findings are summarised below and tabulated such that columns describe the variables (Tables 5.15 and 5.16):

- Column 1, the waste subcategory
- Column 2, the recycling process and disposal options
- Column 3, the Ecopoint estimated for each recycling procedure
- Columns 4 and 5, the Ecopoint and cost per tonne for each disposal option
- Columns 6 and 7, the ranking system to guide choice of BPEO, in which
    - A ranking of 1 is deemed the most desirable BPEO waste disposal option.
    - A ranking of 4 or 5 is deemed the least desirable waste disposal option.

The determination of BPEOs for managing demolition waste allows stakeholders to consider, in an easy-to-use format, the environmental impact of input energy and resources and output pollution of alternative disposal routes. Environmental impact estimations (using Ecopoints) complement an examination of cost and a legislation checklist, allowing comparison of disposal options in an objective and comprehensible way.

Resulting from the discussion above, opportunities to review a full range of construction activities and building types in terms of their respective environmental impact and cost variables might now be deemed relevant. Next, Section 5.2.5.4 reviews opportunities to assess the life cycle considerations of historic structures towards their deconstruction and subsequent reuse of building material arisings.

5.2.5.4 Life-cycle assessment (LCA), waste reuse and recycling

This section reviews the extent to which deconstruction and a subsequent reuse of waste-arising building material might be relevant to historic structures.*

The planned deconstruction of built assets (literally taking buildings apart for maintenance, refurbishment, retrofitting or demolition purposes) has taken on increasing importance in recent years, due to the finite nature of raw materials, respective cost increases and a desire to recognise environmental imperatives. This section discusses the extent to which best practicable environmental (and financial) options for the waste management of building maintenance and demolition arisings can be identified.

Potential deconstruction of listed/historic buildings and historic town centre maintenance is reviewed. Material scarcity juxtaposed with a reduced availability of skilled tradespeople is likely to have an effect on the condition

---

* Input from Richard Laing,[111,119,124,125,127,128,130,132–154] received with thanks.

Table 5.15 Waste management options

Waste	Waste management option		Process value	Disposal value		Ranking	
	Recycling process	Disposal option	Ecopoint	Ecopoint	$/t	Environment	Cost
Concrete, masonry arisings	Sand/fine-aggregate production; primary		0.6110				
	Rock/crushed aggregate production; primary		0.6189				
	Sand/fine-aggregate production; secondary		0.4110				
	Rock/aggregate crushed production; secondary		0.0189				
	Transportation of material, t/km		0.2283				
		Landfill + primary material		11.8056	28	4	5
		On-site material recycling		0.8676	12	2	3
		Off-site material recycling		11.8260	15	5	4
		Product salvage: on-site		[a]	6	1	1
		Product salvage: off-site		11.1867	11	3	2
Glass	Glass manufacture: primary		8.24×10^6				
	Glass manufacture: secondary cullet		7.61×10^6				
	Glass reprocessing aggregate: secondary 20 mm		1.88×10^{-2}				
	Glass reprocessing aggregate: secondary 10 mm		2.89×10^{-2}				
	Glass reprocessing as aggregate: secondary 5 mm		4.31×10^{-2}				
	Glass reprocessing as aggregate: secondary 0.063		4.31×10^{-1}				
		Off-site recycling: container glass		7.61×10^6	[b]	1	1
		Off-site recycling: glass fines		7.61×10^6	57	1	3
		Landfill + primary material		8.24×10^6	46	3	2
		Product salvage[b]					

(continued)

Table 5.15 Waste management options (Continued)

| Waste | Waste management option | | Process value | Disposal value | | Ranking | |
	Recycling process	Disposal option	Ecopoint	Ecopoint	$/t	Environment	Cost
Insulation materials	Stone wool:	Low-grade recycling/reuse	1.5973	109.597	10	1	1
	Paper wool:	Low-grade recycling	1.4133	109.413	10	3	1
		Incineration: energy recovery	1.2978	109.298	52	1	3
		Landfill	1.4102	109.410	28	2	2
	Flax:	Low-grade recycling	2.6353	110.635	10	2	2
		Incineration: energy recovery	2.5331	110.533	52	1	4
		Landfill	2.6377	110.638	28	3	3
		Compost	2.6412	110.641	9	4	1
Gypsum-based products	Manufacture with primary resources		2.1300				
	Manufacture with recycled resources		1.6400				
	Incineration for energy recovery		1.2300				
	Soil enhancement		0.1000				
		Landfill + primary material		13.3167	28	4	3
		Manufacture: recycled feedstock		12.8267	15	3	2
		Incineration: energy recovery		12.4167	52	2	4
		Recycling: soil enhancement		11.2867	9	1	1
		Proprietary product salvage†					

Bituminous mixes/asphalt pavement				
Cold-mix bitumen emulsion: primary	7.0700			
Hot-mix limestone filler: primary	0.8800			
Hot-mix bitumen: primary	2.3100			
Cold bituminous mix paving production: primary	0.0120			
Hot bituminous mix paving production: primary	0.0588			
Processing bituminous material: secondary	0.0150			
Remixing of bituminous pavements: secondary	0.0405			
Landfill + primary material	18.2567	60	5	3
Recycle off-site: hot mix	13.9555	63	3	4
Recycle off-site: cold mix	17.2087	70	4	5
Recycle on-site: process/tread	10.9106	59	1	1
Recycle on-site: remix/repave	10.9361	59	2	1

Note: A summary of preferred options for demolition waste (part 1 of 2).

a Manual salvage: nominal impact.
b Established reuse/recycling route: intrinsic value.

Table 5.16 Waste management options

Waste	Waste management option		Process value	Disposal value		Rank	
	Recycling process	Disposal option	Ecopoint	Ecopoint	$/t	Environment	Cost
Timber/particle-board	Timber landfill		0.8392				
	Timber incineration		0.1182				
	Timber grinder/processor chip manufacture		0.0105				
	Particleboard manufacture: secondary		2.8634				
		Landfill + primary material		12.0364	28	4	2
		Product salvage: reuse off/(on)-site		11.1867	26	1	1
		Recycle: grind/chip off-site		11.1972	35	2	3
		Recycle: board manufacture off-site		14.0501	46	5	4
		Incinerate product		11.3154	52	3	5
Plastics	PVC landfill		4.6496				
	PU landfill		1.4798				
	PVC incinerate		2.4437				
	PU incinerate		2.5589				
	Thermoplastic mechanical recycle		0.1939				
		Landfill PU		21.7700	60	2	2
		Off-site mechanical recycling PU		17.3200	3	1	1
		Incinerate PU		19.5700	83	3	3
		Landfill PVC		18.6000	60	3	2
		Off-site mechanical recycling PVC		17.3200	3	1	1
		Incinerate PVC		19.6800	83	2	3
		Product salvage[a]					

Ferrous metals (steel)	Steel manufacture	5.6410		
	Steel recycling: secondary feedstock at 8% of total	0.4513		
	Supply material			1000
	Reclamation			200
	Manufacture: primary		11.1202	300 [a]
	Recycle		5.9302	220
Nonferrous metals aluminium, copper	Aluminium manufacture	15.3800		
	Copper manufacture from concentrate	7.9500		
	Copper manufacture from scrap metal feedstock	2.4800		
	Copper manufacture from scrap cable feedstock	2.8871		
	Supply material: aluminium		20.8592	1700 [a]
	Supply material: copper		13.4292	3000
	Manufacture: aluminium		15.3800	1200
	Manufacture: copper		7.9500	1500
	Recycle: aluminium		0.7–4.16	2000
	Recycle: copper		2.48–0.89	3000
Cables, PVC, PU, Cu	PVC cable landfill	3.4062		
	PU cable landfill	1.2018		
	PVC cable incineration	1.5129		
	PU cable incineration	1.5631		
	Cable component: copper recycling	2.8871		
	Cable component: thermoplastic recycling	0.8470		

(continued)

Table 5.16 Waste management options (Continued)

| Waste | Waste management option | | Process value | Disposal value | | Rank | |
	Recycling process	Disposal option	Ecopoint	Ecopoint	$/t	Environment	Cost
		Manufacture primary			2100		
		Landfill PVC		9.1137	60	3	2
		Recycle scrap PVC		6.5545	−60	1	1
		Incinerate PVC		7.2204	80	2	3
		Landfill PU		6.9093	60	2	2
		Recycle scrap PU		6.5545	−60	1	1
		Incinerate PU		7.2706	80	3	3
		Product salvage[a]					
Soil		Transportation of material, t/km	0.2283				
		Composting: upper limit	2.9957				
		Composting: typical constituent	1.7055				
		Crushing stone/rock	0.6189				
		Landfill + primary material		11.8056	50	4	4
		Landscape on-site		0.4566	30	1	1
		Landscape off-site		5.9358	35	2	2
		Recycle compost/agricultural enhancement		7.6413	40	3	3

Note: Demolition waste preferred options (part 2 of 2).

a Established reuse/recycling route: intrinsic value.

and appearance of urban centres in the future. If regional variation in material quality, characteristic and aesthetic is deemed important, then the development of a sound pragmatic approach to deconstruction within historic locations is needed to ensure a continuity between, on the one hand, the architecture of the past and, on the other hand, new development essential to the continued vitality of urban areas. This section describes deconstruction processes.

> The deconstruction process of historically valuable built assets must integrate the design of new work with specifications that stipulate both the use of recyclates and indicate where these recycled materials may be obtained.

In this way, perhaps such a structured approach for historic buildings/ historic areas may help contribute to the upkeep of a location's built heritage.

A perennial problem in construction is that projects seek to alter, rather than work within, an environment; new assets should perhaps seek to recognise, catalogue, process and rework same-location (demolition) materials for future reuse. Existing high-quality natural materials (sandstone limestone and granite) emanating from what might be termed the existing local built heritage(s), require sustainable reconditioning detailing and skilled (re)application.

This type of relationship, between economic performance, social sustainability and environmental conditions, has been somewhat central to many urban policy-makers in recent years. What might be termed an environmental pillar is to specifically recognise the need for economic growth to *not* be at the expense of natural resources. Of relevance, too, is to work within the various location-specific economic recovery plans which, in addition to noting the importance of accelerating key construction projects to facilitate an economic recovery across a region, specifically identify an improvement in the energy efficiency of buildings as being central to a region's economic recovery.

Towards sustainable policy, it is noted that buildings benefit considerably from planned preventative maintenance, the extent to which this is prevalent within the existing built-heritage has been identified as requiring a step change in organisation and practice; the following discussion seeks to identify and address the skills and management of deconstruction (literally, as mentioned above, taking buildings apart for maintenance, refurbishment, retrofitting or demolition purposes).

There is a clear relationship between the user or owner of a building, which must be recognised and explored, and the manner in which that building will perform over time. This inevitably also demands a consideration of the manner in which one addresses a longer term consideration of building operations, including disposal of whole or part of a structure as it reaches the end of its useful life.

Life cycle costing analysis (LCCA) and life-cycle assessment (LCA) have been developed as methods to understand the costs, resources and demands of the built environment over the building's whole life cycle, including the planning, design, construction and occupancy stages.

Most LCCA studies have tended to conclude that the vast majority of resources, perhaps as much as 90%, are likely to be consumed after construction has taken place, yet the embodied energy required to replace a building simply because it fails to meet current functional requirements is often overlooked; older (listed) buildings require considerable thought beyond residual considerations.

Although a significant amount of work over many years has demonstrated the applicability of life cycle costing in construction, a failure to engage with LCCA across the building sector is noted. On the other hand (noting the UK perspective), processes developed since the 1990s including private finance initiative (PFI) and public private partnership (PPP) do try to place an emphasis on the management of the building and its fabric, as part of a management contract.

Tendering arrangements aside, the choice of the BPEO to manage construction maintenance, retrofitting or (partial) demolition and its resultant waste arisings requires EIA in terms of the constituent resources and energy associated with product inputs, as well as the pollution and waste from product outputs.

As mentioned previously (in Section 5.2.5.3, which discusses sustainable specification options), LCA provides a basis on which to compare the environmental impacts of building materials by identification and quantification of the energy and raw materials used and emissions and wastes consequently released. Indeed, weightings (representing a snapshot consensus of stakeholders) can be prepared for environmental impact categories such as climate change, ozone depletion, low-level ozone creation, ecotoxicity and the like, such that environmental impact can be expressed in terms of a single unit to assist comprehension.

> It is to be recognised in many urban centres around the world that most buildings that will exist in 25 years' time have already been built:
>
> - It is vital that stakeholders address sustainability effectively, either refurbish and 'retrofit' existing buildings so that they meet current and future standards for energy performance or change the building function or user behaviour within the space.

In recent years, numerous historic bodies have completed studies related to 'change works', suggesting that specific and quite detailed technical solutions can be applied to historic properties, including those which may lie within World Heritage sites.

Studies over the past 5 to 10 years represent a fairly comprehensive series of technical guidelines for owners/users of historic buildings through the process of being able to retrofit those buildings for better energy performance, without necessarily destroying the fabric that formed a basis for conservation taking place in the first instance.

The materials and skills for satisfactorily salvaging and reusing materials to support maintenance, repair and new builds using traditionally common building materials and techniques requires a planned and systematic approach that takes into account materials and skill scarcity and waste management, as outlined below

Materials and skills scarcity

Improved guidance is needed to allow stakeholders to recognise the full potential of construction demolition arisings as a building resource. The introduction of a landfill tax across many of the world's centre's ensures that the construction and demolition industry is becoming more aware of material usage and the need to recycle and reuse waste materials.

Concerns regarding the physical and practical suitability of the material must be addressed, including material strength, durability and the original quarry source. This is quite different, of course, from the more usual recycling of masonry in which the emphasis would be on reuse as an aggregate, rather than as a high-grade building material.

It can perhaps be suggested that some of the main reasons for reluctance amongst clients to use recycled stone materials includes the limited knowledge of availability, lack of standards covering their use and concerns about an increased risk of liability.

Associated with risk is a wider concern over the lack of a skilled workforce in many countries equipped to deal with the materials resulting from reclamation; attempts to reskill need to be addressed.

There remains widespread confusion about the importance of using traditional materials within the industry, which perhaps leads to inadequate specification; only a small proportion of materials used in the maintenance and (re)construction of traditional buildings were of local origin, with the figure standing at less than 25% in many areas.

In many towns and cities, the traditional building material (to refurbish with a view to maintaining continuity) would have been sourced from nearby; such sources are now seldom available. Thus, given the unavailability of original specifications, salvaged materials present the only real alternative for historical retrofitting. The salvage supply chain is therefore important. High levels of sorting, reuse and recycling are possible for demolition waste, and empirical comparisons are now possible to allow informed decisions regarding reuse and recycling, to further encourage an uptake of recyclates and material reuse (as discussed in Section 5.2.5.3, sustainable specification options).

Waste management

Technical expertise exists to utilise waste arisings as new building products. European specifications BS/EN-12620 coupled with BS 8500 as an addendum to EN206, encourage greater certification and specification of recycled materials. Alongside new standards for materials specification, revised codes of practice reexamine demolition (as discussed previously in Section 5.2.5.2, demolition protocol) to take account of advances in new technology as well as construction health and safety regulations and environmental protection (as discussed previously).

Costs represent a major determinant in the uptake and utilisation of demolition arisings by industry and propensity towards the disposal alternatives of salvage and reuse, low-level processing as fill, high-level processing as feedstock for reconstituted building products, landfill and, where appropriate, incineration for energy recovery.

Waste management alternatives for each of the categories of the demolition waste stream can be described depicting the nature of the demolition material, disposal options and the range of processes that contribute to its overall environmental impact. A methodology to predict and subsequently compare the environmental impact of the range of waste disposal options builds has been presented above (Section 5.2.5.3.1, stepwise guide to determine demolition arisings).

The findings show that after salvage and reuse locally, on-site crushing and recycling as low-level fill gives the next most desirable environmental impact compared with other options. Off-site salvage provides a desirable BPEO depending on the nature of the material involved, whereas recycling off-site is found to be comparable to the combined environmental impact(s) of landfill coupled with the supply of replacement primary material.

With regards to practice within the construction industry, care quite clearly needs to be taken when addressing the issues described above. The 'built heritage', almost by definition, concerns built artefacts that society may consider as holding sufficient value, both intrinsic and functional, that they should be passed from one generation to the next. Therefore, the concept of a buildings' useful life requires careful consideration in relation to the built heritage, and wider concepts of waste recycling, material sourcing and design for deconstruction do not so easily relate to the subject matter of heritage studies. Indeed, the topic of planned deconstruction or demolition of the built heritage is, perhaps intrinsically, a complex and difficult subject with which to engage.

Nevertheless, where buildings constructed using traditional materials are demolished, for whatever reason, there is good reason to ensure that materials will be sympathetically deconstructed, reconditioned, catalogued and stored. Ultimately, the themes of waste reuse within construction require a corresponding need to address a lack of primary materials and skills within the built heritage and demand a coordinated strategy. Policies have been developed in response to the growing awareness of these issues, and practice must follow suit.

Having discussed at length deconstruction and recycling construction and demolition materials, along with the need for specification choices that are sympathetic towards the incorporation of recyclates in design solutions, there is need perhaps to now discuss the extent to which it is not just civil engineers that must play a part but the full range of design team participants.

Cross-disciplinary specification for construction activities, whether using recycled materials or virgin materials, increasingly require an approach that is able to integrate the full range of specialist design input; one such integrated approach, that of building information modelling (BIM), is discussed next in Section 5.2.6. Section 5.2.6 reviews specifications and the extent to which it is affected by BIM.

5.2.6 Specifications and BIM

Having developed specialist design specifications, a need arises to integrate such data. BIM or built-environment digital information modelling integrates interdisciplinary specialist designs and element-by-element specifications (alongside knowledge-management processes for a collective technical database). BIM incorporation of life cycle data facilitates virtual prototyping of development proposals and can be used to evaluate (in the long term) a design engineer's specification choices before committing to construction.

Digital technologies are available (in Australasia) and are increasingly being promoted by clients and, not least, design package suppliers/vendors,[155] albeit explicit applied model awareness of BIM in civil and construction engineering remains less widespread.

Professional bodies are addressing, through structured continuing professional development courses, any perceived lack of knowledge across the supply chain of the parties involved in design and construction. Training is beginning to look at communication disparities amongst consultant expert-engineering disciplines, and the historical reluctance to share BIM information with contractors (not least as a result of concerns about risk-sharing and liability).

BIM training courses are continuing to address

- A variance of opinions on BIM standards and BIM guidelines
- Confidence disparities in the trustworthiness of all of the data entered

- Potential misunderstandings of the extent to which life cycle analysis might, practically, allow design to go beyond (traditional two-dimensional drawings) and
  - Current three-dimensional (3-D) approaches for the construction phase
  - Help with fourth-dimensional (4-D) time-sequencing for on-site activities
  - Give practical fifth-dimension (5-D) life cycle cost analyses and address
  - Sixth-dimensional (6-D) facility management and maintenance throughout an asset's entire usable life towards subsequent decommissioning/recycling

A key regional specification system supplier, NATSPEC,[156] believes that digital information, including 3-D modelling and BIM, will enhance methods of specification for engineering design; NATSPEC supports what they call 'open global systems' towards efficiency gains through a sharing of specifications.

The NBS national BIM library[157] features a range of generic and proprietary construction elements suitable for BIM including

- Five hundred and fifty proprietary and preconfigured generic objects covering all major building fabric systems for walls, ceilings, roofs and floors as a primary source of free-to-use platform-neutral UK BIM objects.

Although BIM is increasingly presented as having the potential to generate greater efficiency and knock-on cost-saving benefits,[158] it remains somewhat early in the day to determine the full extent to which significant tangible benefits accrue (and can be shared equitably) from managing knowledge bases cumulatively by the use of an all-in collective information (BIM) model for every and all available design decisions. The discussion in Section 5.2.6.1 reviews any current tangible benefits in BIM uptake.

5.2.6.1 BIM, the construction industry and specifications standards

This section reviews BIM and the extent to which integrated project delivery (IPD) is being embraced in the workplace.*

The following discussion reviews the uptake of BIM and the extent to which it may be enhanced by both public and private client support, as well as by increased company training initiatives to combat the concern felt by

---

* Input from Mark Luca,[159–176] received with thanks.

many local industry professionals regarding software application capabilities, standard compliances and legislative responsibilities.

This discussion explores the extent to which the large upfront cost of purchasing suitable software, alongside increased IT and technical support, is a potentially detracting factor in BIM system implementation locally. The increase in drafting costs by up to 25% (to address BIM) must be seen by stakeholders to produce knock-on tangible improvements in the productivity of the construction process and a reduction in change orders and information requests.

Potentially improved collaboration within design and construction team(s) resulting in faster, less-expensive fit-for-purpose (WA mining infrastructure) projects, can justify the increase in BIM system drafting costs, with staff training as one of the larger cost variables contributing to introducing a BIM system. Expenditure from companies is perceived to be compounded by a lack of any real practical support from training bodies and industry associations. Although industry recognises that BIM has the potential to reduce interparty disputes by the provision of clear and concise information, a need remains for industry and professional associations to address legal/contractual issues and the associated risks with information-sharing on a national (and international) scale.

Building information modelling having grown from earlier initiatives towards computer-integrated construction and ultimately seeks to integrate information bases. In the light of a construction industry that has become more specialised, leading to fragmentation resulting in data duplication, knowledge management inefficiencies and uncoordinated data source integration.

As each specialist produces a package of work, industry strives to integrate these for the client and respective stakeholders; a historical problem is the lack of coordination of the different specialist activities. With the adoption of an integrated system such as BIM, the whole process may arguably become less fragmented, more streamlined and positively address communication and quality control.

Alongside an implementation of BIM are the perceived challenges of modelling related costs and training, with many small to medium-sized enterprises and companies hesitant to embrace BIM because of the daunting and costly process of retraining staff. The Australian Government Department of Innovation, Industry, Science and Research suggests that there should be limited knowledge and training across the entire supply chain (which is acting) as a barrier to greater adoption in the industry, and further emphasises that training is critical to increasing BIM implementation. BIM uptake might be discussed in terms of ongoing and prospective perception, software, education, liability productivity, costs and communication, as follows:

Perceptions

The perception of BIM in industry varies between different user groups. Contractors seem to have the most positive perception of BIM; currently, the lightest users, they are expected to have the greatest uptake in the future. Whether involved in informational modelling collectives or not, most industry members have experienced IPD approaches to projects.

Software

Industry needs to develop greater expertise in software systems in creating and managing modelling; a barrier to IPD is the limited ability of different BIM platforms to share data and support supply chain workflows. Although 64% of users experience gains, 14% argue that the costs of BIM do not balance its benefits; standards require development in both industry and software.

Education

Industry argues that training staff with new software is a cost to business, not a long-term investment; however, amongst explicit users of BIM, half report that offering new BIM services significantly benefits their business. Pressure from new graduates for training, as well those undergraduate educators, is likely to drive instruction.

Liability

Current legal frameworks hinder BIM adoption with no articulation of embedded data nor (whole-life) risk and audit; new legal frameworks are required to address data risk, intellectual property, fees, responsibilities, legal liability and insurance. Confusion exists over supply chains, procurement, model ownership and dispute resolution processes.

Productivity

BIM technology designs can help construction/fabrication efficiencies. Two-thirds of BIM users report a positive return of investment overall. Although drafting costs seem to increase by 22% at the outset, over time, there are likely to be fewer change orders and requests for information, faster collaboration and time savings in construction sequence options.

Costs

BIM can result in reductions of approximately 7% in manufacturing and 9% in installation time across all project costs. Government reports are upbeat and suggest that a cost reduction through BIM use in Australia (during the next 10 years) might result in an estimated one-of (beneficial) increase in GDP of Aus$5 billion.

Collaboration and communication

An even greater opportunity to integrate work and work together is suggested to result from IPD/BIM. Changes to communication systems may reduce several independent interdisciplinary paths to one, allowing faster and more effective information sharing.

The discussion below builds on previous perceptions related to costs, training, liability, productivity effects and collaboration levels by seeking educational and industry opinions related to stakeholder opinion in terms of uptake and ongoing rollout of BIM in the local (Western Australian) civil engineering and construction sector.

Educational online student survey

A straw poll was conducted and might be used to suggest that only 15% of student respondents had any real understanding of the underlying approaches to BIM (other than a feeling that it might be 'new software'), implying that universities are, thus far, not managing to educate students about 'explicit' BIM relevancies. Somewhat aligned with this, although the majority (80%) indicated that they would like more instruction about computer information modelling, the survey did however reveal that most final-year students already felt competent across the full range of software perceived to be currently used by their local industry (60% were very confident of their software knowledge, only 15% deemed themselves not competent in industry software, with approximately the others undecided).

Perhaps it might be suggested that the 'implications' of IPD through BIM, rather than any particular software needs additional emphasis at this stage.

Industry/experienced practitioner survey

A snap online industry survey yielded a reasonable response rate from a fair distribution of disciplines (civil/structural engineering, architecture, building and drafting) with professional experience levels ranging from 5 to more than 25 years. The survey results showed that 65% of respondents had some understanding of the concept, although 20% of respondents had had no exposure to BIM in their roles thus far. Respondents with 5 to 10 years of industry experience reported the most exposure and use of BIM in day-to-day activities.

The industry survey results showed that most of the respondents (80%) were unsure about the current education systems put in place by their company for BIM. Respective in-house training systems were largely unknown to staff; opportunities then, for companies to develop and promote training more holistically perhaps.

The general opinions of the respondents (85%) were positive to the benefits and uptake of BIM systems. No respondents thought using a BIM system would be unable to yield improvements in productivity of their work group; indeed, 95% of respondents would choose to use a BIM software system if they had the choice. Based on these industry results, it seems that a very high percentage of industry representatives believe that BIM would be an effective tool that should be implemented into companies towards increases in productivity and staff satisfaction.

Somewhat in contrast with previous perceptions regarding training opportunities, most of the respondents (75%) perceived that management was trying to implement BIM into their workgroup. This indicates that leadership places high value on implementing BIM into the company and is trying to communicate this attitude to their staff, albeit with limited availability of explicit software training. From these results, it might be implied that firms have a positive attitude towards BIM and implicitly embrace its benefits. It can be seen that management is actively preparing employees for BIM, although with no system in which this knowledge or promotion of training is being delivered.

Industry interviews

A representative industrial panel of professionals were identified and asked to comment on broad topics surrounding BIM. Among the respondents, 25% had been directly involved with using BIM, although all professionals interviewed were aware of it and made statements related to its growth in popularity, its potential theoretical benefits and its expected indispensability in the future.

There was a general consensus that BIM is likely to be a beneficial system and is deemed essential for the future of their respective firms. Most of the professionals were able to give an example of an instance in which they would have increased the quality of the product, saved time or reduced the cost on a project, had some aspect of BIM been used.

> Positive industry perceptions towards BIM centre on its 3-D modelling capabilities, clash detection abilities and automation of tasks; for these aspects alone, BIM is seen as an increasingly useful tool.

There was also a consensus that, in a consultancy, smaller jobs do not necessarily require BIM. Traditional approaches and less detailed software systems are deemed as continuing to have a role in the industry for less complex projects. BIM is perceived to be most effective when dealing with specific large-scale complex projects, which require extensive documentation and detail.

A common perception gleaned from the interviews was that government is encouraging and asking for companies to use BIM in their projects, and that whereas respondents suggest government to be currently trailing the large mining organisations in Western Australia in using BIM, it is hoped that government may ultimately create incentives for companies to uptake the use of a BIM software package.

In regards to in-house BIM training and staff development, interview respondents (similar to survey respondents) were largely

unsure of respective company training. The most common perception was that the best way to learn was to use BIM through
practical application on a project with the help of a mentor. Some
respondents thought an online training module with targeted
tutorials would be helpful in learning the basics of the software
package, before undertaking a 'real' BIM project. A very senior
manager within the interview sample did mention the availability
of industry training but that this had not yet been officially communicated as available to staff. Indeed, staffing respondents also
identified that although education systems were perceived to exist,
they hadn't been encouraged by management to extend their skills
and knowledge.

Respondent consensus is that IPD systems improve the documentation process by eliminating incorrect interpretations of drawings,
and all acknowledge that (in the future) BIM will be developed to
become easier to use, streamlining processes and eliminating much of
the drafting work currently involved.

> BIM is perceived to be most useful if a clear scope is available at
> the outset, but changing concepts midproject would be difficult;
> this is not regarded as overtly negative, rather it implies a require-
> ment to negotiate projects in a different way.

Essentially, the BIM system is limited by industry's explicit knowledge of it, and from the interviews, it seems the system is acknowledged
as essential for the future, but is not yet essential at present—especially amongst small to medium-sized enterprises. The positive perception of BIM, combined with a call for good training programmes,
as well as public sector/government uptake should, the respondents
feel, facilitate change in the industry; making information modelling
even more prevalent in the future.

Building upon the further perceptions of integrated modelling of project
information discussed above, a case study review of an ongoing construction job is presented next.

Case study

The review presents an initial preliminary review of a BIM (design)
case study, the construction of which is underway. This project,
conducted using BIM, is an $18M contract to develop a residential/
retail/office project in Perth, Australia. The project design team was
approached for information related to their (early stage) perception of
project-specific BIM factors such as rationale for use, training, cost
implications, communication and liability concerns.

Rationale for using BIM

The design consultancy proposed their services to the client emphasising their use of BIM, which is perceived as contributing in part to their winning of the competitive tender. The firm decided to incorporate BIM explicitly within their tender submission because senior management involved were highly motivated to use a BIM system.

Processes used to train staff

The interviewees involved in this project were already familiar with BIM and did not require additional educational programmes. The other (specialist design) service subcontractors involved in the project did/do, however, require training, which is proving to be a 'substantial' cost for this project (the value for which is still in calculation). According to senior management, where there is a need for (subcontractor) training, external industry schemes rather than team leader mentoring is being utilised.

Design cost implications using BIM

A greater design cost is involved with the project, as more documentation is required when using a BIM system. In the context of the leading consultancy, cost reduction is proving to be somewhat more of a theoretical benefit rather that an actual benefit when using the BIM system; this is argued since two-dimensional systems allow for interpolation of the drawings; hence, less specification of detail. Thus, the cost of (preparing additional 3-D) design is increasing on this BIM project although it does/is result(ing) in a much better quality of understanding for the client; respondents argue that the difficulty then lies in selling/passing-on that benefit in a tangible way to the client.

Even though there is a cost increase in design, the advantage of using BIM is (beginning to show) much less uncertainty at the construction phase. In addition to this, once contractors learn that documentation is more accurate and has fewer issues on-site, they (are beginning to) reduce the contingency in their lump sum construction cost (submissions). Essentially, the design team suggests that this is taking costs off the builder and onto the designers—it is (being) seen as a more effective expenditure of budget.

> BIM systems give less probability of variation (and claim); overall building costs may be (are being) lowered.

Communication (improvements as a result of) using BIM

It was a common opinion expressed by all that using BIM, communication has become more effective and reliable; and respondents suggest that the amount of communication needed is reduced as the BIM system gives a better project overview.

Liability

Intellectual property and liability are not an issue in this particular case study as only drawings were/are given to the client—the digital model is not given. Respondents suggest that a (positive) legal procedural working structure of BIM projects evolve from implementation and ongoing working culture. The liability associated with changes made to the model (in this particular scenario) is easily controllable as different people have different levels of access and all edits are associated with the person/computer that made them.

The challenges of changing to BIM might be suggested as being offset with government and major private client support, alongside educational systems, both at tertiary and continuing professional development levels, to address software (standardisation) support concerns. The upfront cost of purchasing, maintaining and operating software is seen as a detracting factor of a BIM system implementation.

Staff training is one of the larger costs to introducing a BIM system; a more than 20% increase in drafting costs is anticipated, although the overall productivity of the construction process can be improved with fewer change orders and information requests.

Collaboration within the design/construction team may, through BIM, result in a higher quality, faster, less-expensive project justifying drafting costs. Although industry associations must continue to strive to allay legal and contractual concerns, BIM may reduce disputes through a greater availability of clear and concise information.

The survey results found that most of the respondents were unsure of the current BIM education systems put in place by their companies. This was reinforced by structured interviews, which showed an inconsistent process; education systems were largely unknown, requiring management to work harder to communicate BIM training options to their staff.

From the online survey, 85% believed that BIM would improve the productivity and effectiveness of their work group and 95% of respondents would choose to use a BIM software system if they had the choice. The interviewees demonstrated a positive attitude towards the use and implementation of a BIM system, which is also associated with increases in productivity and an essential part of a company's development.

The case study demonstrated that greater design cost is involved in BIM; however, this extra design cost does provide better understanding for clients, with improved (somewhat intangible) working relationships and reputations, albeit with no improvements in a firm's explicit productivity. It is interesting that BIM is 'perceived' to increase productivity. This disparity may be due to the fact that the theoretical benefits of increases in productivity are very much a long-term, next-project projection.

The online survey showed that 95% of the respondents would choose to use a BIM-enabled software system if they had the choice; interviewee's opinions mirrored these responses, with staff generally motivated to learn and use BIM systems. Case study respondents contend that even though BIM increases design costs, the money is well spent as it results in a higher quality product, has less on-site error and improves reputation. In Western Australia, as perhaps with locations nationally and internationally, BIM seems to be perceived as a positive investment.

Whether in a cumulative model or not, *specifications* will continue to describe *how* design decisions are applied on-site and subsequently utilised. Specifications providers (such as NATSPEC and NBS, as discussed previously) are indeed keen to retain their established roles as providing standard systems and BIM-friendly specification statements for the construction industry.

In conjunction with specifications, whether specifying traditional virgin materials or indeed C&D waste arisings (in BIM models or not), there remains a need to quantify the materials and work task requirements of a civil engineering undertaking. Section 5.3 describes the quantification of design (in BQs) towards clarification of *how much* work is required. Section 5.3 describes civil engineering BQs.

5.3 DESIGN MEASUREMENT AND MENSURATION: CIVIL ENGINEERING BILLS OF QUANTITIES

Civil engineering design solutions are represented typically by four key tender documents:

- Drawings (describe *what* to build)
- Specifications (show *how* to prepare and install the built elements)
- General conditions of contract (describe site *responsibilities*)
- Bills-of-quantities (clarify *how much* work and resources are required)

Bills-of-quantities (BQ) provide a structured unambiguous account of all the materials and work to be carried out under the contract. A number of highly informative texts provide practitioners with guidance and reference notes in the use and application of standard methods of measurement for construction and civil engineering.[177–182]

BQs are traditionally prepared by the client's representatives. An unpriced BQ with item descriptions and quantities but without unit-rate prices is distributed to potential builders. Builders keen to submit a tender-bid, input unit-rate prices (albeit submissions apt to be weighted to assist revenue flow returns) and forward documents for client consideration, which in turn may assist in the selection of an appropriate contractor to construct the design solution.

Measurement of the design in BQs defines agreement of what construction work is to be done and helps communicate the scope of the work and forecast the cost. A standard method of measurement and mensuration for a BQ defines parameters for itemisation and systematic pricing of elements and subelements.

International,[183] national,[184] and state authority[185] variations of standard methods of measurement for construction work exist and cover somewhat similar ground, to encompass the broad scope of building operation quantification whilst remaining flexible enough to cater to less tangible on-site activities. Software options also exist.[186]

Standard methods of measurement to prepare BQs for civil engineering work include

- Civil Engineering Standard Method of Measurement, fourth edition CESMM4 (or the superseded third edition CESMM3) by the Institution of Civil Engineers

Australia-wide standard method(s) of measurement such as

- AS1181-1982, Method of measurement of civil engineering works and associated building works (albeit somewhat overly textual and lacking ease of tabular reference and cross-reference)
- Australian Standard Method of Measurement of Building Works, fifth edition by the Australian Institute of Quantity Surveyors and the Master Builders (albeit somewhat less inclined to cover civil engineering subelements in detail)
- State Standard Method(s) of Measurement for Highways Construction are available in, for example, Western Australia's public infrastructure authority, Main-Roads
- Numerous software packages to facilitate preparation of bulk BQ

One comprehensive, highly structured system to quantify all WUC in civil engineering projects is the 'Civil Engineering Standard Method of Measurement', currently in its fourth edition (CESMM4).[187]

Following a systematic tabulated format, a civil engineer may progress through the guide and, in essence, tick-off the measurable items of all design(ed) components described in the project with reference to the available drawings and specifications. Measurement of the works using CESMM4 helps to checklist and quantify design decisions already made, and indeed can be useful in the identification of any design oversights and specification ambiguities.

	Standard method of measurement civil engineering work	
Item		CESMM4 section class
Definitions		1
General principles		2
Work measurement rules		3
General items		A
Ground (geotechnical) testing		B–C
Demolition and site works		D
Earthworks		E
Concrete		F–H
Pipework		I–L
Metalwork		M–N
Timber		O
Piles		P–Q
Roads and rail-track and tunnels		R–T
Brick, block and masonry		U
Painting and waterproofing		V–W
Sewers and simple building works		X–Z

Section 5.3.1 provides measurement examples for civil engineering work.

5.3.1 BQ preparation

Chapter 2 (detailed estimating, Section 2.2.3) introduced nomenclature for a typical BQ item and how to build up a unit-rate cost for such an item. The illustration used previously is repeated below and depicts a first column standard method of measurement (SMM) reference (where 'E' signifies earthworks, '4' identifies the first division classification of general excavation, '2' describes a second division subcategory of soil type and '4' describes the third division depth category of less than 2 m), the second column describing this item in words, alongside columns that then identify the unit, quantity, unit-rate, and a running total in monetary units.

Typical BQ item		Unit	Quantity	Unit rate	Total
	Earthworks			Aus$	Aus$
E424	General excavation Material other than topsoil, rock or artificial hard material Maximum depth: 1–2 m	m³	25	6.50	162.50

Prior to detailed cost estimation (Section 2.2.3) of a civil engineering job, reference must be made to all available drawings and specifications to allow a detailed measurement of the works and a systematic description of the items.

Section 5.3.1.1 (acknowledging texts by Ivor Seeley and others[188]) provides a structured method for 'taking-off' quantities (traditionally take-offs from drawings) in preparing project BQs.

5.3.1.1 BQ preparation: Practice example (site clearance)

After the preparation of their design drawings (and related specifications), the civil engineer needs to *take-off* and itemise the quantities of the work under the contract. Four steps to prepare the project's BQs are as follows:

Step 1: Collate and review the available design drawings and specifications

Step 2: Identify a standard method of measurement consistent with requirements; refer to SMM in checklist fashion, to itemise/measure all parts of the design

Step 3: Prepare to 'take off' (manually or by Excel spreadsheet/software[186]) to input
- SMM references
- Works descriptions in words
- Unit quantities (and respective dimensions) generated by both default volumes/quantities from CAD design software and remote measurement where default quantification is unavailable
- Retain 'take off draft' for future cost-control cross-check of dimensions

Step 4: Transpose take-off draft items and finalise unpriced BQs

A scenario might involve the need to measure and prepare a BQ item for a future land site clearance.

Step 1
The land to be measured for a future general site clearance is a nonuniform rectangle depicted on a general arrangement drawing of approximately less than 85 × 50 m; with no discernable trees, stumps, buildings, other structures or pipelines.

Step 2
The standard method of measurement to be used in this case is CESMM3; specifically, class D, demolition and site clearance (as shown in Table 5.17).

CESMM3

Class D: Demolition and site clearance

Includes:	Demolition and removal					
Excludes:	Removal below the original surface					

First division	Second division	Third division	Measurement rules	Definition rules	Coverage rules	Additional rules
					C1...	A1...
I Site clearance ha				D1...	C2...	A2...
2 Trees nr	I girth: 500–I m 2 I–2 m...			D2...	C3...	A3...
3...	...	...	...	...	...	...

Adapted from CESMM3 published by Thomas Telford, 1991.

Step 3

Take-off 'paper' is prepared to facilitate description (Figure 5.4) and is divided into six columns:

- 'SMM reference' column on the left
- A large second 'item description' column, followed by
- A columns on the right to allow the input of units (m, m², m³ or ha)
- Additions (for the input of area dimensions length × width)
- Deductions (to allow omission of voids)
- A final quantities column

- The CESMM3 reference is D100
- The works description in words:
  - Class D: Demolition and Site Clearance
  - General Clearance
- The unit of measurement is ha (hectares)
- Remote measurement add dimensions:
  - 78 × 45 (shown linearly)
  - 3362 m² (with take-off diagram)
- Quantification in the final column:
  - 0.34 ha
- Side note rules state C2 and D1 apply to hedges and fences to be included in this case.

Step 4

The finalised transposed BQ item is shown below alongside, for clarity, unit-rate and total; albeit rates and totals usually remain blank to facilitate bids by prospective tenderers.

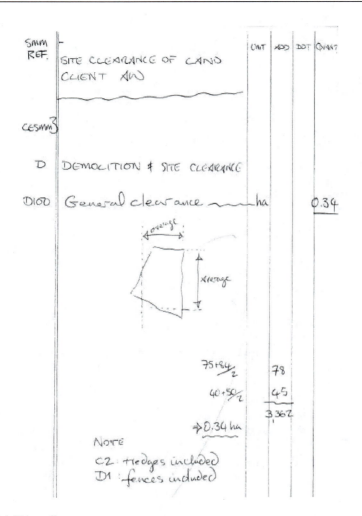

Figure 5.4 Take-off.

	Demolition and site clearance	Unit	Quantity	Unit rate	Total
D100	General clearance	ha	0.34	$1800	$612

5.3.1.2 BQ practice example (concrete strip foundation)

Measurement and itemisation of a concrete strip foundation.

Step 1: Drawings show strip foundation of grade 25 of some 30 m × 2.4 m × 0.9 m.

Step 2:   CESMM3 class F, *in situ* concrete is shown next:

Class F: In situ *concrete*

Excludes:	In situ *concrete for boreholes, etc.*					
First division	*Second division*	*Third division*	*Measurement rules: M1–2*	*Definition rules*	*Coverage rules*	*Additional rules*
Provision						
I Standard m³	I...	1.10 mm		D1–D4		A1...
2 Designed m³	I C7.5	aggregate				
	...	...				
	5 C25	7. 20 mm				
		aggregate				
...	...	...	...	...	...	...
Placing						
5 Mass m³	I	I thick				
	blinding	<150 mm				
	2 base	...				
	footings	4 thick				
		>500 mm				

Source: Adapted from CESMM3 published by Thomas Telford, 1991.

Step 3a: Prepare quantities to 'take off' (Figure 5.5). Take-off 'paper' from six columns

- The CESMM3 reference is F257 (Figure 5.5).
- The work's description in words: Class F: *In situ* Concrete, Provision of concrete Designed mix Grade 25; as AS1379; 20 mm aggregate.
- Unit m³.
- Remote measurement dimensions are $30 \times 2.4 \times 0.9$ linearly = 64.80 m².

with take-off-diagram of strip-founds supporting a grade C30 retaining wall

Step 3b:

- The CESMM3 reference is F524 (Figure 5.6).
- Works description: Class F: *In situ* Concrete.
  - Placing of concrete.
  - Mass, bases, footings.
  - Thickness, >500 mm.
- Unit is m³.
- Dimensions as previous item F257 (Figure 5.5).

Step 4:   Prepare finalised BQ item. The finalised BQ items are shown below with unit-rate and total to be completed by the various tenderers bidding for the job.

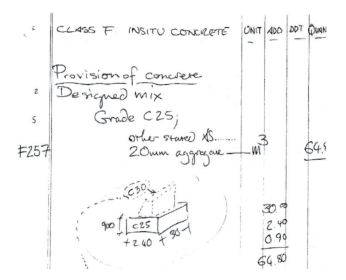

Figure 5.5 Placing 'take-off'.

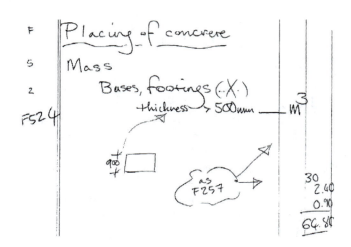

Figure 5.6 Final item.

In situ *concrete*	Unit	Quantity	Unit rate	Total
F257 Provision of concrete Designed mix, Grade C25, AS1379, 20 mm aggregate	m³	64.80		
F524 Placing of concrete (mass, bases and footings; thickness >500 mm)	m³	64.80		

More detailed examples of BQs that use the CESMM3/4 structure are presented in Chapter 7.

Having discussed the preparation of BQs in a structured universal format (in the case above, by adopting CESMM3/4), there is now a need to look at the existing guidelines in place to structure the presentation of (arguably) the most important of the four documents deemed to make up the contract documents, the design drawings. Section 5.4 discusses existing conventions towards the preparation of design drawings for civil engineering solutions.

5.4 DESIGN DRAWINGS

The key to communicating a civil engineering design solution is the set of design drawings. Design drawings are a fundamental component of a contract and, alongside the specification, the schedule of rates/BQ and the standard form of contract, clarify explicitly what work is to be done under the contract.

Although somewhat widely assumed to take precedence over the other contract documents, design drawings are deemed to rank equally with the specifications, the BQ/schedule of rates and the contract (unless otherwise stated as dominant in special clauses/inclusions) because standard forms of contract, such as Australian Standards AS4000, do *not* indicate a hierarchy of precedence of documents. In summary,

- Drawings, specifications, BQ and contract, all rank equally in Australian Standards.
- Equity of documentation importance is deemed to apply, unless special clauses in the contract indicate a precedence that goes beyond standard clauses such as AS4000 clause 2: the Nature of the Contract, and clause 8: Contract Documents.

Design drawing conventions are covered in a wide range of national standards. In Australia, for example, the following standards and guidance texts might be deemed to apply to civil engineering drawings:[189-195]

- AS ISO 128.21-2005 Technical Drawings—General Principles of Presentation—Preparation of Lines by CAD Systems; AS13567.1/2-1999 Technical Product Documentation—Organisation and Naming of Layers for CAD—Overview and Principles; and AS3883-1991 Computer graphics—Computer Aided Design (CAD)—Guide for structuring of computer graphic information

- AS1100.101-1992 Technical Drawing—General Principles
- AS1100.501-1985 Technical Drawing—Structural Engineering Drawing
  - This structural engineering drawing practice compliments AS1100, Part 101 with the types of structures covered including:
    - AS1250 SAA Steel Structures Code
    - AS1475 SAA Block-work Code
    - AS1480 SAA Concrete Structures Code
    - AS1481 SAA Prestressed Concrete Code
    - AS1640 SAA Brickwork Code
    - AS1720 SAA Timber Engineering Code
    - NAA SRA Bridge Design Specification
    - AREA Manual for Railway Engineering
- HB 7-1993 Engineering drawing handbook
- HB 20-1996 Graphical symbols for fire protection drawings
- HB 1:1994 Technical drawing for students
- AS1100.301-2008 Technical drawing—Architectural drawing
- AS1100.301-2008/Amdt 1-2011 Technical drawing—Architectural drawing
- AS1100.401-1984 Technical drawing—Engineering survey and engineering survey design drawing
- AS1100.401-1984/Amdt 1-1984 Technical drawing—Engineering survey and engineering survey design drawing
- AS/NZS 1100.501:2002 Technical drawing—Structural engineering drawing
- AS 1203.4-1996 Microfilming of engineering documents—Microfilming of drawings of special and exceptional elongated sizes
- AS ISO 128.1-2005 Technical drawings—General principles of presentation: Introduction and index
- AS ISO 128.20-2005 Technical drawings—General principles: basic conventions for lines
- AS ISO 128.21-2005 Technical drawings—General principles: preparation of lines by CAD systems
- AS ISO 128.22-2005 Technical drawings—General principles of presentation—Basic conventions and applications for leader lines and reference lines
- AS ISO 128.23-2005 Technical drawings—General principles: Lines on construction drawings
- AS3643.2-1992 Computer graphics—Initial graphics exchange specification (IGES) for digital exchange of product definition data—Subset of AS3643.1—Two-dimensional drawings for architectural, engineering and construction (AEC) industries

5.4.1 Design criteria: Toward graphical representation

The design phase necessitates the preparation of finalised working drawings that confirm a preferred solution that is able to fully satisfy a client's brief.

Detailed design drawings must

- Comply with existing standards/conventions as above
- Extrapolate/extend the client's brief towards an illustration of practicable solution(s)
- Represent consideration and detailed analysis of design criteria

Design criteria might be summarised as in the Civil engineering design criteria and as detailed below.

CIVIL ENGINEERING DESIGN CRITERIA

- Address user group/worker OH&S
- Assess critical sizes and key detailing
- Acknowledge aesthetics and weathering
- Analyse gross/net floor ratios
- Address proportionality/scale ratios
- Provide structural/services clash detection
- Review site conditions and limitations
- Give a flexible life cycle design
- Monitor/control mechanical and electrical (M&E) services
- Compare/choose from fit-for-use specs
- Seek innovative application of concepts

Civil engineering development design criteria might be expected to include

- A detailed examination of current and future user group(s) for civil engineering amenities and infrastructure developments, alongside full recognition of all user (and construction worker) occupational health and safety requirements
- Design of all critical sizes and key details supported by calculations of dead and live loads and the like, set out neatly and well annotated to facilitate ease of double-checking, alongside considerations of appropriate envelope/footprint shape
- Aesthetic considerations able to compliment structural integrity and appease public perception of large structural members, using finishes, textures and colours, not just appropriate at the time of construction but over prolonged exposure to the elements

- Economic considerations such as the gross/net floor ratio
- Proportionality and scale of superstructure items such as access features to cladding items, satisfactory circulation and the like
- Services considerations, with particular attention paid to the potential for service-line clashes and best-lie impediments from structural members
- Existing site conditions and limitations considered and mitigated at the early stages
- Flexibility of design to accommodate changing user needs and changing fuel sources, and allowances for upgrading/refurbishment and future access to allow re-fitting, and not least, acknowledge the need for flexibility to accommodate future changes in aesthetic taste, design fashion and surface finish trends
- The use of mechanical/electrical (M&E) controls to monitor and adapt energy use
- Full consideration of fit-for-purpose specification alternatives (structural sub/elements as well as finishes and equipment) to acknowledge life cycle operation, maintenance and durability as well as decommissioning/recycling
- Not least, civil engineers should seek innovative creative design solutions that acknowledge materials science advances within the constraints of the laws of science, economic limitations and user expectations

Representation/drawing provision of design decisions is a key activity of a civil engineer. Traditional design is characterised by progression and iterative reassessment through a number of steps towards issuing final installation drawings, namely by

- Problem identification and need clarification
- Preliminary ideas identification of a range of potential technical solutions
- Choice justification and refinement of a preferred design solution
- Detailed analysis of the preferred (element/subelement) design solution
- Integration of sub/elemental design into the full concept design
- Drawings produced in the required format (hard copy and/or soft copy)
- Virtual sharing of documentation via BIM and the like where appropriate
- practical implementation and design drawing updates with reference to life cycle(s)

Three-dimensional modelling of design solutions and incorporation of all aspects of specialist input into a collective BIM environment, although not without risk and liability concerns, is increasingly expected by a range of industry stakeholders.

5.4.2 Selected design guides for civil engineers

The preparation of design drawings might be expected to refer to the full range of local, national and internationally applicable design guidance texts. Review and application of suitable texts prior to design (drawings) preparation might assess references such as

Road design guidance texts for Australia/Western Australia might include[196]
- *Austroads Guide to Road Design*
- MRWA Supplement to *Austroads Guide to Road Design*—Part 3

Treatment of sewage and disposal of effluent and liquid waste:[197] sewage/sewerage design texts might include, for example, in Western Australia
- Heath (treatment of sewage and disposal of effluent and liquid waste) regulations 1974 Part 4a regulation 42 (and associated updates), made explicit in documentation such as Code of Practice for the Design, Manufacture, Installation and Operation of Aerobic Treatment Units (ATUs) by the Department of Health, WA Government; commercial hybrid toilet systems ranges endorsed by Green-Smart; commercial wastewater recycling systems such as BioMax and the like.

Disposal of storm-water runoff[198]
- Storm-water management at industrial sites might be gleamed from Australia/WA-based texts such as Water Quality Protection Note by the Government of Western Australia, Department of Water; Disposal of Storm-water Runoff Soakage and the like.

Retaining walls passive anchorages and the like[199,200]
- Internationally applicable piling and anchorage guidance texts might include: Piling Handbook by Arcelor Mittal; Sheet Piling Handbooks Design.

Geomechanics design and rock properties[201,202]
- Design guidance for geomechanics and rock properties applicable in Australia and further afield including: geomechanics design handbooks, *Geomechanics Principles in the Design of Tunnels and Caverns in Rocks* and texts related to developments in geo-technical engineering.

Structural steelwork: standards for structural steel work in Australia might be expected to include
- AS4100 Steel Structures Code
- AS4600 Cold-Formed Steel Structures Code
- AS1397 Hot-Dipped Zinc-Coated or Aluminium/Zinc-Coated Steel Sheet and Strip
- AS1595 Cold-Rolled Unalloyed Steel Sheet and Strip
- AS1627 Metal Finishing—Preparation and Pretreatment of Surfaces

- AS4860 Hot Dip–Galvanised (Zinc) Coatings on Fabricated Ferrous Articles

Guidelines for design and specification of industrial floors and pavements: pavement guidance in Australia includes

- Industrial floors and pavements guidelines for design construction and specification by the Cement and Concrete Association of Australia

Concrete wall panels and tilt-up technology design related to Australia might include

- Recommended practice design of tilt-up concrete wall panels by the Concrete Institute of Australia

Representative civil engineering drawing examples that utilise and apply the listed examples of design codes can be found in Chapter 7. Design drawings (Section 5.4) alongside the contract (Section 5.1), specifications (Section 5.2) and rates–schedules/BQs (Section 5.3) represent the work under the contract. The civil engineering contract documentation (and related activities in preparation, dissemination, tendering and review) must satisfy inherently an ethical imperative to provide-for and protect society.

Next, Chapter 6 discusses ethical application(s) in detail.

Chapter 6

Engineering ethics and professional development

Civil engineers and, by extension, their activities and design decisions, must demonstrate an appreciation of the responsibilities of ethical professional practice in addressing quality, cost and time variables in the planning, procurement and contractual administration of sustainable civil engineering projects.

This chapter discusses the evolution of engineering traditions and ethics, and the extent to which an ethical imperative for continuing professional development must embrace leadership, communication, change management and an ability to empathise with multidisciplinary design team members:

- Engineering traditions and (natural) philosophy (Section 6.1)
- Professional engineering ethics and institution membership (Section 6.2)
- Leadership theory (Section 6.3)
- Communication (Section 6.3.4)
- Change management (Section 6.3.5)
- Professional integration in a multidisciplinary design team (Section 6.4)

An understanding of engineering ethics is essential in attaining chartered membership to the national[1-3] and international[4] learned professional bodies of engineers.

Chartered membership of any of the national professional engineering institutions certifies that the civil engineer practises in a competent, independent and ethical manner; and indicates that they are leaders in their field nationally and internationally.

- The International Engineering Alliance is more commonly known as the 'Washington Accord'. Six international agreements govern mutual recognition of engineering qualifications/professional competence. Participating countries/economies become signatories to the agreement; the body submitting the application verifies that it is the appropriate national representative and stipulates that graduates must adhere to respective ethical codes for the engineering practitioner.

- Engineers Australia: National Chartered Membership strives 'to represent the highest standards of professionalism, leadership, up-to-date expertise, quality and safety and the ability to undertake independent practice'.
- Institution of Professional Engineers New Zealand is the professional body that represents professional engineers from all of its national disciplines.
- Institution of Civil Engineers represents practitioners in the United Kingdom and frames its (ICE) code of professional conduct.

All international professional institutions of (civil) engineering (as well as all professional engineering body signatories to the Washington Accord) adhere strictly to published codes of ethical professional conduct. Consideration of professional ethics requires an appreciation of tradition and the development of professional values; thus, engineering traditions and engineering philosophy[5] represent a foundation for engineering ethics and are discussed in Section 6.1.

6.1 ENGINEERING TRADITIONS

Today's civil engineers strive to address the needs of an increasingly demanding society and seek innovative, creative design solutions that acknowledge materials science advances within the constraints of the laws of science, economic limitations and user expectations. Indeed, this has always been the case; early engineering and applied mechanics, physics and natural sciences courses under the umbrella term *natural philosophy* similarly sought solutions within practicable acceptable limits.*

Engineers are increasingly charged to communicate their decisions to a wide range of stakeholders. Recently, the ability to communicate effectively was brought into question because of the traditional reliance of engineering courses on technical, mathematical and scientific reasoning at the expense perhaps of sociopolitical/environmental issues.

Concerns by the professional bodies themselves, as well as educators (and requisite validation boards), over perceived levels of social detachment have brought about undergraduate curricula changes (over the past decade or so) that now seek engineering courses to have a reasonable proportion of course content based on management studies, economics, supply-chain

---

* Available at http://www.universitystory.gla.ac.uk/chair-and-lectureship/?id=790. Natural philosophy being part of the oldest UK course in Civil Engineering; offered at Glasgow University continually since 1840.

considerations, law, risk appreciation, creativity, structured reporting, team-working, as well as the embedding of communication skills into all/most undergraduate engineering activities.

The public largely regards engineers positively albeit somewhat aloof:[6]

- Public attitudes to and perceptions of engineering and engineers, sought by the Royal Academy of Engineering and the Engineering and Technology Board, found that ≈70% of the public are positive about engineers; overall, the public say that 'engineers fix things', but there is a need for more effective public engagement.

For engineers (and graduates) to play an effective role, the ability to communicate technical problems is needed, alongside social/political awareness, so that solutions are shared sensitively with all stakeholders. Although as mentioned the public perceives engineers as *fixers*, engineers themselves must seek a more explicit explanation of how their activities affect society, be able to measure such impact and seek to ensure that activities remain ethical by building upon (an explicit knowledge of) how the solid engineering traditions of natural philosophy must continue to provide social benefit.

6.1.1 Engineering (natural) philosophy

Towards a more explicit explanation of engineering activity, beyond the truism that *engineers translate science into useful design*, it can be argued that engineering design seeks new concepts and leaps of imagination (metaphysics in action, perhaps) towards novel applications of existing knowledge (applied empirical evidence) where civil engineers examine the link between cause and effect (causality) in their mathematical calculation of (abstract objects of) dead and live loads, with subsequent incorporation into what are currently vacant spaces/gap sites (through abstract conceptualisation).

Civil engineering design, therefore, might be termed a philosophical examination of:

- Metaphysics (seeking new concepts and leaps of imagination for design solutions)
- Natural empiricism (novel applications of existing scientific knowledge)
- Causality (examining cause and effect in the application of structures to locations)
- Abstract conceptualisation of objects (loading calculations for vacant areas/sites)

These approaches to design realisation seem to sit easily with the term natural philosophy as applied originally/historically to all early science applications, mechanics, thermodynamics and materials courses; and indeed it was, until somewhat recently, the label of choice for modern-day units/modules concerning foundations studies in physics and mechanics.[7]

There is perhaps an empiricist tradition of philosophy embedded in engineering given that engineers search for an understanding of the external world, look for meaning and truth in terms of physical entities, investigate reality and also question then develop knowledge bases towards (civil engineering) applications that will be used by society day in, day out.[6]

The links between engineering and philosophy might perhaps be typified by the pursuit of undergraduate final-year projects, and not least by higher research degrees in civil engineering in which the investigation of reality—questioning and then developing the knowledge base, and the examination and validation of experiences for future application in the world—determine success in contributing to civil engineering knowledge.

If society is to be provided for, then engineers must work within the cultural expectations of that society. In other words, engineers must acknowledge an accepted (collectively agreed upon) right conduct to provide society with (infrastructure) solutions.

This collectively agreed upon right conduct forms the basis of the tenets of professional ethics applicable to all practitioners making up the building design team (with Section 6.4 reviewing the interaction). Teams commonly include a range of players: architects, civil engineers, structural engineers, mechanical and electrical services engineers, quantity surveyors, project managers and superintending officers; building practitioners such as contract managers, construction managers, resident engineers, quality control professionals, building surveyors, heating and ventilation engineers, environmental planners, asset and facility management professionals, landscape architects, estate managers, interior designers, specialist subcontracting builders and building materials, fittings and fixtures suppliers. All interdisciplinary team members must work within professional ethical codes.

6.2 PROFESSIONAL ENGINEERING ETHICS

Morals might be termed the rules of right conduct for individuals although *ethics* might be argued to extend individual morality towards an accepted collective, group behaviour. In the case of civil engineers, the group might be termed as a professional body of specialist practitioners offering solutions to the community based on existing norms and innovation. The professional body of engineers is then charged to self-regulate its legitimate acceptable group behaviour towards the community; in other words, develop and apply acceptable behaviour and codes of ethics to help serve society: effectively, an engineering code of ethics.

Engineering codes of ethics guide practitioners in seeking a balance between society, the employer and fellow engineers by ultimately scrutinising individual decisions in terms of something that may be held to account by the other members of the profession. By having a code that requires the individual to judge themselves against the action of a body of other reasonable engineers, an engineering code of ethics helps to establish shared standards and avoid negligent deeds. If the reasonable engineer would have their loading calculations checked and rechecked, then this action should set the standard.

Each national body of professional engineers have their own code of engineering ethics,[8] which are largely similar in expectation (see Engineering code of ethics).

ENGINEERING CODE OF ETHICS

- Demonstrate integrity
- Practice competently
- Exercise leadership
- Promote sustainability

Engineers commit to practising in accordance with an engineering code of ethics and agree to be held accountable for conduct under engineering's disciplinary regulations.

Practicing engineers are required, by their respective code of ethics to

- Demonstrate integrity
  - Act on well-informed conscience
  - Be honest and trustworthy
  - Respect the dignity of all
- Practise competently
  - Maintain/develop knowledge/skill
  - Represent competence objectively
  - Act on adequate knowledge
- Exercise leadership
  - Uphold trustworthiness of engineering
  - Support/encourage diversity
  - Know of the reliance of others on engineering expertise
- Promote sustainability
  - Be responsible in the community
  - Foster health/safety/well-being of the community
  - Balance the needs of the present with the needs of the future

Sustainability in civil engineering needs to be categorised explicitly.

The triple bottom line of a sustainable design solution must fully address the three E's of sustainability; namely, environment, economics and everyone.

- *Environment*: An acknowledgement of all environmental impact assessment (EIA) legislation and environmental management standards and codes of practice
  - Appropriate application of life cycle assessment (LCA) methodology to compare fit-for-purpose design specification options.[9]
- *Economics*: Consideration of the full financial whole-cost of a proposed facility, with acknowledgement of all/any potential intangible (improved access) benefits
  - Appropriate application of value management/value engineering principles and tools such as life cycle cost analysis (LCCA).[10,11]
- *Everyone*: Full compliance with all social safeguard legalisation and codes of practice and national standards relating to noise, particulate matter and the like.

Generally, engineers are required to recognise that their actions must be ethically defensible to peers and society. A rule of thumb might be to suggest that an engineer assume that the facility designed and built, shall be used by their own Aunt, financed by that Aunt's tax dollars (see Civil engineering ethics: Rule of thumb guidance).

Civil engineering ethics: Rule of thumb guidance

Recall that all solutions designed and built are

- For use by an Aunt and her friends
- Financed in full by the Aunt's tax dollars

Methodologies to assist engineers to address difficult decision-making and ethical dilemmas[12] might include the acknowledgment of

- Ethical frameworks that progress from 'does the law give an answer' to 'formal guidelines of corporate policy' to 'informal guidelines asking if one would tell a TV reporter, or Tweet or blog the data' to '(non) consequential duties and rights analyses' towards 'categorical imperatives of consistency and respect'
- Problem-solving line-drawings, placing issues along a long line with negative paradigms at one end and positive paradigms at the other, towards the identification of the cumulative best choice

In more prescriptive terms, it might be suggested that actions cannot be morally right even though they might be approved by law or custom because low standards and moral wrongs cannot be justified in civil engineering (ethical relativism is seldom applicable). Conversely, it is also perhaps unacceptable to maintain exactly the same practices both at home and overseas because civil engineers must try to take into account external and cultural variables on-site (ethical absolutism is somewhat similarly to be avoided).

On the other hand, moral (group) judgements might be argued to be largely contextual and to be made in relation to many factors of custom and culture (therefore, ethical relationalism might perhaps be one way forward?).

Given the discussion in this section of the progression from engineering tradition, to community expectation, to a code of ethics to guide civil engineering activity, Section 6.2.1 seeks to assist in the more practical exercise of preparing the engineer for ethics assessment as part of the process of attainment of professional chartered status.

6.2.1 Ethics assessment: Chartered professional membership practice example

An understanding of engineering ethics, alongside an express ability to apply the tenets of the code of ethics, is essential in attaining chartered membership to the respective national professional body(s) of engineers. Civil engineers with a reasonable level of postgraduate participation in the construction industry will subsequently be assessed on their knowledge of ethics. The following step-by-step procedure is deemed applicable.

Step-by-step guide to addressing civil engineering, professional chartered membership assessment, ethical questions:

Step 1: Summarise and review the key issues of any question posed, and seek to prioritise the range of issues

Step 2: Seek to identify a known/real project suitable for use as case study reference

Step 3: Identify the current relevant local code of ethics and seek to explicitly reference specific applicable regulation items

Step 4: Ensure that the coverage of issues encompasses reflection on the three E's of sustainability, with (a somewhat typical) prioritisation in order of (i) everyone, (ii) the environment and then (iii) economic considerations

The following practice example scenario seeks to illustrate the application of a relevant ethical approach.

A scenario might involve a design-and-build/turnkey project in which the contractor requests you (the client's expert administrator) to accept minor irregular (i.e. nonconforming to recent engineering practice) proposals to maximise project profits.

Step 1
Clarification of the key issues might first require an explanation that

- Design-and-build/turnkey projects place responsibility and obligation on the main contractor to get all the plans and drawings approved. Major items like the development plan, building plan, infrastructures plan for earthworks, drainage, roads, water and sewer reticulation, streetlight plan, as well as electrical and mechanical (M&E) drawings need to be approved before any work commences on-site. Other drawings and somewhat more minor approvals/consent are also often required from the authorities related to detailed land beautification and road-naming.
- Although an expert administrator might be approached to approve or assist with nonconforming work methods or approaches, it might be argued that current measures of approval and authorisation will eventually uncover even minor noncompliant aspects, alongside scrutiny of all technical compliances in quality management systems.

Step 2
Identification of a known/similar case study is somewhat expected to allow reflection upon the issues. If a case study project is not within the candidate's actual range of experience of projects covered prior to the chartered status exam, reference to a real or historical engineering failure may be applicable.

- Engineering failure case studies relating to compliance waivers of technical standards and the like might perhaps refer to Sampoong Department Store, South Korea (1995).
- Historic engineering failures are tragically plentiful and include a broad range of technical failure stemming often from interpersonal/management inadequacies
  - Tay Bridge, UK (1879); St. Francis Dam, US (1928); Tacoma Narrows Bridge, US (1940); Fréjus Dam, France (1959); Ronan Point, UK (1968); West Gate Bridge, Australia (1970); Hotel New World, Singapore (1986); Hanshin Expressway, Japan (1995); World Trade Center, US (2001); Charles de Gaulle Airport T:2E, France (2004); Mississippi Bridge, US (2007); Tho Bridge, Vietnam (2007)
    - Quebec Bridge, Canada (1907); Seonsu Bridge, China (1979); Christchurch Grammar, Australia (2008); Silver Bridge, Ohio, US (1967); Injaka Bridge, South Africa (1998); Hyatt Regency Walkway, US (1981); Tropicana Casino Garage, US (2003); Armand Cesari Stadium, France (1992);

Mianus River Bridge, US (1983); College La Promesse, Haiti (2008); Burnaby Supermarket, Canada (1988); Versailles Wedding Hall, Israel (1988); Lotus Riverside Shanghai Complex, China (2009); Schoharie Creek Bridge, US (1987); Guadalajara Sewer Explosion, Mexico (1992); Highlands Tower, Malaysia (1993); Vajont Dam, France (1959); Hartford Civic Center, US (1978); De La Concorde Overpass, Canada (2006); Buncefield Oil Storage Depot, UK (2005); Boston Molasses Disaster, US (1919); Tin Chung Court, Hong Kong (1999); FMG Rail Camp, Port Hedland, Australia (2007); and the infamous Bhopal Gas Tragedy, India (1984).

Step 3
Reference to a local guideline might, for example, identify the Engineers Australia Code of Ethics (July 2010 edition), where tenets applicable in this case might include

- Item 2.3(a), practising in accordance with legal and statutory requirements, and the commonly accepted standards of the day... where standards updates ignorance
- Item 2.1(a), requires continual development of relevant knowledge and expertise... because there is a need to
- Item 1.1(c), act appropriately and in a professional manner when something is perceived as wrong... so that
- Item 3.3(a), clear communication issues such as risk are made... ultimately to
- Item 4.1(b), inform employers or clients of the likely consequences on the community... as a means to
- Item 4.2(a), incorporate social over economic considerations into engineering tasks

Step 4
Overall, a request for a relaxation of compliance, no matter how apparently minor, might simply represent the thin edge of a wedge; maximisation of profitability and *economic* considerations (or indeed any blinkered concentration on environmental impact in isolation perhaps) cannot override the well-being of *everyone* in the community.

Having attained chartered status with a respective national learned professional body of engineering, the practitioner is now ready to not only manage the resources required to realise a civil engineering project but also *lead* the team of professionals charged with bringing that project to fruition. Section 6.3 extends previous review of the tools and techniques of *management* towards a discussion of *leadership*.

6.3 LEADERSHIP

An overview of the integration of civil engineering design (quality) on the one hand, and construction (time, cost and contract variables) on the other hand, might be presented graphically, with recognition of the central role of 'managing' the interface (Figure 6.1).

Similarly, Figure 6.2, which represents the processes required in the preparation, dissemination and subsequent review of offers that result from the tender documents, somewhat underemphasises both the management

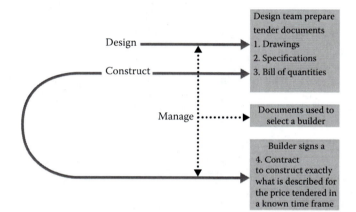

Figure 6.1 Management.

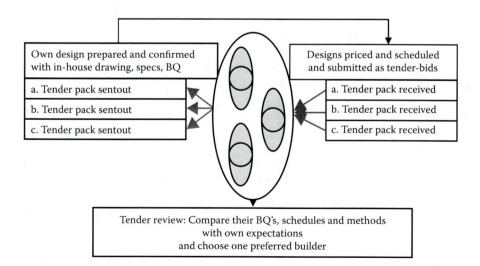

Figure 6.2 Procurement.

of the procedures (in the textboxes) and the leadership of (a bird's-eye view of the heads and shoulders of) the stakeholders.

Next, Section 6.3.1 seeks to present a somewhat more explicit definition of managing and leading, discusses the differences between management and leadership and gives an overview of leadership theory.

6.3.1 Management versus leadership

Management might be stated as the coordination of resources by planning, organising, directing and controlling through the application of a range of tools and techniques such as

> Scheduling, CPM, float analysis, PERT, network diagrams, resource aggregation, crash costing, linear scheduling, decision trees, LCCA, LCA, brainstorming, function analysis system technique, risk analysis, cost–benefit analysis (B/C ratio), stakeholder analysis technique, SCAMPER, SWOT, as well as operational management approaches including check sheets, scatter diagrams, cause-and-effect diagram, Pareto charts, flow process charts, histograms, statistical process control, Taguchi target specs, quality loss function, materials requirement planning, assignment methods, JIT layout, U-shaped shop floor, level schedules Kanban, supply chain location evaluation (such as factor-rating method, location breakeven analysis, centre of gravity and transportation model).

> Leadership is less about the application of product process techniques and more about looking after the stakeholders involved in the process.

Leadership involves describing the task at hand and the necessary direction and actions required to achieve that task, the maintenance of group harmony and team spirit, as well as meeting the needs of the individuals that make-up the group. By accepting leaders, group members define status and acknowledge an unequal power in the group, albeit members are not powerless and can choose en masse to simply not follow direction.

Leadership addresses and somewhat measures and rewards the behaviour of the group in their fulfilment of the tasks at hand. Power to reward and power over members becomes key and, traditionally, power is deemed to stem from opportunities and propensity to not only reward (either financially or in terms of status) but also to apply coercion (and the ability to influence behaviour by threat), an application of legislation, use of simple hierarchal position (as a line manager in an internal structure) to exert influence and to direct events with reference to recognisable expertise (and application of recognised skill-set that builds on experience).

A balance of an attained skill-set is important. An effective manager may be adept at planning, organising and administrating but lack motivational skills whereas effective leaders may indeed inspire and enthuse but lack an ability to structure team input. Indeed, identification of leader's skill-set and their subsequent ability to address engineering tasks and fulfil followers' aspirations, as well as instil a vision, is something that many organisations seek to measure, and is discussed briefly in Section 6.3.2.

6.3.2 Leadership measurement by personality testing

A number of employers across the world, including the large players in construction and engineering, use tools that seek to measure 'leadership' awareness or potential. In the United States, there are approximately 2500 personality tests on the market, with one of the most popular being the Myers–Briggs type indicator or MBTI,[13] which is reportedly being used by 89 of the Fortune 100 companies; MBTI has been translated into 24 languages and adopted by governments and military agencies around the world.[14]

Tests, it might be argued, very broadly speaking, seek to categorise individuals[15] as having either a managerial or leadership slant, and requires the completion of a range of responses to a number of questions about particular scenarios. Respondents might be classified as pertaining to a number of different personality types including categories such as big thinker, counsellor, go-getter, idealist, innovator, leader, mastermind, mentor, nurturer, peacemaker, performer, provider, realist, resolver, strategist or supervisor.

Interesting and somewhat at odds with the personality categorisation, a number of guidance sheets and tips to manipulate responses are increasingly available;[16] if, for example, an engineering leadership role is determined by an applicant to be more useful to a prospective employer, a respondent may seek to identify, note and subdivide the questionnaire into its various components and 'balance' responses accordingly. Thus, over the course of the 150 or so personality test answers, if a respondent charts and keeps note of the various questions concerning planning, idea generation, emotional response and group interaction, they might influence categorisation. If a respondent ever so slightly mentions a preference for being 'spontaneous' over being a 'planner', ever so slightly prefers/favours 'ideas' over 'facts', ever so slightly prefers to use one's 'head' over the 'heart', expresses alignment with both male and female guardians and ever so slightly acknowledges the male influence, then a 'mastermind' type might potentially result from ever so slightly favouring 'introverted' behaviour over 'extroverted' behaviour, whereas on the other hand, a 'leadership' type might potentially result from ever so slightly preferring to be an 'extrovert' over being an 'introvert'.

Techniques seeking to categorise personality and propensity to lead might be suggested as complimenting specific leadership theory, which is discussed in Section 6.3.3.

6.3.3 Leadership theory

A number of theories have been developed towards an explanation of why some individuals exude leadership, including having inherent 'greatness'; displaying certain traits (such as intelligence and self-confidence and even physical traits such as being tall); adapting certain behavioural approaches that may be learnt; adopting contingency approaches in different situations; or by a determined, dynamic engagement with followers that emphasises challenging, inspiring, encouraging, actioning a task and setting an example.

The behavioural theory suggests that behaviours can be learnt (unlike traits where, for example, it is impossible to learn how to be tall). Again, subcategorisation exists within this behavioural approach to leadership such as having a task-oriented style (in which supervision of tasks and getting the job done is key) or an employee-oriented style (in which motivating—not controlling—and trust and respect dominate interaction).

The Blake and Mouton managerial matrix grid[17,18] is one such means to allow respondents to reflect upon their task-oriented or employee-oriented styles. The contingency or situational theory suggests that an effective leadership approach reflects a particular situation, an organisational culture and a certain task requirement, and that no one trait and no one style is deemed optimum in all situations.[19]

A number of approaches might be suggested as falling underneath the umbrella of the contingency theory of leadership including

- A situational readiness and willingness to interact (popularised by the Hersey and Blanchard situational leadership model[20])
- A recognition that perhaps leaders should 'stick-to-their-knitting' and that trying to *change* style is inefficient (as highlighted in the Fiedler model of leadership style[21])
- A leadership approach that emphasises the path/goal in different situations and a leadership approach that reviews both the quality and acceptance of decisions (popularised as the Vroom–Yetton model of leadership[22])

Although many have sought to theorise leadership effectiveness, others shy away and link leadership to dictatorial behaviour, suggesting that those who *want* to lead present a concern, and that the wish to lead represents perhaps an unconscious effort to satisfy unfulfilled needs (in other words, the more psychoanalytic approach to leadership originated by Freud more than 100 years ago that still seems to polarise opinion[23]).

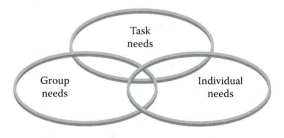

Figure 6.3 Adair's action-centred leadership.

Other viewpoints suggest that leaders are lionised and are placed upon pedestals by followers who just seek to simplify their own roles and actions, or that charisma is key or that leadership should be transactional and seek to instil confidence in followers.

Leaders generally might do well to recognise that a balance is needed between three key variables, namely, task functions of defining and planning and monitoring activities, group needs of forming and also maintaining team spirit and motivation and individual needs addressed through feedback related to current ability and future training needs.[24,25]

There is a need to balance the task; group and individual functions require the leader to recognise their interrelated nature (as shown in Figure 6.3).

The leader (and indeed the manager) needs to communicate effectively with subordinates individually and the team in general about the tasks at hand.

Next, Section 6.3.4 discusses communication paths and techniques.

6.3.4 Communication

Engineering and construction stakeholders in a building design team need to communicate; this may be categorised by verbal messages (alongside body language), written signals and graphical means. This section presents a discussion of communication processes for civil engineers that is somewhat beyond the usual expectation for drawings and written specifications.

There have been a number of schools of thought that have attempted to break down the communication process. An early influential representation of the factors involved in communication was Shannon and Weaver's mathematical theory of communication,[26] which identified the need for an information source, a transmitter, a noise source and a receiver and a destination (Figure 6.4).

Despite perhaps underemphasising feedback, context and alternative signals, this early text allows for the recognition that, if there is a wish to signal and send a message, then there is need for a receptor to be ready and willing to receive that communication.

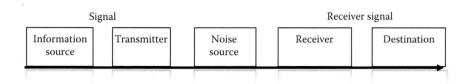

Figure 6.4 Mathematical theory of communication. (Adapted from E. Shannon and W. Weaver, *The Mathematical Theory of Communication*. University of Illinois Press, USA, 1949.)

Effective communication, somewhat implicitly, requires that source and receiver have high empathy. Unfortunately perhaps, there is a truism in the industry that the specialists who make up the civil engineering design team are less empathetic than is required. Communication disparities do seem to exist within the consultant disciplines and, as a result, it might be suggested that the messages may become distorted.

Building design team practitioners come from disparate (educational) backgrounds and often there is a misalignment between 'what-is-said' and 'what-is-heard' in the interdisciplinary design team because the different 'cultures' that might apply, for example, to an architect as opposed to a civil engineer is shaped by educational (and professional) processes that shape profession-specific attitudes, values, beliefs and resistance to change.

An example of such misalignment of the signal-transmitted and the signal-received can be found in the much repeated and much adapted 'tree swing' graphic, perhaps best attributed to John Oakland.[27]

Figure 6.5 presents (yet one more rehash of) the perceived misalignment of communication within the construction design team.

Generally then, a communication of the message must be able to work with the stakeholder's respective attitudes, values, beliefs and resistance to change.

Different theories that have sought to allow for this have included communication theories such as

- The classical approach in which the message is attended to, comprehended and accepted
- The two-way school of thought that promotes participant empathy
- An interpersonal view encouraging support and trust
- Group process thinking, allowing an atmosphere of opportunity to interact[19]

The different schools of thought mentioned above, as a means to help direct and structure approaches to address communication, can be attributed to the following discussion of ways to perhaps assist with the verbal, written and body language communication and listening skills required by civil engineers in day-to-day activities.

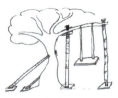

What the CLIENT wanted:
'a nice tree swing'

What the ARCHITECT conceptualised:
'functional space enhancement'

What the STRUCTURAL ENGINEER
thought: 'structural stability'

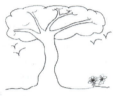

What the QUANTITY SURVEYOR
anticipated: 'low cost'

What the BUILDER thought:
'install around the tree'

What the ENVIRONMENTAL
IMPACT ASSESSORS ruled: 'EIA'

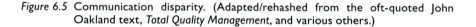

What the occupational health and saftey and risk assessor
thought: 'avoid OHS/risk'

Figure 6.5 Communication disparity. (Adapted/rehashed from the oft-quoted John Oakland text, *Total Quality Management*, and various others.)

Verbal communication aspects to address

Prepare what is going to be said; explain why it is important for the 'audience' to listen to the delivered message; summarise the important points; ensure 'feedback mechanisms' are ongoing such as looking for positive signs such as 'receptors' note-taking, asking questions and general body language signals related to their respective attentiveness, loss of concentration or confusion.

Written communication (to compliment authorise ad hoc texts/SMSs/ e-mails/Tweets)

Concise length of three distinct parts related to: first, the provision of a brief statement summarising the subject matter; second, an amplification and description; and third, an explicit suggestion of what needs to be done with the information.

Listening skills factors to bear in mind

The higher the leader's status, the less important it is to speak and the more important it is to listen; avoid an urge to inject own viewpoints early on, both mentally and verbally; seek to hear all of their messages/facts before seeking to interpret; bear in mind that individuals spend one-third of their waking hours listening but retain only one-quarter of all messages and thus waste 4 hours a day by not listening properly; individuals can be argued to think faster than they speak (700 versus 150 words per minute) thus practice summarising and verifying throughout respective listening processes; look to clarify key questions of what is meant, how is it known and perhaps what is being left out.

Body language[28] *aspects to be aware of (but avoid fixating on) and to be interpreted within (or indeed excluded by) existing cultural and national norms, include*

Defensiveness interpreted as crossing of arms or legs (or both), placement of hands as fists and pointing an index finger; suspicion perhaps expressed by not looking directly at the speaker, sideways glances, presenting the silhouette to the speaker, feet/body pointing towards exits and also perhaps by touching and rubbing the nose; Disagreement argued to be shown by having fingers on lips/hand on jaw; cooperation being expressed by open hands, hand-to-face gestures, a tilted head; and finally, nervousness being represented by clearing the throat, pinching own flesh and tugging own ear.

The communication variables discussed directly above, although noteworthy, are recognised as being much generalised.

A more applicable form of communication for civil engineers perhaps, is the process of organising and transferring graphical/specification messages; in other words, communication transfer that occurs through building information modelling (BIM) and integrated project delivery (IPD) systems (as mentioned in Section 5.2.6, which discusses specifications and BIM).

BIM and IPD initiatives represent a change in the industry and, as mentioned previously, change and resistance-to-change impinges on effective communication.

Next, Section 6.3.5 looks at change management and uses BIM/IPD as an example to place change management into context for the civil engineer.

6.3.5 Change management: BIM, communication and dissemination

Section 6.3.4 discussed how change and resistance to change affects communication. One change to the construction industry procedure is BIM (as mentioned previously in Section 5.2.6, which discussed specifications and BIM); this section briefly examines change management approaches by using BIM as an example.

BIM (or built-environment digital information modelling) integrates (knowledge management of collective) the interdisciplinary specialist design team input not just at the design stage but throughout the life cycle. Indeed, life cycle data is increasingly required to facilitate virtual prototyping of development proposals to assist specification choices prior to committing to construction. Digital technologies are readily available in Australia and is sought by clients. Figure 6.6[29] presents the growth of BIM in Australia.

Largely, BIM seeks to extend three-dimensional (3-D; width/length/height) communications of a design solution towards four-dimensions (4-D; time/sequencing) at a construction phase, into so-called fifth-dimension/sixth-dimension interpretations of life cycle cost analysis and facilities management.

Ongoing issues related to BIM uptake involve communication and a leader's ability to encourage sharing in BIM, where issues include communication disparities between consultant disciplines and a reluctance to share BIM information (with contractors), concerns over risk sharing and liability, a lack of consensus related to agreed standards/guidelines and lack of confidence in trustworthiness of data entered.[30]

To confront some of these issues, it might be suggested that a civil engineering leader acknowledge change management approaches in which there is recognition of

- Recognising organisational change to manage alterations to the norm
- Determining triggers and technology

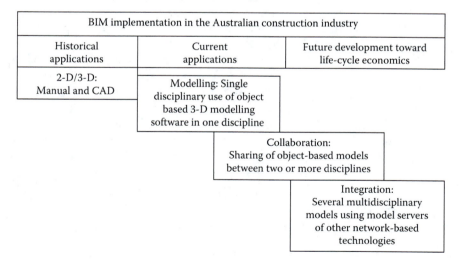

Figure 6.6 BIM implementation (Aus). (Adapted from the Australian Institute of Architects, Cooperative Research Centre, *National Building Information Modelling [Conceptual Framework]*. AIA Publications, Australia, 2009.)

- Addressing and understand resistance
- Controlling process and resource organisation to overcome resistance to change

In the case of BIM and its resultant requirement for change management, it might be suggested that

- Change is somewhat initiated by 'disorganising pressures' outside the traditional firm, where BIM presents a 'change' for a more integrated industry that seeks 'time data' for 'virtual prototyping' of project specifications prior to construction.
- Facets of organisations are linked such that BIM needs 3-D (design) and now 4-D (time) modelling with 'skill input' changes, and where team members now need to address these aspects in lobbying or tendering activities.
- Conflicts and frustrations amongst the needs/aspirations of employees may well result from changes in tendering procedure in which BIM requires (more traditional) engineers to adopt/use the new technologies.
- Time lags are apt to result in uncoordinated interdepartmental adherence where some adapt quickly, others take longer and where
  - Generally, triggers include competitor's activities, customer tastes, technology, materials, legislation, social cultural values and changing economic circumstances.
- Modifications in attitude, motivation, behaviour, skill, knowledge/performance needs and 'changing' job design, product design, office/factory layouts, allocation of responsibilities and, not least, technology and software application.

Change management schemes to address organisational change must address the variables of task, people, technology and structure where change can occur in any one, but affects all. Figure 6.7 identifies the variables to be addressed in a change management process.

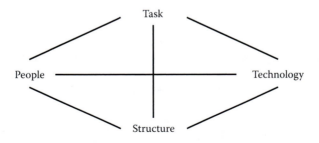

Figure 6.7 Change management variables adapted from Buchanan. (From D. Buchanan and A. Huczynski. *Organisational Behaviour*, 1991.)

Leaders who wish to introduce organisational change in a planned way, to address BIM by way of example, may

- *Modify the task*: Change strategy from design solutions that offer 2-D/3-D representations to 4-D (timed) design specifications
- *Modify the technology*: This popular general strategy affecting knowledge and skill demands on users is somewhat explicit in terms of BIM software uptake
- *Modify the structure*: Another popular general strategy to change individual's jobs and responsibilities, which again is perhaps a way to structure software application
- *Modify the people*: By addressing motivations/values/attitudes, although dealing with experienced civil engineers is unlikely to be straightforward and resistance is likely to be common

Thus, leaders might seek to confront and manage resistance to organisational change.

- Organisational change through addressing four common causes:
  - Parochial self-interest in which individuals seek to protect a status quo
  - Misunderstanding and lack of trust where often the introduction of change is resisted, rather than the change itself
  - Contradictory assessments in which different engineers shall try to interpret in different ways, although perhaps this range of divergent and different interpretations can also lead to constructive criticism in an adoption of BIM
  - Low tolerance to change in which self-doubt and self-concept need to be addressed largely perhaps through retraining in software application
- Manage organisational change as a result of BIM by specifically addressing
  - Nonparticipating groups in which resistance to new standards and software might be met by splitting-up existing groups into new factions
  - Identification of representative groups alongside the reporting of their increased empowerment; aligned with the identification of participative groups early-on
  - Certainly, education and BIM training for trusted employees is necessary, and is linked to facilitation and support to overcoming trepidation over new systems
  - Negotiation and agreed rewards upon stage completions, or conversely and somewhat as an extreme measure (not applicable for design work?), is

- Some form of covert manipulation (although silent compliance is likely to lead to future problems), and as final measure for serious noncompliance of company policy, explicit coercion via downsizing/retrenching/and redundancy

Change management generally then, as a process implemented by a leader, is expected to result in success if that company addresses at an in-house departmental level the variables of participation, consultation and involvement.

As a result of change management approaches in industry moving towards BIM and IPD, it is perhaps noteworthy to reconsider the parties that might be expected to integrate their respective knowledge bases; indeed perhaps discussion of the extent to which the various members of the multidisciplinary design team are predisposed to integrate is worthy of review. Section 6.4 extends the discussion regarding BIM and IPD, and reviews opportunities for professional integration in a multidisciplinary (BIM orientated) design team.

6.4 PROFESSIONAL INTEGRATION IN A MULTIDISCIPLINARY (BIM ORIENTATED) TEAM

This section reviews opportunities towards assisting and integrating the professionals as well as their professional contribution in the multidisciplinary design team.[31-45]

Broadly speaking, the realisation of a client's brief in the construction industry requires the interaction of a range of specialist engineering, architectural and building professionals. If effective participation is sought to improve efficiency levels in multidisciplinary construction design teams, perhaps motivational mechanisms (such as increased identification with disparate colleagues, overarching goal acceptance and trust) might be deemed appropriate perhaps to compliment expert technical input.

This section discusses whether perhaps fulfilment of professional potential may best be found in combining particular skills efficiently in the participative building design team; certainly, this is a goal of BIM systems that integrate the knowledge bases. The section seeks to further discuss team integration.

Do the values and expectations of disparate design team members influence the design process and the final built product and, might addressing these variables at a formative educational stage, through structured cross-disciplinary project work, have a potential to prepare practitioners for a more effectively integrated (BIM orientated) industry?

Modelling interdisciplinary contribution in the creation and maintenance of the built environment is increasingly important. The complex nature of specialist input into the design and construction process, and the importance of integrating specialist subsolutions, has lead to the promotion of BIM as a means to facilitate appropriate virtual prototyping of development proposals to assist specification choices before committing to construction, both in new build projects and life cycle maintenance programmes.

A raft of digital technologies are available in Australia; however, the Built Environment Digital Modelling Working Group calls for further research in this area, arguing that there is still much more work to be done in the development, promotion and utilisation of digital models that involve the integration of technically disparate disciplines. If productivity and environmental gains are to be realised in the Australian construction industry through BIM, practitioner motivation to contribute must be addressed.

Although 3-D modelling and the concept of BIM is known and accepted in the Australian building and engineering industry, the Cooperative Research Centre for Construction Innovation finds that there remains communication anomalies, apt to negatively affect the interaction between consultant disciplines charged to realise the client's design brief for a project. It might be suggested perhaps that there is somewhat of a reluctance by design team specialists to share information in the medium of BIM. This reluctance stems from both risk management worries as well as a lack of trust in the input from other, contractually procured, 'team' members. There is a perceived lack of agreed standards and guidelines governing all input and this seems to have resulted in a lack of confidence (amongst engineers and architects) regarding the trustworthiness and reliability of some areas of specialist input data.

BIM systems do seek to aid design team integrated decision making; however, it might also be suggested that computer-integrated construction mechanisms, although procedurally beneficial, have thus far failed to fully address requirement gaps in information sharing.

Although physical modelling shows how the components of a building project relate to each other (allowing a virtual environment for interdisciplinary professional interaction), perhaps this process must also acknowledge the attitudes, values and distinct professional cultures of the specialist disciplines expected to use such models. This section seeks to highlight any attitude differences amongst building design team practitioners and how these might affect communication generally and the willingness to embrace BIM specifically.

Today's construction teams, charged to realise a client's brief, are made up of a wide range of professions and disparate design and building specialists brought together (traditionally, by competitive tendering procurement mechanisms) for one-of, short-term building projects; parties are contractually bound to design and then construct the required asset.

Teams (as mentioned previously in a review of tradition in Section 6.1) are made up of a range of professional disciplines.

Design teams include not only the commonly held principal players such as

- Architects, civil engineers, structural engineers, mechanical and electrical services engineers, quantity surveyors, project managers and superintending-officers, but also
- Building practitioners such as contracts managers, construction managers, resident engineers, quality control professionals, building surveyors, heating and ventilation engineers, and environmental planners, as well as asset and facility management professionals, landscape architects, estate managers, interior designers and specialist subcontracting builders, along with building materials, fittings and fixtures suppliers

Input from the various specialists in the team was traditionally (and perhaps probably is still) somewhat of a linear process, starting with a civil engineering or architectural designer who then passes forward their respective overall concept designs for subsequent input by the other specialist practitioners. However, an increasing appreciation of the life cycle requirements of complex building has lead to recognition of the importance of a more integrated collective contribution at the outset. Building and construction design teams increasingly recognise the importance of allowing all members the opportunity for early, initial stage input, although such early input into the specifications of design is procedurally more difficult to organise. The BIM system seeks to primarily address this difficulty.

It might be suggested perhaps that information and communication technology techniques, although able to order and make available specialist data sets, must primarily seek to complement the interaction of the multidisciplinary design team members. Computer-integrated construction and BIM require recognition of the working practices and professional cultures to justify the levels of success forecasting for this growing cross-disciplinary integration medium.

> The decision processes of design are still likely to reflect the overall psychology of design team members, with software application unlikely (on its own) to enable optimum multidisciplinary team interaction.

Problems do often arise from the fragmentation of the design process and might perhaps be argued to affect performance negatively, although fragmentation is difficult to address because its roots lie in the historical development of the disparate construction professionals. Fragmentation can be argued to create professional dissonance, which can be bad for the practical realisation of innovative design. It might be suggested that if professionals are to grow in understanding and work better together, they need to know the pattern underlying any data set in their own and other's terms. In other words, empathy with other professionals and acknowledgement of the potential of specialist databases, irrespective of the mode of communication and sharing adopted.

Some studies that have examined technology with group processes and communication show no significant improvements in equity of participation or decision consensus, and have found slower group decision speeds and a negative relationship to exist between quality of solution and consensus. Although building information modelling/computer-integrated construction is markedly different from group decision-making systems, group consensus can be argued to contribute much to the effectiveness and efficiency of innovation construction. A somewhat prevalent opinion might be suggested that overt reliance on technology (in isolation) limits the use of social cues in meetings, and that this is perhaps problematic for groups forging new relationships. The limitation of social cues poses an obvious problem for groups like the traditionally procured building design team, which is continually required to forge new project-specific relationships.

It might be suggested that a building that is functionally (and aesthetically) successful must necessarily extend the initial creative art of invention (as discussed in Section 6.1.1, dealing with engineering philosophy) towards a practicability for the user. In other words, the specialist members of the building design team must collectively go beyond the single designer's engineering/architectural, philosophical and metaphysical creative leap of imagination that initiated the innovative design concept; the next stage must necessarily move into the realms of the empirical practical application of existing knowledge and causality, represented by input from the other disparate construction specialist knowledge bases able to provide solutions that link technical cause and effect to allow (innovative) built assets to function.

Beyond the first stage of the individual metaphysical creative spark, and after the second level of a collective application of empiricism and the addressing of causality, the building design decision making process might be argued to then require, what Aristotle has termed, the controlling hand of a person of practical decisions with the ability to lead decision making; such an individual must seek to avoid a situation in which a single specialist is given the opportunity to settle a debate over other disparate specialist practitioners. Again, it might be suggested that this is somewhat beyond the remit of BIM. Generally, although a designer is charged with the initial creative art of invention, it is the specialist skills of the building design team (reasonably lead) that should be responsible for the realisation of innovative design; again, this process must occur irrespective of the technology that facilitates a sharing of information.

Integrating design teams by more than just BIM

Group members will make choices about which features of BIM systems they are prepared to use. These choices are often linked to group dynamics, and it might be suggested perhaps that information technologies that focus primarily on task completion without supporting personal relationships will be somewhat less successful.

To understand the technological effects, research should not only focus on the packages of hardware and software but also on the specific task and social variables that are inherent in specialist interaction.

In other words, the role of social relations remains important in collaboration and neglect of these relations may lead to the failure of existing information technology systems to adequately support technological specialist interaction. This is not a new concept. Groundbreaking research undertaken in the Tavistock Report of the mid-1960s, through to work by Faulkner and Day 20 years later, then Latham in the mid-1990s as well as recent initiatives such as the UK Government's raft of guides and documentation to encourage 'achieving excellence in construction', all indicate a relationship between performance of the construction design team and (perceived hierarchal) orientation of the professions that make up the team.

Given moves toward BIM in Australia, there remains a need to perhaps reexamine possible communication anomalies, negative consultant interaction, a reluctance to share information and any perceived lack of trust between the construction design team members; the following case study discussion attempts to review some of the main issues.

Attitude study

The examination presented below looks at interdisciplinary values held by building design team members at their formative stages, towards analysis of the professional relationships that require collaborative effort to fulfil their respective objectives. The role of social relations in the collaborative process and communication in the building design team are seen as a function of disparate professional cultures; these need to be addressed if local tools such as the Australian BIM system are to integrate successfully different specialist knowledge bases.

Digital representations of innovative ideas require reaction and response prior to incorporation. Specialist data files may be shared and integrated into the whole only after consensus has been reached over their positive contribution to the scheme. Although design teams may well use new technology to relay scheme design ideas to colleagues, it might be argued that the facility to empathise with disparate specialist suggestions remains rooted in professional culture.

The study below was conducted to explore the effect of attitudes and values held about peer groups on interdisciplinary processes, irrespective of the means of communication. The subsequent discussion highlights the major findings from a multitude of responses elicited at various times from a sample of several hundred selected as representative professions that make up the integrated design team.

Case study sample

The experimental case study sample group was made up of more than 500 respondents from five different universities pertaining to eight undergraduate degree courses that encompassed: first, architectural/engineering design–orientated course structures; second, construction economics and management courses that concentrate less on design and more on the cost-engineering and building organisation and control of the process; and third, building surveying and building engineering courses that seek to develop knowledge of minor building design work and the effect on maintenance and life cycle asset refurbishment and retrofitting. These university disciplines and their respective course make-ups were held to be representative of education for today's construction industry. Respondents were categorised by university discipline/course, stage of study, as well as pertaining to on the one hand, disciplines taught in isolation and segregated from the other courses tested and, on the other hand, courses that participate in educational initiatives designed to encourage multidisciplinary building design activities (periodic integrated workshops involving many courses together in role-play design projects).

Findings: Attitudes influenced by course enrolled

A two dozen–point attitude scale sought respondents' attitudes towards disciplines other than their own. The attitude scale highlighted attitudes towards creative motivation, orientation towards other people, mental habits, purpose and responsibility, information handling, social status, level of training received, level of education received, contribution to the building process, usefulness of information provided and leadership.

A discriminant function analysis was carried out on a data set of 534 attitude scale questionnaire responses. This allowed an analysis and identification of 'groupings' on the basis of independent variables.

- Potential predictors of attitude scale scoring were thought to include perhaps: age, construction industry experience, course undertaken, curriculum structure full or part-time, parent/guardian occupation, gender, university attended and current stage of studies. Discriminant analyses were ordered by size of correlation within the function.

'Course' was identified as contributing most substantially to *differences* (different attitudes held about team members) in group scores

- The variable of 'course' was a significant indicator of attitude score.
- Respondents from, on the one hand, 'design courses' (such as architecture) and, on the other hand, (construction) 'management courses' were linked by their *mutual unfavourable attitude towards their disparate team members.*

The findings suggest a need to address, at the formative educational stage, the traditionally accepted truism that suspicion exists between the 'design element' and the 'building management element' of the traditionally procured construction design team. Adherence to traditional vocational courses seems to influence attitudes held towards professional peers; this in turn might be argued as negatively influencing the communication process of the building design team.

Findings: Addressing negative attitudes by cross-course project workshops
Given that the educators at the universities chosen for study were inclined to address (what was proven empirically by the application of this attitude scale to be) negative attitudes held by certain courses about other courses through small, integrated cross-course project workshops, an analysis and direct comparison of attitudes measured before and then after participation in (and completion-of) small cross-disciplinary project work was undertaken.

Further examination of approximately one-third of the sample was carried out. Nonparametric tests allowed the assessment of differences (before and after changes) in attitudes towards other courses resulting largely from participation in small interdisciplinary workshops (shared undergraduate project work).

Results found the 'staging' of workshop projects in either the first, intermediate or final years of a student's progression through the course was a key factor.
- Early stage integrated projects did not produce a statistically significant change in attitudes.
  - In other words, first-year students did not change their (relatively positive) attitudes towards students from different courses as a result of participation in interdisciplinary projects.
- Intermediate staging of cross-disciplinary project work was found to be the most effective in improving the somewhat negatively held attitudes towards those from courses different from the respondent's

own (attitudes which had, interestingly, grown somewhat more negative after the first year of study).

- In other words, students in the middle years of their undergraduate studies did, after participation in cross-disciplinary workshops that sought an element of specialist input to fulfil overall goals, see students from other courses in a more positive light.
- Final-year students of undergraduate courses who participated in cross-course project work did not change attitude after participation in cross-disciplinary workshops.
  - If anything, project work at a final stage of courses reinforced (albeit in degrees not statistically significant) previously held (relatively negative) attitudes towards those from courses other than their own.

Generally, this further analysis found that a respondent's stage in a course (first year, intermediate years or final year), as well as the content of interdisciplinary initiatives, play an important part in addressing attitudes displayed by students from specific courses towards disparate peers. The educators in this study were informed that the 'best' results might be expected to come from interdisciplinary projects at the intermediate year of study, where disparate team members were able to contribute to group efforts in small specialist, discipline-specific ways.

Findings: Attitudes differ but may be addressed

Results show that the particular vocational course of study undertaken by a professional during formative development is a key contributor to design team attitude difference. Courses, and by extension, belonging to a particular profession, influence group differences in attitudes.

Attitudes differ but may be addressed:

- Mutual enmity between courses (professions) was found to increase after the initial stages of vocational education.
- Attitudes displayed toward disparate professionals became significantly less favourable and less empathetic as the student progressed through their full-time course of study.
- Final-stage students were found to display little affinity towards those from different courses; indeed, interdisciplinary project work, conducted at the final stage to encourage interdisciplinary integration resulted instead in increased enmity to those from other built-environment courses.
- It might be argued that different patterns of thinking can lead to a reduction in respect for, or lack of trust in professional colleagues and that, consequently, the building process suffers. Educators

may be able to address this through cross-disciplinary project work, but the staging and content of these workshops require staging at the intermediate years with requirement for small specialist contributions.

The research conducted suggests that vocational education instils attitude differences amongst building design team members, instilling professional cultural differences that in turn affect communication and integration in the multidisciplinary building design team. Participation requires high levels of integration to achieve organisational innovation solutions. The social process of communication and, in particular, the individual variables of the innovative process of an organisational team provide perhaps a key to effective integration.

Moves towards BIM techniques require going beyond a somewhat simplistic assessment of how to input and integrate stand-alone specialist databases to improve the (often somewhat) strained social processes of communication in the building design team.

In the main, BIM systems seeking to integrate building design knowledge bases must perhaps first understand the integration process of the professionals who will use the packages. Systems must acknowledge the building blocks of professional integration. Communication is seen as a function of a cultural system, where culture is derived from attached values and selected ideas and where values are applied to the objects of need desire and attitude, and the extent to which attitude is open to modification from interdisciplinary initiatives during tertiary education. Based on the discussion above, it might be suggested here that BIM software support is unlikely, in rude isolation perhaps, to instil in the user empathy with disparate professional objectives.

Innovative designs must acknowledge holistic specialist input to address sustainable innovation design but the opportunities for doing so are restricted, to a large extent by interdisciplinary disharmony. BIM presents a means to address interdisciplinary specialist information discord (such as, at a very minor level, clash detection) but must acknowledge the attitude, values and distinct professional cultures of the specialist disciplines expected to use such technologies.

BIM is a way forward but the case study findings discussed above seem to show that building design team professionals at the formative stages might be somewhat disinclined toward the integrative dictates of a software package (in isolation).

Construction industry practitioner attitudes, arising during the vocational educational years of professional development, might somewhat be expected to persist no matter the method of integration. Physical and mathematical modelling by industry through BIM may begin to address the integration requirements of knowledge bases, but it is perhaps unlikely to improve the integration of the specialist themselves who have thus far registered disquiet over the BIM input process related to risk and liability.

Although calls for users to empathise and trust the value of respective specialist input by other disparate professionals towards the realisation of optimum technologically complex solutions to clients' building needs might be somewhat idealistic, current levels of disquiet over risk and liability do seem real. Educators and the professional bodies that validate courses might, therefore, be required to provide greater support in addressing any latent attitudes of mistrust that might surround integrated modelling systems.

Having previously discussed, somewhat philosophically, the integration of specialist knowledge through BIM, perhaps a return to a more grounded discussion of the means and methods and approaches to structure a civil engineering design solution is timeous.

The following chapter (Chapter 7) presents first, a method to structure a civil engineering design solution that explicitly integrates cost and time variables, and then reviews three practice examples towards an illustration of the integrated approach.

Chapter 7

Integrated design and cost management solutions

Generally, a client's key indicators of brief realisation might be deemed to include

- Drawings and calculations incorporating materials specifications for supply and installation (made explicit in AS4000 clause 8.2, item 15, and clause 8.2)
- Method and timeline and related work-task sequence and a critical path of activities (made explicit in AS4000 clause 32)
- Priced bills of quantities (structured with the civil engineering standard method of measurement CESMM, 3rd or 4th edition)
- Risk analysis, safety concerns, project constraints, supply and installation and testing of all elements and components within a quality management system (expressly requested in AS4000 clause 29) and
- Tender-offer expectations (framed with AS2125 Tender form offer, and AS4000 page 41, annexure A, B and C, where such appendices are completed and issued as part of a tender document and incorporated into the contract).

Advice to progress design realisation, related to tendering procedure, might be taken from documents such as the Australian Constructors Association,[1] which requests

- Ethical practice and no improper advantage/collusive practice
- An evaluation of tenders based on the conditions of tendering and selection criteria related to bid conformity, capability, innovation, price and building period
- Tender preparation periods of a reasonable time (of 2 weeks plus, alongside a similar evaluation period of 2 weeks or so) based on scope and complexity
- Client's notifications

- Other aspects such as opportunities for partnering/alliances/direct negotiation, and expectations for building information modelling and the like

Local authority agencies, such as Main Roads[2] in Western Australia, also provide guidelines for tender document preparation, consisting of the main sections for major works (or alternatively minor works) utilising AS2124, with suggestions for road and bridge design consultancy (AS4122), specifications preparation and authority-issued methods of measurement, in which such tender submission requires a suitable skill-set (to prepare: book 1, tender submission; book 2, conditions of contract; book 3, general and management requirements; book 4, roadworks technical specification; book 5, bridgeworks technical specification; books 6 and 7, drawings; and book 8, information for tenderers). Previous chapters have sought to review such applicable knowledge.

The previous discussions have addressed, in relative isolation, civil engineering approaches to

- Cost planning, estimation and life cycle analysis
- Scheduling project timelines and risk
- Quality management systems
- Contract documents: contracts/specifications/BQs/drawings
- Ethical (procurement) processes and sustainability

Section 7.1 builds on these skill-sets to address a client's brief towards integrated design and cost management solutions. The practice example scenarios presented in Sections 7.1.1 through 7.1.3, although expansive, are not presented as definitive solutions; rather design quality, cost and time explanations might be regarded as representative of the approach of a civil engineer in a formative brief development.

7.1   INTEGRATED DESIGN PRACTICE EXAMPLES

A step-by-step guide to addressing a client's civil engineering detailed brief (beyond the approximate and preliminary estimating stages, which are deemed to have been given their respective phase 'go-aheads') is as follows:

Step 1:   Review the client's brief; clarify the problem/need and scope of the proposal (as Section 5.4.1).

Step 2:   After further client briefing meeting(s) and site visit(s), address a working concept for the design that considers fit-for-purpose design options (as Section 5.4.2).

Step 3:   Generate working reference specifications and design drawings (as Sections 2.2.2, 2.2.3, 4.1, 5.2 and 5.4).

Step 4:   Clarify appropriate detailed specifications for the project (as Sections 2.2.2, 2.2.3, 4.1 and 5.2).

Step 5:   Establish the construction methodology/method statement (as Sections 2.2.3, 2.2.4, 3.1, 3.1.1 and 3.2.2).

Step 6:   Prepare the project timeline schedule (as Sections 3.1, 3.1.1 and 3.2.2).

Step 7:   Assess risk, confirm quality management systems and occupational health and safety compliance andaddress environmental impact (as Sections 3.3, 4.1 and 5.1.7).

Step 8:   Prepare Bills of Quantities (BQ) from drawing/details take-offs (as Section 5.3).

Step 9:   Generate a detailed priced BQ; input unit rates into the final BQ (as Sections 2.2.3 and 5.3).

Step 10: Reflect upon the final project value and seek elemental cost comparisons (as Sections 2.2.2, 2.2.3 and 3.6).

Step 11: Method-statement resources review towards future cost-monitoring (as Sections 2.2.3, 2.2.4, 3.1, 3.1.1, 3.2.2 and 3.5).

Step 12: Value engineering: alternative specification options review (as Sections 2.2.2, 2.2.2.1 and 3.6).

Step 13: Extrapolation of cash flow/revenue 'S-curve' analysis (as Section 2.3.1).

Step 14: Preparation of a tender-pack toward selection of a contractor (Chapter 5).

The following scenarios adopt the 14 steps above for 'client's brief' development.

7.1.1   Practice example: Design solution for siteworks for a new school

Step 1: Review the client's brief; clarify problems/needs/scope (as Section 5.4.1)
A charity organisation proposes to build a new private primary school (to cater to more than one thousand students) in a relatively rural shire, just west of a state's main city.*

---

* This hypothetical/academic/educational example acknowledges with thanks contributions by Cong Bui, Nicholas Loke, Chen Shok Yin and Sun Yini; all designs, specs, timings and costings are provided for illustrative purposes only.

Permanent access to the complex is from an adjacent principal road. Damage to existing bushland to be minimised to only trees and vegetation within 10 m of the proposed siteworks/centreline of road alignment or in car park areas. The school requires an outdoor recreation area/oval, five basketball courts, 30 classrooms, an assembly hall, gymnasium and library. Universal access required between buildings.

The client seeks a suitable design for the site-layout of the proposal, including site-earthworks, water supply and sewerage to complement a suitable geometry for all roads, along with adequate earthworks and drainage provision(s).

It is hoped that the earthworks can commence in early October and be completed, within 4 months or so, by January of the next year. The client would like to then start construction of the new school buildings in January and have the classrooms ready 12 months later. Design of the building structure is *not* required.

Due to nearby residences, sitework is prohibited between 6 p.m. and 7 a.m. Monday to Saturday, and all-day Sunday.

Step 2: Generate a working concept for the design solution (as Sections 5.4.1 and 5.4.2)
The site is a 6-ha greenfield with an undulating terrain. The design includes earthworks, water supply system, sewerage system and storm-water system, with earthworks designed to accommodate current slope levels in three (cut and fill) sections to reduce the need (for expensive) removal of soil from the site.

The slope level is significant, therefore, a retaining wall is recommended with a maximum height of 10 m. There are currently no existing utilities at the site's location.

A water supply system consisting of three existing pipes of 100 mm diameter connected to gather and divert water to the site was found to provide sufficient pressure. Because the nearest sewerage pipe is located approximately 25 km from the site, the use of an on-site sewerage treatment system by proprietary suppliers Biolytix, alongside two 3,000-L BioPods, was deemed appropriate for the project, with treated water released back into the ground using a proprietary leach drain from RainSmart. For the storm-water system, a (large-volume) subsurface tank system (in lieu of an open storage system to alleviate student access/danger) shall store and dissipate rain water with soak well(s) for roads to reduce diversion piping.

Site inspection: soil is brown gravelly loam overlaying clay. The type found on-site is classified as Class A with little or no ground movement in accordance with Australian Standard 2870. Testing was done according to Australian Standard 1726.

There are no gas/telecommunications/underground power services to the site; high-voltage power lines are available nearby for future connection. All calculations considered:

- Contour map (Figure 7.1) shows that the site area is a hill in which the difference between high and low levels is 35 m.
- The site near Rd-L is much more lower than the corner close to Rd-G, where a steep drop of ground level is noted.
- Water Corporation (Dial Before You Dig) records existing services where pressure checks confirm sufficiency.
- No existing sewerage systems noted.
- Existing storm-water via inspection designed for road run-off.

Step 3: Generate working/reference specifications and design drawings (as Sections 2.2.2, 2.2.3, 4.1, 5.2 and 5.4)
Specifications and design drawings are deemed necessary for

- Site layout and excavation:
  - 5.76-ha basketball courts (32 × 19 m); recreational area 135 × 110 m); library 1,200 m²; classroom 130 m²; three design levels for earthmoving: 217, 230 and 238 m; excess excavated material estimate ≈1,600 m³
- Retaining wall(s):
  - Maximum height 10 m, 10 kPa imposed load; cribwall design to AS3600 in lieu of conventional cantilever… due to lower carbon footprint and reduced imported fill cost savings of ≈63% and 50%, respectively, and for aesthetic improvement

Figure 7.1 Contour.

- Road(s):
  - Two-way access road (Road 1-1) enters existing road Rd.W and first deviates into Road 2-2, then diverges into disabled parking to provide access to most school buildings and lastly ends with entry and exit off oval's visitor parking. First deviation (Road 2-2) is a one-way road for drop-off point as well as entry of staff and student parking, then exit to existing Rd.W.
  - Total road width reserved for Road 1-1 and Road 2-2 is 9 m and 8.7 m, respectively. Road 1-1 has mountable concrete kerbing on both sides as there is a pedestrian walkway along. For Road 2-2, only some parts, which are not connected with the parking, have the kerbing. All the road geometric and design specs are in accordance with Main Roads WA Standards and Guidelines.
  - AustRoad pavement design guidelines give different design thickness for granular asphalt wearing surfaces. Effective pavement thickness is 50 mm asphalt with 300 mm granular subbase. Other options such as granular pavement with surface seal and full-depth asphalt/asphalt surface pavement on cemented base available but design rejected due to materials costs and durability.
- Car park areas:
  - Three separate car parks are proposed with overall dimensions of 96.9×20, 15×56, and 11.4×22.4 m² to accommodate different users: staff (46), Yr-12's (14), visitors (40), universal access (3).
  - Surface finish to be asphalt work as road pavement to ease construction as well as the compaction criteria. Barrier kerbing of 150 mm height is designed along the parking edge to prevent vehicle encroachment on the side.
  - Parking space with bay angle at 90° and shared space or area markings comply with AS2890.1 and AS2890.6 pavement markings.
- Water supply services:
  - Water supply system calculation records for pressure checks (Bernoulli equation) of changes in elevation, major loss in pipes and minor loss in pipe recorded for off-site and on-site. To obtain sufficient pressure, three different existing outlets of 100 mm diameter by Water Corporation require connection and linkage to site, made of appropriate materials addressing cost and the roughness of the pipe.

- Initial flow and pressure to AS2419.1-2005, ranging from 200 kPa min residual pressure to 20 L/m min flow rate for Water Corporation
- External pipeworks deemed to need 10 m overlapping coverage between fire hydrants; therefore, two fire hydrants should be added to external pipeworks
- Site pipeworks: fire hydrants allocated as Australian Standard 2419.1-2005
- Sewerage systems:
  - Wastewater design flow rates link with sewerage design, with wastewater mainly directed by gravity fall to the connection point. Proprietary products selected include: polypropylene pipe, septic tank (BioPot) and leach drain for the library, recreation centre and classrooms. Peak dry weather density goes towards pipe size and gradient.
- Storm-water services:
  - A new storm-water service is designed according to available rainfall intensity charts. The school is divided into different sections by function. Variables include discharge, concentration time and velocity. Services include gully, polypropylene pipe, soak well and subsurface tank system.
  - Gullies will be used along the road to collect storm-water and will be connected by polypropylene pipes.
  - Pipes to AS/NZS5065 with full discharge and velocity to the minimum diameter for the pipe of 300 mm.
  - Storm-water collection is achieved with the help of the soak well and subsurface tank system. The system consists of a number of enviro-modules from Ausdrain stacked together to reach high compressive strength. Chosen for its durability, light weight, low maintenance and environmentally friendly features. Capacity/arrangements of modules given in proprietary product specs. Quantity of the soak wells achieved by using storage volume from different segments and volume of the selected soak wells with position arranged with respect to the site layout and water pipe connection.

Design drawings as follows in Examples 7.1.1a through 7.1.1m (excluding typical details for kerbs/services/proprietary systems via easy access standard drawings downloads from Local Authority Co.):

Example 7.1.1a: Plan layout (not to scale)

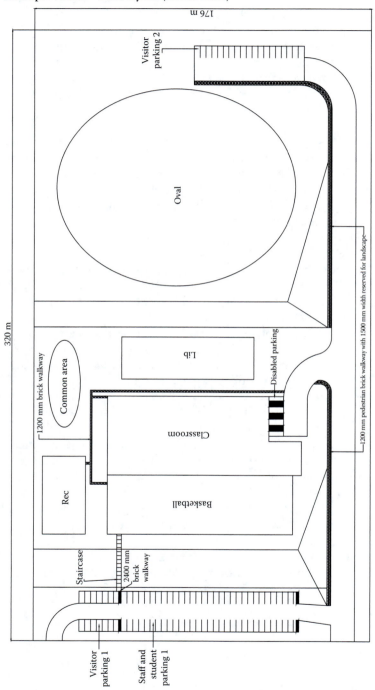

Example 7.1.1b: Road 1-1 longitudinal section (not to scale)

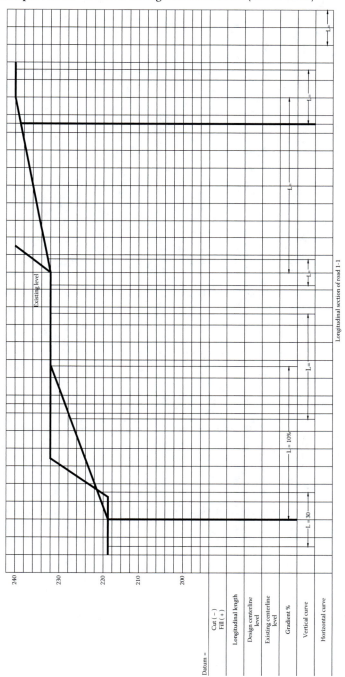

Example 7.1.1c: Road 2-2 plan and longitudinal section (not to scale)

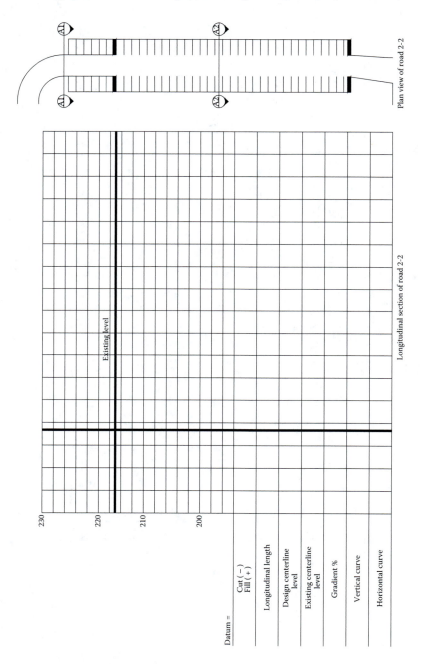

Plan view of road 2-2

Longitudinal section of road 2-2

Existing level

230		220		210		200	

Datum =

Cut (–)
Fill (+)

Longitudinal length

Design centerline
level

Existing centerline
level

Gradient %

Vertical curve

Horizontal curve

Example 7.1.1d: Cross-section and kerb detail (scale as shown)

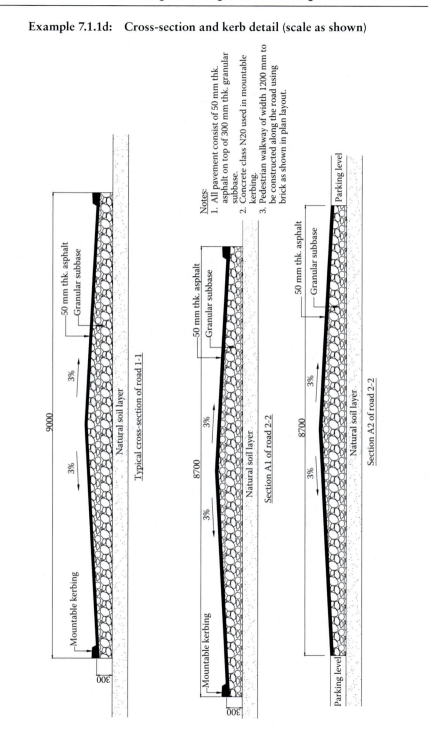

Notes:
1. All pavement consist of 50 mm thk. asphalt on top of 300 mm thk. granular subbase.
2. Concrete class N20 used in mountable kerbing.
3. Pedestrian walkway of width 1200 mm to be constructed along the road using brick as shown in plan layout.

9000

50 mm thk. asphalt
Granular subbase

3%   3%

Mountable kerbing

300

Natural soil layer.

Typical cross-section of road 1-1

8700

50 mm thk. asphalt
Granular subbase

3%   3%

Mountable kerbing

300

Natural soil layer.

Section A1 of road 2-2

8700

50 mm thk. asphalt
Granular subbase

3%   3%

Parking level

Parking level

Natural soil layer.

Section A2 of road 2-2

Example 7.1.1e: Parking bay layout (not to scale)

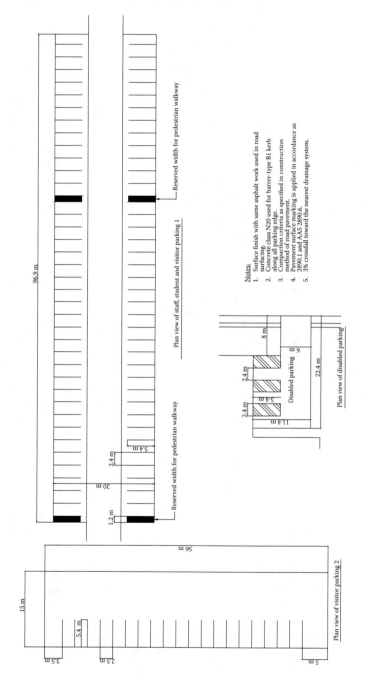

96.9 m

Plan view of staff, student and visitor parking 1

Reserved width for pedestrian walkway

Reserved width for pedestrian walkway

2.4 m

5.4 m

20 m

1.2 m

Notes:
1. Surface finish with same asphalt work used in road surfacing.
2. Concrete class N20 used for barrer-type B1 kerb along all parking edge.
3. Compaction criteria as specified in construction method of road pavement.
4. Pavement surface marking is applied in accordance as 2890.1 and AAS 2890.6.
5. 3% crossfall toward the nearest drainage system.

8 m

2.4 m

6 m

2.4 m

5.4 m

22.4 m

11.4 m

Disabled parking

Plan view of disabled parking

56 m

15 m

5.4 m

3.5 m

2.5 m

5 m

Plan view of visitor parking 2

Example 7.1.1f: Location of new external pipes and fire hydrant (water supply)

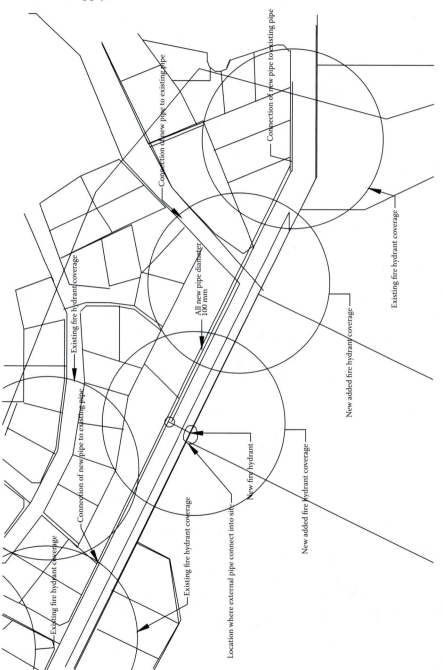

Example 7.1.1g: Location of on-site pipes and fire hydrant (water supply)

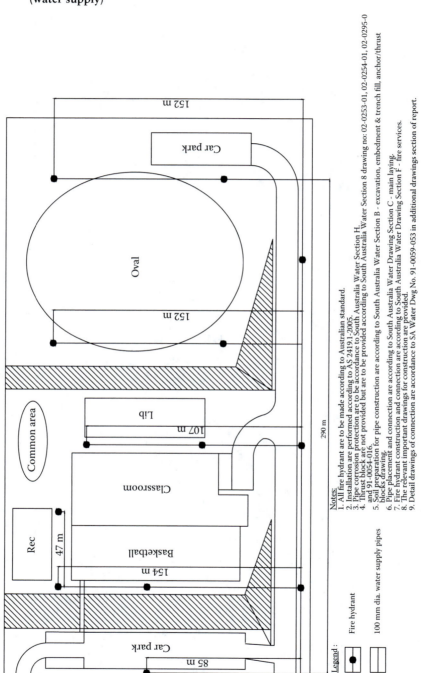

Notes:
1. All fire hydrant are to be made according to Australian standard.
2. Installation are performed according to AS 2419.1-2005.
3. Pipe corrosion protection are to be accordance to South Australia Water Section H.
4. Thrust block are not provided but are to be provided according to South Australia Water Section 8 drawing no: 02-0253-01, 02-0254-01, 02-0295-0 and 91-0054-016.
5. Soil preparation for pipe construction are according to South Australia Water Section B - excavation, embedment & trench fill, anchor/thrust blocks drawing.
6. Pipe placement and connection are according to South Australia Water Drawing Section C - main laying.
7. Fire hydrant construction and connection are according to South Australia Water Drawing Section F - fire services.
8. The relevant important drawings for construction are provided.
9. Detail drawings of connection are accordance to SA Water Dwg No. 91-0059-053 in additional drawings section of report.

Legend :

●   Fire hydrant

☐   100 mm dia. water supply pipes

Example 7.1.1h: Existing and design ground level on the east side (not to scale)

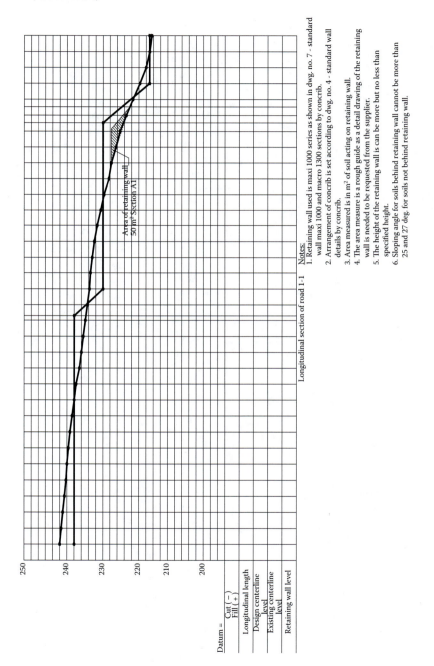

Longitudinal section of road 1-1

Datum =
Cut (–)
Fill (+)
Longitudinal length
Design centerline level
Existing centerline level
Retaining wall level

Area of retaining wall
50 m² Section A1

Notes:
1. Retaining wall used is maxi 1000 series as shown in dwg. no. 7 - standard wall maxi 1000 and macro 1300 sections by concrib.
2. Arrangement of concrib is set according to dwg. no. 4 - standard wall details by concrib.
3. Area measured is in m² of soil acting on retaining wall.
4. The area measure is a rough guide as a detail drawing of the retaining wall is needed to be requested from the supplier.
5. The height of the retaining wall is can be more but no less than specified height.
6. Sloping angle for soils behind retaining wall cannot be more than 25 and 27 deg. for soils not behind retaining wall.

Example 7.1.1i: Existing and design ground level on west side (not to scale)

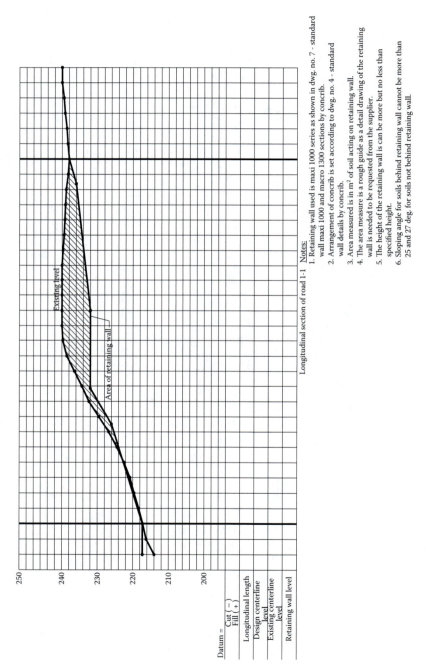

Longitudinal section of road 1-1

Notes:
1. Retaining wall used is maxi 1000 series as shown in dwg. no. 7 - standard wall maxi 1000 and macro 1300 sections by concrib.
2. Arrangement of concrib is set according to dwg. no. 4 - standard wall details by concrib.
3. Area measured is in m² of soil acting on retaining wall.
4. The area measure is a rough guide as a detail drawing of the retaining wall is needed to be requested from the supplier.
5. The height of the retaining wall is can be more but no less than specified height.
6. Sloping angle for soils behind retaining wall cannot be more than 25 and 27 deg. for soils not behind retaining wall.

Example 7.1.1j: Existing and design ground level on south side (not to scale)

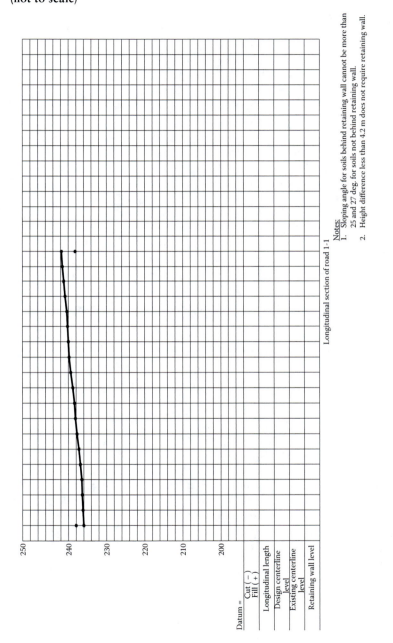

Longitudinal section of road 1-1

Notes:
1. Sloping angle for soils behind retaining wall cannot be more than 25 and 27 deg. for soils not behind retaining wall.
2. Height difference less than 4.2 m does not require retaining wall.

250   240   230   220   210   200

Datum =
Cut (−)
Fill (+)
Longitudinal length
Design centerline level
Existing centerline level
Retaining wall level

Example 7.1.1k: Cross-section of retaining wall (not to scale)

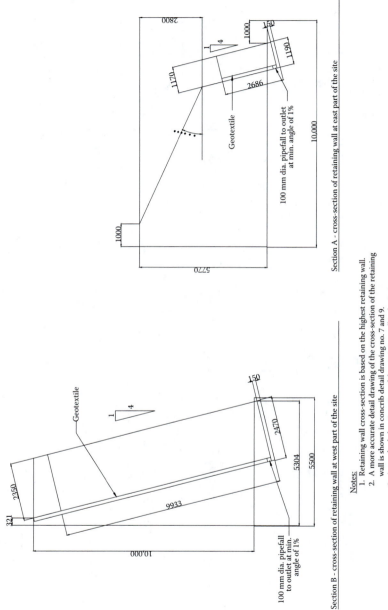

Section A – cross-section of retaining wall at east part of the site

Section B – cross-section of retaining wall at west part of the site

Notes:
1. Retaining wall cross-section is based on the highest retaining wall.
2. A more accurate detail drawing of the cross-section of the retaining wall is shown in concrib detail drawing no. 7 and 9.
3. A standard elevation view of the retaining wall is shown in concrib dwg. no. 1 - standard retaining wall.
4. Wall details are shown in concrib dwg. no. 4 for maxi 1000 header and dwg. no. 5 for macro/macro 2350 double header.

Example 7.1.1l: Location of sewerage treatment system and pipes (not to scale)

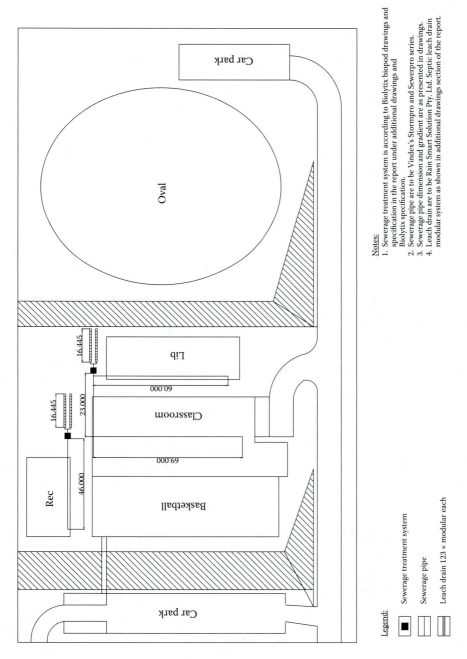

Notes:
1. Sewerage treatment system is according to Biolytix biopod drawings and specification in the report under additional drawings and Biolytix specification.
2. Sewerage pipe are to be Vindex's Stormpro and Sewerpro series.
3. Sewerage pipe dimension and gradient are as presented in drawings.
4. Leach drain are to be Rain Smart Solution Pty. Ltd. Septic leach drain modular system as shown in additional drawings section of the report.

Legend:

◼ Sewerage treatment system

▭ Sewerage pipe

▦ Leach drain 123 × modular each

Example 7.1.1m: Location of storm-water retention system gully and pipes

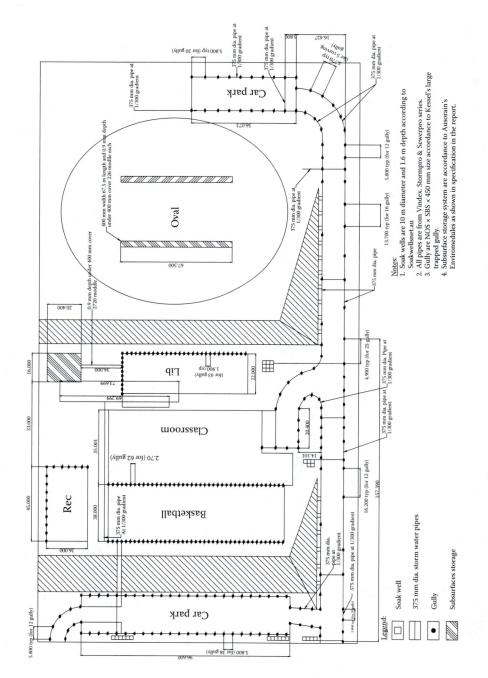

Step 4: Clarify appropriate detailed specifications (as Sections 2.2.2, 2.2.3, 4.1, 5.2 and 5.4)
Reference specifications are somewhat included in the sidenotes of the drawings presented above. Reference specifications are also given in the sidenotes of the other inferred local authority/proprietary supplier (easy access) typical design detail drawings (not given here).

A detailed specification for this proposal is required to access, review and then adapt (as necessary) the detailed specification pro forma provided by Main Roads WA as well as the proprietary systems pro forma given in appropriate website links such as

- Clearing Specification 301
  - https://www.mainroads.wa.gov.au/Documents/Spec%20301%20 14%20Jun%202007.u_2571350r_1n_D09%5E23288647.PDF
- Earthworks Specification 302
  - https://www.mainroads.wa.gov.au/Documents/Specification%20 302%5E2C%20Earthworks%5E2C%2026%20August%20 2011.u_3326799r_1n_D11%5E23208480.pdf
- Revegetation and Landscaping Specifications
  - https://www.mainroads.wa.gov.au/Pages/default.aspx
- Retaining Wall Specification
  - http://www.concrib.com.au/pdf/specifications.pdf
- Pavement Specifications 501 and 503
  - https://www.mainroads.wa.gov.au/Pages/default.aspx
- Pavement Marking Specifications 604
  - https://www.mainroads.wa.gov.au/Pages/default.aspx
- Drainage specifications 402, 403, 407
  - https://www.mainroads.wa.gov.au/Pages/default.aspx
- Water Supply Specification
  - http://www.lismore.nsw.gov.au/content/planning/manuals/ D11WaterSupplyNRLG.pdf
- Storage Tanks Specification
  - http://www.biolytix.com/biopod-specifications/
- Storm-water System Specifications
  - http://www.ausdrain.com/downloads/ausdrain_enviromodule2_ tanks.pdf

Step 5: Establish construction methodology towards method-statement(s) compilation (as Sections 2.2.3, 2.2.4, 3.1, 3.1.1 and 3.2.2)
Siteworks for the project shall take approximately 3 months, after which the construction of utility systems and buildings will run concurrently for a total duration of 1 year. The project is roughly estimated to take about 15

months. The total duration incorporates 1 month float-time, transferrable to the critical path if need be.

The method-statement for the work is abridged as text as follows:

- Site establishment:
  - Locate water supply/suitable site access/pending existing lane closure(s)
- Site clearance:
  - Locate water supply point and suitable site access, and pending existing single-lane closure(s).
  - Site clearance, stockpiling and processing of vegetation: trees that are more than 150 mm are processed/sold, whereas trees that are less than 150 mm are ground/mulched using a Whole Tree Chipper (Morbark 2000) with a capability of 100 tons per hour.
  - Three zones for site clearance (Figure 7.2): once Zone 1 is cleared, it is then used as a stockpiling and processing area for the cleared vegetation and as the location for the site office. Zones 2 and 3 will be the zones for earthworks for cutting, filling and compaction of the soil for levelling and construction commencement. Zones 1 and 2 will be 90 × 160 m, and Zone 3 will be 180 × 17 m via Caterpillar. Once the trees have been cut down, a Caterpillar track-type tractor D7R series (Caterpillar 2010) is equipped to push vegetation from the ground. Track

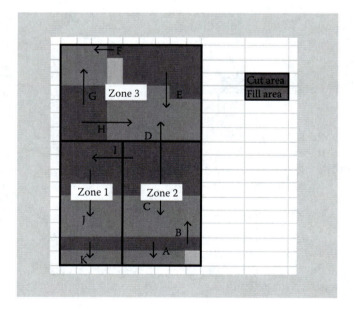

Figure 7.2 Zones.

Forest Machine transports vegetation to the tree chipper and stock tree logs.
- Cut, fill and compaction:
  - Zones 2 and 3 cleared of vegetation to Zone 1. With the top soil removed the cut and fill of the soil to the design level begins.
  - Once soil levelling is completed at Zone 2, construction of the retaining wall will begin alongside soil compaction.
  - Formwork for the footing is then built on lean concrete poured and placement of bar chair, steel rebar and concrete poured.
- Pavement:
  - Spread/compact pavement materials
  - Bituminous/asphaltic surfacing
- Water supply:
  - Water supply system built in two sections
- Storm-water:
  - Soak wells and gullies located to pits
- Sewerage installation:
  - Trench excavation for sewerage pipes

These items then lend themselves to being transposed to text/logic statements in the form of a method statement (as Section 2.2.3.1 and later in Example 7.1.2). The abridged work method statement above is transposed as a tabulated representation of labour, and plant efficiencies are then used in a Gantt chart of a finalised programme. See step 11, method-statement(s) continuity towards necessary reflection/mitigation.

Step 6: Prepare the project timeline schedule (as Sections 3.1, 3.1.1 and 3.2.2)
The type of Gantt chart of the programme is shown somewhat illustratively in Figure 7.3.

Step 7: Assess risk: Confirm quality management systems and occupational health and safety compliance, and address environmental impact (all as Sections 3.3, 4.1 and 5.1.7)
See Example 7.1.2 for an explicit, specific illustration of document preparation.

Step 8: Prepare BQ from drawings/detail take-offs (as Section 5.3)
An example of a BQ take-off using CESMM3 for the project is shown in Figure 7.4a.

An example of a finalised BQ item (CESMM3 or CESMM4) from the take-off is shown in Figure 7.4b.

Once measured and itemised, the finalised BQ is priced as in Figure 7.5.

(a)

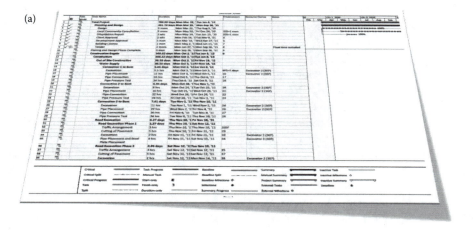

(b)

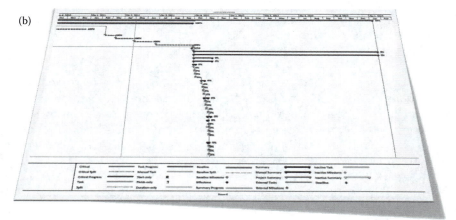

Figure 7.3 Illustrative schedules. (a) Activity-list and (b) duration-bars.

Step 9: Generate a detailed estimate; input unit rates into the final BQ (as Section 2.2.3)
Parts 1 through 5 of step 9 are illustrated in Figure 7.5. Once the BQ has been priced, reflection upon the breakdown can occur as below in Step 10.

Step 10: Project value reflection and elemental/class cost comparison (as Sections 2.2.2 and 2.2.3)
The total cost of siteworks for the new school development is $3,355,000.

(a)

Reference	Description	Unit	Add	Sub	Qty
CESMM3					
R	Roads and Pavings				
	Kerbs, channels and edgings				
662	In situ concrete kerbs and edgings, straight or curved to radius not exceeding 12m.	m			25.8
	(Cross-sectional dimensions refer to appendix of DWG No. 04)				
	Ancillaries				1317.5
824	Surface markings, continuous lines – Road 1-1 both sides, 2x313.7m	m	627.4		
	Road 2-2 both sides which is not connect to parking,				
	2x(158.4+64.9+96.9)m		133.1328		
	Parking space line, 20 x 5.4 m		108		
	2x40 x 5.4m		432		
	5 x 5.4m		27		

(b)

Reference	Description	Units	Quantity
R	**CLASS R: ROADS AND PAVINGS**		
	Kerbs, channels and edgings		
R661	In situ concrete kerb and edgings, Straight or curved to radius exceeding 12m	m	1141.13
R662	In situ concrete kerb and edgings, Curved to radius not exceeding 12m	m	25

Figure 7.4 CESMM3 take-off exemplar (a); CESMM3 item exemplar (b).

- Total cost: $3,355,000 (Figure 7.6).
- Earthworks constitutes 45% of the total cost due in large part to the estimated volume of 160,450 m³ of soil to be excavated/retained to obtain the design level and acceptable grade of the road: balanced cut and fill with excess topsoil removal of 1,298 m³.
- Supply and installation of the proprietary retaining wall constitutes $373K.

Reference	Job: Site Works for New School-Lesmurdie Description	Units	Quantity	Rate, $	Total
A	CLASS A: GENERAL ITEMS				
	Contractual requirements				
A120	Insurance of works (TCV $2,500,000)	item			$4,038
	Premium – 0.15% TVC		1	3,750	
	Deductible		1	500	
	Fire Safety Levy		1	0	
	GST – 10% Premium		1	375	
	Stamp Duty – 10%(Premium+ GST)		1	412.5	
A130	Third party insurance (TCV $2,500,000)	item			$8,075
	Premium – 0.30% TCV		1	7,500	
	Deductible		1	1,000	
	GST – 10% Premium		1	750	
	Stamp Duty – 10%(Premium+ GST)		1	825	
	Specified requirements				
A211	Accommodations for the Engineer's staff Offices	weeks	47	80	$3,760
A231	Equipment for use by the Engineer's staff Office equipment	months	11	20	$220
	Method related charges				
A315	Accommodation and buildings Canteen and messrooms	weeks	47	100	$4,700
A321	Services, Electricity	item	1	600	$600
A322	Services, Water (TCV $2,500,000)	item	1	2,750	$2,750
A332	Plant, Cranes	hours	10	160	$1,600
				Total C/Fwd	$25,743

Figure 7.5 Class A method-related charges and unit rates given above are illustrative. Class D–Z unit rates usually deemed to encompass labour + plant + materials supply/installation + overheads + profit, unless otherwise stated.

Reference	Job: Site Works for New School-Lesmurdie Description	Units	Quantity	Rate, $	Total
				Total B/Fwd	*$25,743*
B	CLASS B: GROUND INVESTIGATION				
	Geotechnical Report (drilled borehole and sample testing, 3 site condition testing, 3 laboratory testing)	nr	7	450	$3,150
D	CLASS D: DEMOLITION AND SITE CLEARANCE				
D100	General Clearance	ha	5.632	2,910	$16,389
D210	Trees, Girth: 500 mm-1 m	nr	13	242	$3,146
D220	Trees, Girth: 1-2 m	nr	6	495	$2,970
E	CLASS E: EARTHWORKS				
	General excavation				
E412	Topsoil, Maximum depth: 0.25-5 m	m3	10947	4.45	$48,714
E421	Materials other than topsoil, rock or artificial hard material, Maximum depth: not exceeding 0.25 m	m3	150	4.40	$660
E422	Materials other than topsoil, rock or artificial hard material, Maximum depth: not exceeding 0.25-0.5 m	m3	1250	4.40	$5,500
E423	Materials other than topsoil, rock or artificial hard material, Maximum depth: not exceeding 0. 5-1 m	m3	4450	4.40	$19,580
E424	Materials other than topsoil, rock or artificial hard material, Maximum depth: not exceeding 1-2 m	m3	7946	4.40	$34,962
E425	Materials other than topsoil, rock or artificial hard material, Maximum depth: not exceeding 2-5 m	m3	34050	4.40	$149,820
				Total C/Fwd	*$310,634*

Figure 7.5 (Continued) Class A method-related charges and unit rates given above are illustrative. Class D–Z unit rates usually deemed to encompass labour + plant + materials supply/installation + overheads + profit, unless otherwise stated.

Reference	Job: Site Works for New School-Lesmurdie Description	Units	Quantity	Rate	Total
				Total B/Fwd	$310,634
E426	Materials other than topsoil, rock or artificial hard material, Maximum depth: not exceeding 5-10 m	m3	47319	4.40	$208,204
E427	Materials other than topsoil, rock or artificial hard material, Maximum depth: not exceeding 10-15 m	m3	4050	4.40	$17,820
	Excavation ancillaries				
E531	Disposal of excavated material, Topsoil	m3	1298	13.10	$17,004
	Filling				
E621	Embankments, Excavated topsoil	m3	9649	6.20	$59,824
E623	Embankments, Non selected excavated material other than topsoil or rock	m3	11888	6.20	$73,705
E633	General, Non selected excavated material other than topsoil or rock	m3	85781	6.20	$531,842
E635	General, Imported natural soil other than topsoil or rock - Permeable coarse sand	m3	477	12.5	$5,963
	Landscaping				
E810	Turfing, 100 mm depth	m3	44.364	80	$3,549
				350	
	PRECAST CONCRETE RETAINING WALL				
MC.01	CONCRIB Segmental Retaining Cribwall	m2	1066	650	$373,100
				Total C/Fwd	$1,601,645

Figure 7.5 (Continued) Class A method-related charges and unit rates given above are illustrative. Class D–Z unit rates usually deemed to encompass labour + plant + materials supply/installation + overheads + profit, unless otherwise stated.

Reference	Job: Site Works for New School-Lesmurdie Description	Units	Quantity	Rate	Total
				Total B/Fwd	$1,601,645
I	CLASS I: PIPEWORK - PIPES				
I332	Iron pipes, Nominal bore: not exceeding 200 mm, In trenches, depth: not exceeding 1.5 m	m	1746	144	$251,424
I531	Polyvinyl chloride pipes, nominal bore: 300-600 mm, In trenches, depth: not exceeding 1.5 m				
	- Leach drain	m	66	355	$23,430
	- 450 mm Φ Sewerage pipes	m	152.6	355	$54,173
	- 375 mm Φ Sewerage pipes	m	46	375	$17,250
	- 375 mm Φ Stormwater pipes	m	1600	166	$265,600
J	CLASS J: PIPEWORK – FITTINGS AND VALVES				
J311	Iron pipe fittings, bends, nominal bore: not exceeding 200 mm	nr	5	131	$655
J321	Iron pipe fittings, junctions and branches, nominal bore: not exceeding 200 mm	nr	6	186	$1116
K	CLASS K: PIPEWORK - MANHOLES AND PIPEWORK ANCILLARIES				
K250	Other stated chambers, Precast Concrete	nr	45	1,005	$45,225
K350	Gullies, Precast Concrete	nr	301	920	$276,920
Z	CLASS Z: SIMPLE BUILDING WORKS INCIDENTAL TO CIVIL ENGINEEIRNG WORKS				
Z530	Piped building services, sanitary appliances and fitting	nr	2	4,035	8,070
	- Sewerage treatment tank				
				Total C/Fwd	$2,545,478

Figure 7.5 (Continued) Class A method-related charges and unit rates given above are illustrative. Class D–Z unit rates usually deemed to encompass labour + plant + materials supply/installation + overheads + profit, unless otherwise stated.

Reference	Job: Site Works for New School-Lesmurdie Description	Units	Quantity	Rate	Total
				Total B/Fwd	$2,545,478
R	CLASS R: ROADS AND PAVINGS				
	Sub-bases, flexible road bases and sub surfacing				
R117	Granular material DTp Specified type 1, depth 250- 300 mm	m2	6352.4	14.20	$90,204
R312	Cold asphalt wearing course, depth 30-60 mm	m2	6352.4	9.55	$60,665
	Kerbs, channels and edgings				
R661	In situ concrete kerb and edgings, Straight or curved to radius exceeding 12 m	m	1141.13	48.60	$55,459
R662	In situ concrete kerb and edgings, Curved to radius not exceeding 12 m	m	25	48.60	$1,221
	Ancillaries				
R824	Surface markings, Continuous lines	m	1317.53	1.35	$1,779
R825	Surface markings, Intermittent lines	m	665.97	1.35	$899
X	CLASS X: MISCELLANEOUS WORK				
X124	Fences, Metal post and wire, Height: 1.5-2 m	m	996	35	$34,860
X234	Gates and stiles, Metal field gates, Width: exceeding 3 m	nr	2	1,140	$2,280
Y	CLASS Y: SEWER AND WATER MAIN RENOVATION AND ANCILLARY WORKS				
Y602	New manholes, Depth 1.5-2 m	nr	2	1,515	$3,030
				TOTAL	$2,795,905
	10% GST				$279,590.5
	10% OVERHEADS				$279,590.5
	TOTAL PROJECT COST		$ 3,355,086		

Figure 7.5 (Continued) Class A method-related charges and unit rates given above are illustrative. Class D–Z unit rates usually deemed to encompass labour + plant + materials supply/installation + overheads + profit, unless otherwise stated.

School site-works: Cost breakdown $3,355,000

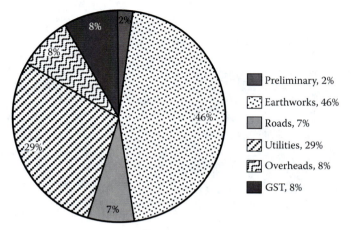

Preliminary, 2%

Earthworks, 46%

Roads, 7%

Utilities, 29%

Overheads, 8%

GST, 8%

Figure 7.6 Cost pie chart.

- Utilities water supply pipeworks and storm-water systems constitute 29% of the total.
  - A relatively expensive piped storm-water drainage system was selected (as shire council requirements) as opposed to less expensive/traditional rural roads open drain systems used in less progressive shire councils.
- Roads and parking pavement comprises 7% of the total project cost.
  - Construction of pavement requiring much compaction alongside ancillaries, such as mountable concrete kerbing, contributes to cost somewhat significantly/necessarily due to pedestrian road/access safety concerns.
- Other items at ≈2% include preliminaries, related insurances, site services/office setup, landscaping, fencing/gating and government service tax (GST) (2% + 8%).

Cost breakdown by work class:

- Almost half of the total cost of the project is allocated to class E: earthworks.
- Three other classes include C-precast concrete, I-pipework pipes and K-pipework manhole/pipework ancillaries make up most of the remainder (Figure 7.7)

Step 11: Method-statement resources review for future cost-monitoring and future feedback towards method-statement continuity (as Sections 2.2.3, 2.2.4, 3.1, 3.1.1 and 3.2.2)

School site-works: $3,355,000 Classification breakdown

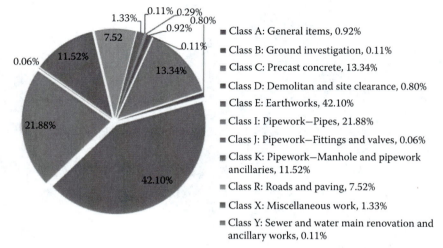

- Class A: General items, 0.92%
- Class B: Ground investigation, 0.11%
- Class C: Precast concrete, 13.34%
- Class D: Demolitan and site clearance, 0.80%
- Class E: Earthworks, 42.10%
- Class I: Pipework—Pipes, 21.88%
- Class J: Pipework—Fittings and valves, 0.06%
- Class K: Pipework—Manhole and pipework ancillaries, 11.52%
- Class R: Roads and paving, 7.52%
- Class X: Miscellaneous work, 1.33%
- Class Y: Sewer and water main renovation and ancillary works, 0.11%

Figure 7.7 Works pie chart.

Other than the retaining wall (step 12), a major cost item is the pipework, which contributes to 22% of the total construction cost. A representative resource (labour/plant/materials) analysis is performed for this pipework item towards an illustrative future cost-monitoring exercise of existing pre-contract method-statement(s) estimates.

To ensure that the total cost of the project does not run overbudget, and to assist remedial action in the event of an incongruity between actual versus predicted costs, a number of key item method-statement continuity exercises are recommended.

Table 7.1 Method statement pipes

Description	Unit	Quantity	Labour	Plant	Material	Total rate
Polypropylene pipe at 375 mm diameter for a length of 1,600 m						
Material	cost/m	1			156	156
Excavator at $75/h, duration of 8.09 days for 24 h	cost/m	1		9.1		9.1
Tradesman	cost/m	1	0.5			0.5
Labour	cost/m	2	0.17			0.34
Total	cost/m					165.94
After round off	cost/m					166

The storm-water pipe element is presented in Table 7.1. It is suggested that material (the key variable) supply chain requires monitoring to ensure material conformance upon arrival; plant efficiency contributes to a lesser extent.

If the supply or installation duration of the element is delayed by 20%, the total time needed to complete the project will be 233 hours, which would result in a plant rate of $11, producing an element rate of $167.84 and resulting in the total cost of the element being $268,544 instead of $265,600 (a difference of $2,944, excluding the extra cost from tradesmen/general labour). Although somewhat inconsequential in isolation for this hypothetical example, the potential knock-on cumulative effect is great.

Method-statement continuity is recommended for a full range of key cost contributors.

Step 12: Value engineering review: Alternative specification options for the retaining wall (as Sections 2.2.2 and 2.2.2.1)
Alternative specification review was done for the retaining wall because it is a major cost of the project (supply and installation of the proprietary retaining wall constitutes $373K). Benefits/disadvantages of a selected specification versus alternatives ensure that a preferred option is indeed the best fit-for-purpose selection.

To compare the differences, specifications for the retaining wall factors have been identified including design life span, cost per square metre and aesthetics:

- Retaining wall design life remains a factor.
- Options include (random rock/boulder) minimum-cost solutions of $140/m^2 versus a more aesthetic option (proprietary concrete crib) of $350/m^2.
- At a maximum height of 10 m, the wall not only dominates its surroundings but also would require the cheaper rocks/boulder option of having a 10-m clearance of vegetation, proving somewhat site-restrictive.
- Alternative interlocking concrete (unreinforced) blocks may be considered at $150, but this retaining wall will not be able to reach the required height for the small width unless the interlocking concrete block was reinforced with a geo-grid, or if mass concrete was used to increase the minimum cost of the retaining wall to $300, comparable with proprietary (concrete crib) systems.
- The concrete crib (proprietary system) retaining wall is deemed justified.

A breakdown of retaining wall alternatives are shown in Table 7.2.

Table 7.2 Value management (VM) walls

Description	Design life (years)	Cost ($)	Aesthetic
Concrete crib	>40	250–450	Yes
Rocks/boulders	>40		
Random		140–250	No
B grade cut sandstone		170–280	Yes
A grade cut sandstone		280–500	Yes
Interlocking concrete blocks	>40		
Unreinforced		150–450	No
Reinforced with geogrids or mass concrete		300–650	No
Concrete sleepers	>15	270–450	Yes
Core-filled 'Besser block'	>40	400–600	No

Step 13: Cash flow/revenue 'S-curve' analysis (as Section 2.3.1)

S-curves allow progress tracking of the project at the precontract stage as well as the all-important on-site construction stage. Major activity durations and costs are tabulated Table 7.3, showing the extent to which preplanning dominates the timeline of the project and somewhat understandably, generates no resultant tangible revenue. Provisional items are somewhat timeous with limited return. Once on-site, excavation activities contribute much

Table 7.3 Site durations

Project duration				Cost		
Days	In %	Cumulative (%)	Activity	$	In %	Cumulative (%)
310	41	0	Planning	28,892.50	1	0
20	3	41	Site clearance	22,505.12	1	1
50	7	43	Excavation	1,138,089.47	41	2
14	2	50	Retaining wall	373,100.00	13	43
16	2	52	Water supply	253,195.00	9	56
14	2	54	Sewerage	106,539	4	65
67	9	56	Storm-water	599,924	21	69
29	4	65	Road and car park	236,520.26	8	90
240	32	68	Provisional	37,140.00	1	99
		100				100

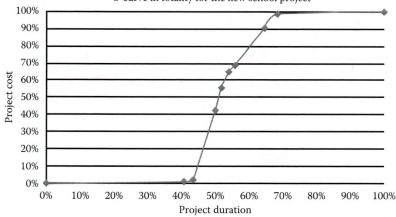

Figure 7.8 S-curve school.

to the project duration and expected costs. Table 7.3 shows the sitework's project activities alongside durations and costs.

The resultant S-curve is somewhat flat at the initial stages following the timeous nature of the planning stages, but is seen to steepen sharply upon on-site commencement and throughout earthworks activities.

- Provisional items and potential provisional sums for the defects liability periods and any resultant snagging activities, as well as the lag-time representing school construction proper, represents a (somewhat skewed) flattening out at the project's end.
- Amplification of the middle section and indeed the preparation of element-by-element 'S-curves' would be appropriate for time to cost/revenue charting of all activities.
- The S-curve in totality for the new school project is presented in Figure 7.8.

Step 14: Tender-pack preparation towards contractor bid and selection (as Chapter 5)

Having presented the client with the range of information discussed in the 13 steps above, an integrated design and cost management solution to the perceived problem and need should now be well within grasp. In a traditionally procured project, selection of a builder to do the job follows a tender-bid; a contract is signed between the client and the contractor to cement responsibilities and obligations in the work under the contract. Tender-bids are usually based on tender-packs distributed by the client.

Traditionally, the tender-pack represents all the information needed by prospective builders to make an offer. Clearly, it is ill-advised for clients, seeking competitive tenders, to make available the full range of detailed (cost) information prepared by the design consultants. Therefore, the tender-pack provided to potential builders is somewhat of an abridged version of the detailed (financial) solution.

Constituents of the tender-pack might include

- Cover of tender-pack contents and rationale for documentation inclusion(s)
- Key dates for the processing of tender submissions
- General scope of the work under the contract
- Design drawings in totality
- Specifications and codes of practice explicitly stated and in totality
- Unpriced BQs (with item descriptions and quantities but no prices)
- Standard form of contract applicable (such as AS2124 with an expectation for the completion of all relevant appendices and information as required)
    - Tendering compliance regulations and codes (such as AS4120)
- Applicable general conditions of tendering and form of tender (such as AS2125, alongside, if deemed appropriate, the AS2127 form of formal instrument of agreement) requesting completion of all sections and acknowledgement of documents to the tender which might include a request that the tenderer address specifically (with no general suffices, rather) a detailed description of the following:
    - Construction methodology request
    - Site layout needs
    - BQs full unit-rate inclusion(s) request
    - Quality management plan offered as part of the work under the contract
        - Material supply/testing methods
        - Risk management plan
        - Occupational Health and Safety Compliances
        - Environmental management plan
    - Project schedule with critical path clarifications, if applicable
    - Project constraints clarifications, if appropriate
    - Selection criteria and criteria compliance, if deemed appropriate
    - Caveats as necessary, such as the fact that the client/principal is not bound to accept the lowest or indeed any tender received
        - Clearly signed/dated monetary offer to complete the work under the contract
- Timeline of tender selection decisions

Once in receipt of the documentation related to the work under the contract, the builder makes an offer. If accepted by the client, a contract is signed and work commences.

7.1.2 Practice example: Design solution for a temporary barge landing facility

Step 1: Review the client's brief; clarify problems/needs/scope (as Section 5.4.1)
This project involves the design and construction of a short-term (up to 3 years) barge landing near a coastal town.* A provision of the facility is to meet the needs for delivery of materials and equipment for construction of a nearby jetty, as part of a larger project commissioned by a mining company to service the production and export of iron ore.

Step 2: Generate a working concept for the design solution (as Sections 5.4.1 and 5.4.2)
The barge landing considered is to be made up of the following major elements:

- Sheet piling wall (11 m high from seabed, and 36 m2), to be backfilled to subsequently allow for the mooring of a delivery barge and unloading by a 200-tonne crawler crane, which is driven atop the landing structure. The structure should be able to withstand all live loading mentioned in addition to the permanent actions provided by adjacent soil and water effects.
- Access to the facility for SM1600 trucks, according to AS5100.
- Sufficient storage capacity and lay-down for materials not immediately transported to the jetty site.
- Retaining structures of all exposed slopes, including beach adjacent to the structure.
- Ultimately, there is a responsibility for the design phase of this project to be completed internally prior to being put out to tender (for construction).

Step 3: Generate working/performance specifications and design drawings (as Sections 2.2.2, 2.2.3, 4.1, 5.2 and 5.4)
Specifications involve dissipating the horizontal large tensile forces and simply connected elements with reference to the manufacturer's sheet piling

---

* This hypothetical/academic/educational example acknowledges with thanks contributions by Cong Bui, Jarrad Coffey, Nicholas Teraci and Kai Teraci; all designs, specs, timings and costings are provided for illustrative purposes only.

systems (loading here deemed much too great to use lightweight rod tie options such as aluminium), thus steel rods with a maximum diameter of 25 to 38 mm were proposed; sheet piling was filled in with a series of layers of select material with compaction characteristics to hold design loadings. Specifications for the road include additional features such as security gates and fencing. For many elements, proprietary equipment may provide guidance. Component specifications in brief include

- *Sheet piling wall*: Retaining vertical plane of granular material for mooring and subsequent unloading of the delivery barge; equivalent-strength fibre-reinforced plastic (FRP) sheets or steel sheet piling (or both).
- *Sheet piling*: 15-m sections of CMI GG-95 FRP sheets or equivalent, embedded to a depth of 4 m at all locations; 144 m (laterally) required, square with side lengths of 36 m, joined and waterproofed, according to the manufacturer's suggestions.
- *Walers*: For application of tension from the tie rods and lateral stiffness to the sheets (mainly at joint locations), five spaced PFC 180 beams to be provided and connected appropriately to the tie rod diameter.
- *Tie rods*: Various sized rods ranging from 25 to 38 mm in diameter, steel yield to be greater than or equal to 500 MPa; five-layer placement in both directions, restraining wall movement.
- *Railings*: For purposes of vehicle (mainly crane) exclusion zone, 4 m from the edge of the sheet piling and safety provided a handrail (1 m high) at the edge of the facility.
- *Backfill of landing area*: Within sheet piling.
- *Backfill of the sheet piling*: Sufficient bearing capacity for movement of the 200-tonne crawler crane/trucks at a (total) depth of 11 m: including rock ballast to varying diameters of up to 950 mm as indicated in the drawings (density, 1.2 t/m³; effective elastic modulus, 200 MPa); natural gravel, maximum diameter of 36 mm as indicated in the drawings (density, 2 t/m³; effective elastic modulus, 200 MPa); crushed rock road base as indicated in the drawings (density, 2.4 t/m³; effective elastic modulus, 1500 MPa); placing with the installation of tie rods and the installation of two 30-tonne capacity bollards; suitably compacted.
- *Foundations*: Geotechnical data indicates that sand present on the site has a bearing capacity of 500 kPa with a density of 2.6 t/m³ and an internal angle of friction of 40°. Thus, sound (and competent when wet) rock exists at 12 m: excavation deemed to require nominal foundation/falsework to this level; maximum particle size to be 100 mm.
- *Road access to the site*: 9 × 115 m long, as indicated in AS5100 for SM1600 trucks; pavement (CIRCLY 5.0 by Mincad Systems). California bearing ratio (CBR) of 20% (~200 MPa resilient modulus) composed of 300 mm crushed limestone as sub-base, 200 mm

crushed rock as road base and 20 mm bituminous seal with a maximum aggregate size of 7 mm.

- *Crane*: Liebherr 200 t crawler crane to be made available according to the manufacturer's info.
- *Lay-down area*: 80 × 50 m for storage of surplus material/plant.
- *Gate/access*: Security at entry location according to ASA6000 (Boom-Gate Australia).
- *On-site offices*: To be located on compacted surface and serviced accordingly.
- *Bollards*: Adjacent to the road of 100 t capacity (allowing for barge tie-off of 'DBB1-80 PM&I' or equivalent), details according to manufacturer's info; footings of steel-reinforced concrete poured *in situ*; 50 mm of 8 N40... reinforcing bars 3 m wide × 4 m high, provided by Onesteel or equivalent, with an elastic yield strength of 500 MPa.
- *Rockwall/rock armour protection*: Approximately 50 m on either side of the barge landing.
- *Landing impact protection*: Protection from impact by the barge being moved into a mooring position is to be provided by Tian Rubber or equivalent.

Design drawings as follows (excluding typical details for kerbs/services/ proprietary systems via easy-access standard drawings downloads from Local Authority Co.).

Example 7.1.2:   Drawing 1: General arrangement

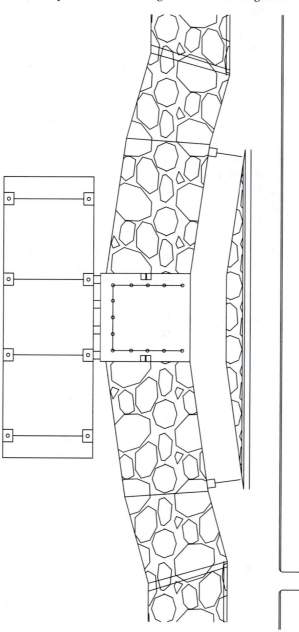

Example 7.1.2:   Drawing 2: Cross-section B-B

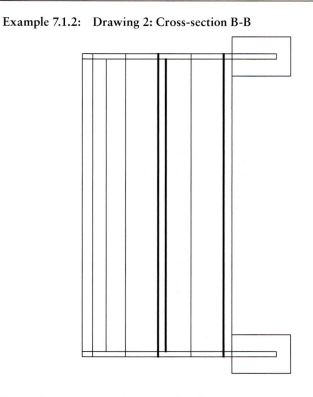

Example 7.1.2:   Drawing 3: Steel rod system

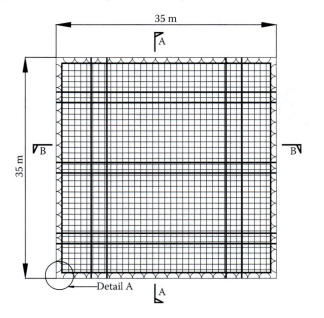

Example 7.1.2:   Drawing 4: Cross-section A-A

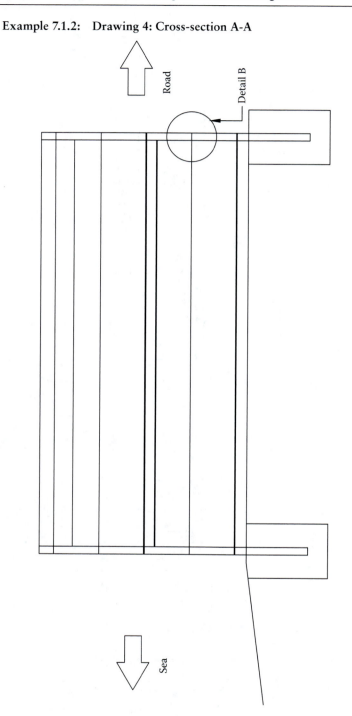

Example 7.1.2:   Drawing 5: Detail A

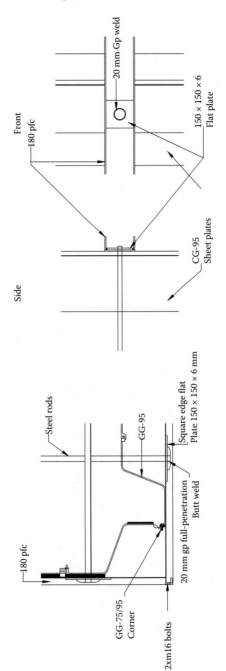

Example 7.1.2:   Drawing 6: Detail B (Section A-A)

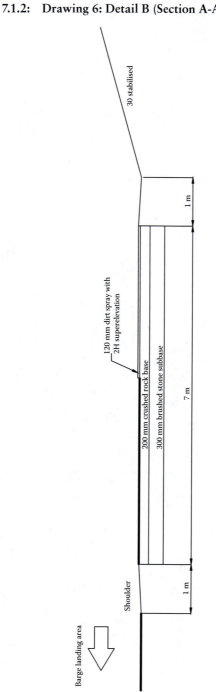

Example 7.1.2:   Drawing 7: Road cross-section

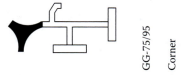

GG-75/95

Corner

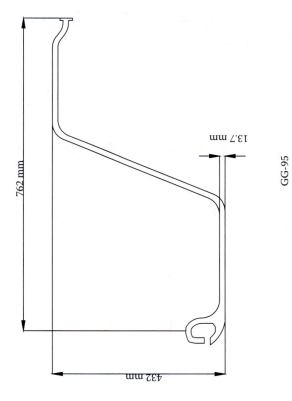

762 mm

432 mm

13.7 mm

GG-95

Example 7.1.2:   Drawing 8: Ultracomposite FRP shore guard sheet piling

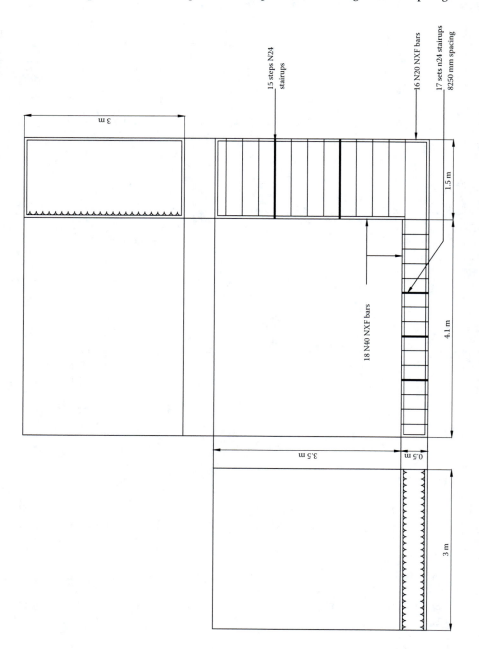

Example 7.1.2:   Drawing 9: Concrete footing (bollard footing)

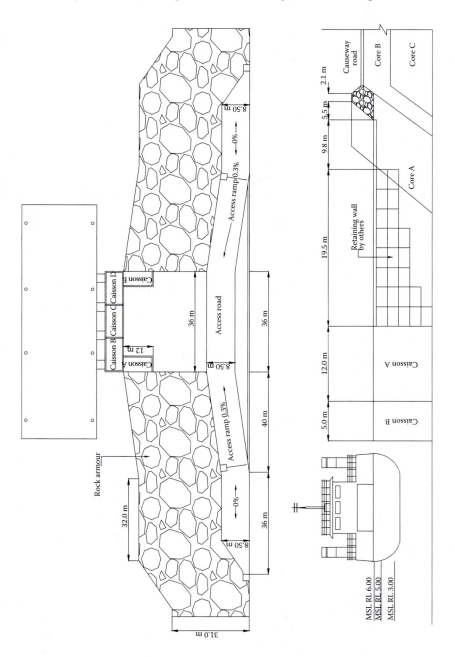

Step 4: Clarify appropriate detailed specifications (as Sections 2.2.2, 2.2.3, 4.1, 5.2 and 5.4)

Reference specifications may be somewhat implicit; see also other inferred Local Authority (easy access) typical design-detail drawings (not given here) and Proprietary Systems, accessed via appropriate website links, as follows:

- Clearing Specification 301
  - https://www.mainroads.wa.gov.au/Documents/Spec%20301%2014%20Jun%202007.u_2571350r_1n_D09%5E23288647.PDF
- Earthworks Specification 302
  - https://www.mainroads.wa.gov.au/Documents/Specification%20302%5E2C%20Earthworks%5E2C%2026%20August%202011.u_3326799r_1n_D11%5E23208480.PDF
- Sheet pile and proprietary CMI sheet piling and associated data
  - http://www.cmisheetpiling.com/sheet-piling.php
  - http://www.cmisheetpiling.com/Privacy.aspx?sflang=en
  - http://www.cmisheetpiling.com/Engineer.aspx?sflang=en
- Marine bollards and proprietary shore-guard/mooring bollards
  - http://www.pacificmarine.net/pierhardwarebollardsDBB1.htm
- Rubber fender details
  - http://www.everychina.com/products/td_z_roller_type_rubber_fender-37074385-zbb6b3d-detail.html
- Retaining wall specification
  - http://www.concrib.com.au/pdf/specifications.pdf
- Pavement Specifications 501 and 503
  - https://www.mainroads.wa.gov.au/Pages/default.aspx
- Pavement Marking Specifications 604
  - https://www.mainroads.wa.gov.au/Pages/default.aspx
- Drainage Specifications 402, 403, 407
  - https://www.mainroads.wa.gov.au/Pages/default.aspx

Step 5: Establish construction methodology (work breakdown structure) towards method-statement(s) compilation (as Sections 2.2.3, 2.2.4, 3.1, 3.1.1 and 3.2.2)

The following work breakdown structure tabulation identifies specific work-task items and task-reference numbers for the barge landing construction, bollard installation and lay-down area construction (Table 7.4).

The on-site work breakdown structure requires resourcing; the following method statement extends the work breakdown structure for siteworks and preaccess, reflecting labour and plant efficiencies by task (Table 7.5).

Table 7.4 Work-breakdown structure barge

Barge landing construction	On-site task ID reference no.	
Site preparation—Clearing 1001.1	Clearing	12
1001.2	Topsoil removal	13
Sheet pile installation 1002.1	Falsework installation	40
1002.2	Sheet pile installation	41
1002.3	Sheet pile waterproofing	42
1002.4	Water pumping	43
1002.5	Survey	39
Earthworks 1003.1	Initial survey	39
1003.2	Dredging	44
1003.3	Cut common	44
1003.5	Cut rock	44
1003.7	Remove unsuitable	44
1003.8	Survey	45
Ballast backfill—Tie rod install 1004.1	Backfill ballast to tie rod level	47, 51, 55, 59, 63
1004.2	Waler installation	48, 52, 56, 60, 65
1004.3	Tie rod installation	49, 53, 57, 61, 66
1004.4	Survey	50, 54, 58, 62, 64, 67
Services placement 1005.1	Survey	69
1005.2	Place services	70
1005.3	Survey	71
Granular material backfill 1006.1	Place material	73
1006.2	Compact material	74
1006.3	Survey	75
1006.4	Final shaping	76
Installation of furniture 1007.1	Side ladder construction	78
1007.2	Rubber stopper installation	79
1007.3	Guardrails	82
1007.4	Painting and signage	83
1007.5	Services	81
1007.6	Bollards	80
Bollard installation		
Site preparation—Clearing 2001.1	Clearing	12
2001.2	Topsoil removal	13
Footing pits 2008.1	Dig pits	87
2008.2	Survey	85
Footing installation 2009.1	Ground concrete	88
2009.2	Formwork	89, 92

(continued)

Table 7.4 Work-breakdown structure barge (Continued)

Barge landing construction	On-site task ID reference no.	
Bollard installation		
2009.3	r/f Installation	90, 93
2009.4	Hold down bolts	94
2009.5	Pour footing	91, 96
2009.6	Strip formwork	97
2009.7	Backfill	98
2009.8	Compaction	98
Bollard installation 2010.1	Survey	100
2010.2	Bracket installation	100
Laydown area construction 1		
Site preparation—Clearing 3001.1	Clearing	12
3001.2	Topsoil removal	13
Subgrade preparation 3011.1	Survey	102
3011.2	Cut and shape subgrade	103
Subbase preparation 3012.1	Survey	106
3012.2	Placement	105
3012.3	Compaction	105
3012.4	Final shaping	107
Base preparation 3013.1	Survey	110
3013.2	Placement	109
3013.3	Compaction	109
3013.4	Final shaping	111
Seal layer 3014.1	Survey	114
3014.2	Placement	113
3014.3	Compaction	113
3014.4	Final shaping	115
Road furniture 3015.1	Signage	134
3015.2	Ground marking	133
Laydown area construction 2		
Site preparation—Clearing 4001.1	Clearing	12
4001.2	Topsoil removal	13
Subgrade preparation 4011.1	Survey	118
4011.2	Cut and shape subgrade	119
Subbase preparation 4012.1	Survey	122
4012.2	Placement	121
4012.3	Compaction	121
4012.4	Final shaping	123
Base preparation 4013.1	Survey	126

(continued)

Table 7.4 Work-breakdown structure barge (Continued)

Barge landing construction		On-site task ID reference no.
Laydown area construction 2		
4013.2	Placement	125
4013.3	Compaction	125
4013.4	Final shaping	127
Seal layer 4014.1	Survey	130
4014.2	Placement	129
4014.3	Compaction	129
4014.4	Final shaping	131
Road furniture 4015.1	Signage	134
4015.2	Ground marking	133
4015.3	Security	135

The method-statement tabulation in Table 7.5 reflects labour and plant efficiencies by task.

The method-statement and work-breakdown structure compilation(s) in Tables 7.4 and 7.5 lead to an opportunity to compile 'resources forecast(s)' such as the labour-forecast in Figure 7.9; this in turn provides an opportunity for resource aggregation/smoothing in the event of excessive peaks and troughs in work-task activity requirements. As below, labour resources, although mostly uniform, require attention at five separate peaks and troughs; these will be addressed by relocation to (and from) the 'rockwall armour installation' as detailed in step 11.

Step 6: Prepare the project timeline schedule (as Sections 3.1, 3.1.1 and 3.2.2)
A Gantt chart programme may be compiled with the input of the information detailed above (using Microsoft Project/Primavera/Excel spreadsheet) towards a bar chart/critical path output similar to Step 6 of Section 7.1.1.

Step 7: Assess (i) risk as part of a quality management system, (ii) Occupational Health and Safety-compliance and address (iii) environmental impact (see Sections 3.3, 4.1 and 5.1.7)
Step 7(i): Risk
Risk analysis is completed by means of quantifying each identified risk by way of a matrix and comparing the consequence and likelihood of each. Guidance from ISO31000:2009 and AS/NZS4360:2004 ensures that risks are addressed acceptably.

Table 7.5 Method-statement barge

Task name	Duration	Start date	Finish date	Predecessors	Resources
Site handover	33.5 days	7/02/*, 8:00	24/03/*, 12:00		
Public notification	24 days	7/02/*, 8:00	10/03/*, 17:00		
Sign information near-site and on-site	1 day	7/02/*, 8:00	7/02/*, 17:00		2× CAT 301.5 excavator and operator, 2× labourer/spotter
Distribution of information to local shires and residents	2 days	7/02/*, 8:00	8/02/*, 17:00		1× site administrator
Meeting with shires	1 day	21/02/*, 8:00	21/02/*, 17:00	4FS + 8 days	2× company reps, 1× company vehicles
Public meeting	1 day	22/02/*, 8:00	22/02/*, 17:00	5	2× company representatives, 1× company vehicles
Meeting with shires	1 day	7/03/*, 8:00	7/03/*, 17:00	5FS + 9 days	2× company representatives, 1× company vehicles
Public meeting	1 day	10/03/*, 8:00	10/03/*, 17:00	7FS + 2 days	2× company representatives, 1× company vehicles
Access road to beach and site laydown area/offices	4.5 days	7/03/*, 8:00	11/03/*, 12:00		
Basic preparation and site tidy-up	2 days	7/03/*, 8:00	8/03/*, 17:00		2× labourers, 1× company vehicle (utility)
Topsoil removal	1 day	9/03/*, 8:00	9/03/*, 17:00	10	1× labourer/spotter, 1× CAT D6R* series 2 bulldozer and operator
Survey access road chainage	1.5 days	9/03/*, 8:00	10/03/*, 12:00	10	1× surveyor, 1× surveyor assistant
Placement of suitable wearing course for access road	0.5 day	10/03/*, 13:00	10/03/*, 17:00	12	1× CAT 267/277 skid steer and operator, 2× labourers
Compaction and shaping of road	1 day	10/03/*, 13:00	11/03/*, 12:00	12	1× CAT CB-334D double drum compactor and operator, 1× labourer/spotter
Laydown area	5.5 days	11/03/*, 13:00	18/03/*, 17:00		

Task	Duration	Start	Finish	Pred.	Resources
Identify and fence intended laydown area	3 days	11/03/*, 13:00	16/03/*, 12:00	14	2× labourers/spotter, 1× CAT 301.5 excavator and operator
Setup access gate for laydown area	1 day	11/03/*, 13:00	14/03/*, 12:00	14	2× labourers/spotter, 1× CAT 301.5 excavator and operator
Level laydown area	1.5 day	15/03/*, 13:00	16/03/*, 17:00	17FS + 1 day	1× labourer/spotter, 1× CAT D6R* series 2 bulldozer and operator
Site office setup for laydown area control	2 days	17/03/*, 8:00	18/03/*, 17:00	16FS + 0.5 days	3× labourers
Site office setup	5 days	17/03/*, 13:00	24/03/*, 12:00		
Identify and fence intended area for site and site offices	5 days	17/03/*, 13:00	24/03/*, 12:00	16FS + 1 day	2× labourers/spotter, 1× CAT 301.5 excavator and operator
Level site office area	1 day	17/03/*, 13:00	18/03/*, 12:00	16FS + 1 day	1× labourer/spotter, 1× CAT D6R* series 2 bulldozer and operator
Set up offices for site	3 days	21/03/*, 13:00	24/03/*, 12:00	22FS + 1 day	4× labourers
Site preparation	38.5 days	7/02/*, 8:00	31/03/*, 12:00		1× laydown area supervisor
Delivery of construction machinery and materials to site	3 days	24/03/*, 13:00	29/03/*, 12:00	23	
Employee training	4 days	7/02/*, 8:00	10/02/*, 17:00		2× Site Administrator
Employee site training	2 days	7/02/*, 8:00	8/02/*, 17:00		
Employee HSE Training	2 days	9/02/*, 8:00	10/02/*, 17:00	27	1× site administrator, 1× HSE supervisor
Initial site survey	1 day	24/03/*, 13:00	25/03/*, 12:00	23	1× surveyor, 1× surveyor assistant
Site clearance	2.5 day	25/03/*, 13:00	29/03/*, 17:00		
Tree and shrub removal	1.5 days	25/03/*, 13:00	28/03/*, 17:00	25FS − 2 days	1× CAT D6R* series II bulldozer and operator, 1× CAT 315C excavator and operator, 2× labourer/spotter

(continued)

Table 7.5 Method-statement barge (Continued)

Task name	Duration	Start date	Finish date	Predecessors	Resources
Boulder and large rock removal	2 days	25/03/*, 13:00	29/03/*, 12:00	25FS – 2 days	1× CAT D6R* series II bulldozer and operator, 1× CAT 315C excavator and operator, 2× labourer/spotter
Top soil layer removal	0.5 day	29/03/*, 13:00	29/03/*, 17:00	32	1× CAT D6R* series II bulldozer and operator, 1× CAT 315C excavator and operator, 2× labourer/spotter
Site survey	1.5 days	30/03/*, 8:00	31/03/*, 12:00	33	1× surveyor, 1× surveyor assistant
Construct barge landing	94 days	31/03/*, 13:00	10/08/*, 12:00		
Sheet piling installation	29.5 days	31/03/*, 13:00	11/05/*, 17:00		
Initial water survey of intended barge landing area	1 day	31/03/*, 13:00	1/04/*, 12:00	34	1× surveyor, 1× surveyor assistant
Placement of support structure for sheet piles	8 days	1/04/*, 13:00	13/04/*, 12:00	37	2× company vessels, 6× scaffolders
Placement of sheet piles	12 days	6/04/*, 13:00	22/04/*, 12:00	38FS – 5 days	2× company vessels, 6× sheet pilers
Temporary waterproofing around sheet piles	8 days	22/04/*, 13:00	4/05/*, 12:00	39	1× company vessel, 2× waterproofing technicians
Removal of water from sheet pile area	2 days	4/05/*, 13:00	6/05/*, 12:00	40	2× water pumps, 2× pumping technicians, 1× pump supervisor
General earthworks within sheet pile area (dredging)	3 days	6/05/*, 13:00	11/05/*, 12:00	41	1× CAT 315C excavator and operator, 1× dredging vessel and operators, 1× Hino tandem axle tipper truck and operator, 1× labourer/spotter
Survey of base levels within sheet pile area	0.5 day	11/05/*, 13:00	11/05/*, 17:00	42	1× surveyor, 1× surveyor assistant
Tie rod installation and ballast backfill	51 days	12/05/*, 8:00	21/07/*, 17:00		

Task	ID	Start	Finish	Duration	Resources
Backfill of ballast to first bar level	43	12/05/*, 8:00	13/05/*, 17:00	2 days	1× Hino tandem axle tipper truck and operator, 1× CAT 315C excavator and operator, 3× labourer/spotter
Installation of walers along base of sheet piles	43	12/05/*, 8:00	17/05/*, 17:00	4 days	3× steel fixers, 1× labourer
Installation of tie rods along base of sheet piles	46	18/05/*, 8:00	24/05/*, 17:00	5 days	3× steel fixers, 1× labourer
Survey of waler and bar tie levels	47	25/05/*, 8:00	25/05/*, 17:00	1 day	1× surveyor, 1× surveyor assistant
Backfill of ballast material up to second bar level	48	26/05/*, 8:00	30/05/*, 17:00	3 days	1× Hino tandem axle tipper truck and operator, 1× CAT 315C excavator and operator, 3× labourer/spotter
Waler installation along second height sheet piles	48	26/05/*, 8:00	31/05/*, 17:00	4 days	3× steel fixers, 1× labourer
Installation of tie rods along second height on sheet piles	50	1/06/*, 8:00	7/06/*, 17:00	5 days	3× steel fixers, 1× labourer
Survey of waler and bar tie levels	51	8/06/*, 8:00	8/06/*, 17:00	1 day	1× surveyor, 1× surveyor assistant
Backfill of ballast material up to third height on sheet piles	52	9/06/*, 8:00	13/06/*, 17:00	3 days	1× Hino tandem axle tipper truck and operator, 1× CAT 315C excavator and operator, 3× labourer/spotter
Waler installation along third height on sheet piles	52	9/06/*, 8:00	14/06/*, 17:00	4 days	3× steel fixers, 1× labourer
Installation of tie rods along third height on sheet piles	54	15/06/*, 8:00	21/06/*, 17:00	5 days	3× steel fixers, 1× labourer
Survey of waler and bar tie levels	55	22/06/*, 8:00	22/06/*, 17:00	1 day	1× surveyor, 1× surveyor assistant

(continued)

Table 7.5 Method-statement barge (Continued)

Task name	Duration	Start date	Finish date	Predecessors	Resources
Backfill of ballast material up to fourth height on sheet piles	3 days	23/06/*, 8:00	27/06/*, 17:00	56	1× Hino tandem axle tipper truck and operator, 1× CAT 315C excavator and operator, 3× labourer/spotter
Waler installation along fourth height on sheet piles	3 days	28/06/*, 8:00	30/06/*, 17:00	57	3× steel fixers, 1× labourer
Installation of tie rods along fourth height on sheet piles	4 days	23/06/*, 8:00	28/06/*, 17:00	56	3× steel fixers, 1× labourer
Survey of waler and bar tie levels	5 days	29/06/*, 8:00	5/07/*, 17:00	59	1× surveyor, 1× surveyor assistant
Final backfill of ballast layer to final ballast level and fifth level of sheet pile	2 days	6/07/*, 8:00	7/07/*, 17:00	60	1× Hino tandem axle tipper truck and operator, 1× CAT 315C excavator and operator, 3× labourer/spotter
Survey final ballast level	1 day	8/07/*, 8:00	8/07/*, 17:00	61	1× surveyor, 1× surveyor assistant
Final waler installation along top level	4 days	8/07/*, 8:00	13/07/*, 17:00	61	3× steel fixers, 1× labourer
Installation of tie rods along fifth height on sheet piles	5 days	14/07/*, 8:00	20/07/*, 17:00	63	3× steel fixers, 1× labourer
Survey of waler and tie rod levels	1 day	21/07/*, 8:00	21/07/*, 17:00	64	1× surveyor, 1× surveyor assistant
Services installation	3 days	22/07/*, 8:00	26/07/*, 17:00	65	1× surveyor, 1× surveyor assistant
Survey locations for placement of services	0.5 day	22/07/*, 8:00	22/07/*, 12:00		1× surveyor, 1× surveyor assistant
Lay services along marked locations	2 days	22/07/*, 13:00	26/07/*, 12:00	67	2× electricians
Survey levels and position of services	0.5 day	26/07/*, 13:00	26/07/*, 17:00	68	1× surveyor, 1× surveyor assistant

Task Name	Duration	Start	Finish	Predecessor	Resource Names
Backfill of granular material	3 days	26/07/*, 13:00	29/07/*, 12:00		1× Hino tandem axle tipper truck and operator, 1× CAT 315C excavator and operator, 3× labourer/spotter
Backfill of granular material/ballast material	2 days	26/07/*, 13:00	28/07/*, 12:00	68	
Compaction of granular material	1 day	27/07/*, 13:00	28/07/*, 12:00	71FS – 1 day	1× CAT CB-334D double drum compactor and operator, 1× labourer/spotter
Final survey of granular material	0.5 day	28/07/*, 13:00	28/07/*, 17:00	72	1× surveyor, 1× surveyor assistant
Fix any discrepancies with granular material level	1 day	28/07/*, 13:00	29/07/*, 12:00	73FS – 0.5 day	1× CAT 267/277B and operator, 1× labourer
Furniture installation on barge landing	7 days	1/08/*, 13:00	10/08/*, 12:00		
Installation of ladders down sides of barge landing	0.5 day	1/08/*, 13:00	1/08/*, 17:00	74FS + 1 day	2× labourers
Installation of rubber stoppers along edge of landing	1 day	1/08/*, 13:00	2/08/*, 12:00	74FS + 1 day	2× labourers
Installation of tie-down brackets on edge of landing	0.5 day	2/08/*, 13:00	2/08/*, 17:00	76FS + 0.5 day	2× labourers
Installation of services along landing	2 days	3/08/*, 13:00	5/08/*, 12:00	77FS + 1 day	4× labourers
Construction of rail guard along edge of landing and to mark safety area	3 days	5/08/*, 13:00	10/08/*, 12:00	79	2× labourers
Marking safety area and information on surface of barge landing	1 day	8/08/*, 13:00	9/08/*, 12:00	79FS + 1 day	2× labourers
Bollard installation	53.5 days	30/03/*, 13:00	13/06/*, 17:00		

(continued)

Table 7.5 Method-statement barge (Continued)

Task name	Duration	Start date	Finish date	Predecessors	Resources
Survey and mark locations for bollard installation	0.5 day	31/03/*, 13:00	31/03/*, 17:00	34	1× surveyor, 1× surveyor assistant
Bollard footing installation	50.5 days	30/03/*, 13:00	8/06/*, 17:00		
Dig pits for bollard installation	1 day	30/03/*, 13:00	31/03/*, 12:00	34FS – 1 day	1× CAT 315C excavator and operator, 1× labourer/spotter
Lay ground concrete	1 day	31/03/*, 13:00	1/04/*, 12:00	85	1× concrete truck and operator, 2× concreters
Formwork construction for base of footing	6 days	4/04/*, 8:00	11/04/*, 17:00	86FS + 0.5 day	3× carpenters
r/f installation for base of footing	4 days	18/04/*, 8:00	21/04/*, 17:00	87FS + 4 days	3× steel fixers
Pour base of footing	1 day	22/04/*, 8:00	22/04/*, 17:00	88	1× concrete truck and operator, 2× concreters
Construct formwork for rest of footing	8 days	25/04/*, 8:00	4/05/*, 17:00	89	3× carpenters
r/f installation for rest of footing	5 days	5/05/*, 8:00	11/05/*, 17:00	90	3× steel fixers
Hold down bolt installation	6 days	12/05/*, 8:00	19/05/*, 17:00	91	3× labourers
Survey locations for hold down bolts	1 day	20/05/*, 8:00	20/05/*, 17:00	92	1× surveyor, 1× surveyor assistant
Pour rest of footing	1 day	23/05/*, 8:00	23/05/*, 17:00	93	1× concrete truck and operator, 2× concreters
Strip footing	1.5 days	2/06/*, 8:00	3/06/*, 12:00	94FS + 7 days	3× carpenters

Task	Duration	Start	Finish	Predecessor	Resources
Backfill and compaction around footing	2 days	7/06/*, 8:00	8/06/*, 17:00	94FS + 10 days	1× CAT 315C excavator and operator, 1× labourer/spotter, 1× Whacker compactor
Actual bollard installation	2 days	10/06/*, 8:00	13/06/*, 17:00		
Installation of bollard bracket on footing	2 days	10/06/*, 8:00	13/06/*, 17:00	96FS + 1 day	2× labourers
Construction of laydown area	42 days	30/03/*, 8:00	27/05/*, 14:20		
Survey intended layout area	1 day	1/04/*, 8:00	1/04/*, 17:00	83	1× surveyor, 1× surveyor assistant
Remove top layer of soil to specified depth	0.5 day	30/03/*, 8:00	30/03/*, 12:00		1× labourer/spotter, 1× CAT D6R* series 2 bulldozer and operator
Sub-base layer	2 days	30/03/*, 13:00	1/04/*, 12:00		
Place and compact sub-base of layout area	1 day	30/03/*, 13:00	31/03/*, 12:00	101	1× CAT 267/277B skid steer and operator, 1× CAT CB-334D double drum compactor and operator, 1× labourer/spotter
Survey level of sub-base layer	0.5 day	31/03/*, 13:00	31/03/*, 17:00	103	1× surveyor, 1× surveyor assistant
Final shaping of sub-base layer	0.5 day	1/04/*, 8:00	1/04/*, 12:00	104	1× CAT 267/277B skid steer and operator, 1× labourer/spotter
Base layer	4 days	20/05/*, 8:00	25/05/*, 17:00		
Place and compact base of layout area	1 day	20/05/*, 8:00	20/05/*, 17:00		1× CAT 267/277B skid steer and operator, 1× cat cb-334d double drum compactor and operator, 1× labourer/spotter
Survey level of base layer	0.5 day	25/05/*, 8:00	25/05/*, 12:00	107FS + 2 days	1× surveyor, 1× surveyor assistant
Final shaping of base layer	0.5 day	25/05/*, 13:00	25/05/*, 17:00	108	1× CAT 267/277B skid steer and operator, 1× labourer/spotter
Seal layer	2.17 days	25/05/*, 13:00	27/05/*, 14:20		

(continued)

Table 7.5 Method-statement barge (Continued)

Task name	Duration	Start date	Finish date	Predecessors	Resources
Place and compact seal of layout area	1 day	25/05/*, 13:00	26/05/*, 12:00	109FS – 0.5 day	1 × CAT 267/277B skid steer and operator, 1 × CAT CB-334D double drum compactor and operator, 1 × labourer/ spotter
Survey level of seal layer	0.5 day	26/05/*, 13:00	26/05/*, 17:00	111	1 × surveyor, 1 × surveyor assistant
Final shaping of seal layer	0.67 days	27/05/*, 8:00	27/05/*, 14:20	112	1 × CAT 267/277B skid steer and operator, 1 × labourer/spotter, 1 × Whacker compactor
Rockwall armour installation along sides of landing	75 days	4/04/*, 8:00	15/07/*, 17:00		1 × Hino tandem axle tipper truck, 1 × CAT 315C excavator and operator, 1 × labourer
Access road	44 days	27/05/*, 14:20	28/07/*, 14:20		
Initial survey of road alignment and levels	1 day	27/05/*, 14:20	30/05/*, 14:20	113	1 × surveyor, 1 × surveyor assistant
General earthworks along road chainage	1 day	8/06/*, 14:20	9/06/*, 14:20	116FS + 7 days	1 × CAT 267/277B skid steer and operator, 1 × labourer/spotter
Subbase layer	11 days	8/06/*, 14:20	23/06/*, 14:20		
Place and compact sub-base of layout area	3 days	8/06/*, 14:20	13/06/*, 14:20	117FS – 1 day	1 × CAT 267/277B skid steer and operator, 1 × CAT CB-334D double drum compactor and operator, 1 × labourer/ spotter
Survey level of sub-base layer	1 day	13/06/*, 14:20	14/06/*, 14:20	119	1 × surveyor, 1 × surveyor assistant
Final shaping of sub-base layer	1 day	22/06/*, 14:20	23/06/*, 14:20	120FS + 6 days	1 × CAT 267/277B skid steer and operator, 1 × labourer/spotter
Base layer	5 days	22/06/*, 14:20	29/06/*, 14:20		

Task	Duration	Start	Finish	ID	Resources
Place and compact base of layout area	3 days	22/06/*, 14:20	27/06/*, 14:20	121FS − 1 day	1× CAT 267/277B skid steer and operator, 1× CAT CB-334D double drum compactor and operator, 1× labourer/spotter
Survey level of base layer	1 day	27/06/*, 14:20	28/06/*, 14:20	123	1× surveyor, 1× surveyor assistant
Final shaping of base layer	1 day	28/06/*, 14:20	29/06/*, 14:20	124	1× CAT 267/277B skid steer and operator, 1× labourer/spotter
Seal layer	5.5 days	29/06/*, 14:20	7/07/*, 9:20	125	
Place and compact seal of layout area	3 days	29/06/*, 14:20	4/07/*, 14:20	125	1× CAT 267/277B skid steer and operator, 1× CAT CB-334D double drum compactor and operator, 1× labourer/spotter
Survey level of seal layer	1 day	4/07/*, 14:20	5/07/*, 14:20	127	1× surveyor, 1× surveyor assistant
Final shaping of seal layer	1.5 days	5/07/*, 14:20	7/07/*, 9:20	128	1× CAT 267/277B skid steer and operator, 1× labourer/spotter
Installation of road furniture	0.5 day	21/07/*, 9:20	21/07/*, 14:20	129FS + 10 days	1× CAT 315c excavator, 2× labourer/spotter
Line and information marking on road	0.5 day	21/07/*, 9:20	21/07/*, 14:20	130FS − 0.5 day	2× labourers
Setup of appropriate sign information	0.5 day	22/07/*, 9:20	22/07/*, 14:20	130FS + 0.5 day	2× labourers
Installation of boom gate for landing access	3 days	25/07/*, 14:20	28/07/*, 14:20	132FS + 1 day	5× labourers

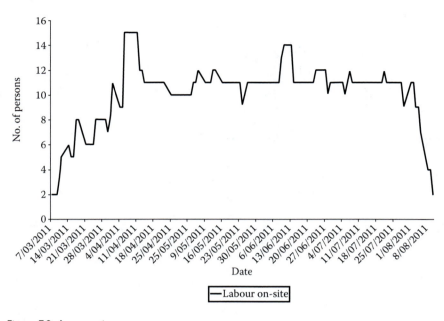

Figure 7.9 Aggregation.

Risk is broken down into four categories: feasibility to commissioning phase, design, construction and handover. From this initial analysis, each risk is defined within five categories, one being the worst (to be avoided at any cost) to five (minor disruption requiring no specific planning); these risks are then quantified in the four categories of environmental impact, capital, operating costs and people costs, with each category ranked from 1 to 5 (1 being the most favourable). A significant result might be the capital cost of a culturally important artefact being uncovered, increasing the project total by more than 10%; a less than 1% increase might arise from worker walk-off. Table 7.6 is indicative of risk analysis.

Step 7(ii): Safety
Several hazards are specific/increased due to the coastal locality of the site. Rigorous consideration of safety risks is aided by consideration of key documents:

- 'The National Standard for Construction Work', compiled by WorkSafe Australia, attention to both the designers and contractors versions and
- WorkSafe Australia's so-called 'subcontractors/subby-pack'

Building up these guidance documents, a methodology is prepared as a matrix (similar to the risk assessment) relating to likelihood (nominal, limited, common and high) and consequence (minor, notable, major and catastrophic).

Table 7.6 Risk examples

			Feasibility to commissioning (sample)							
Event	Explanation	Impact	Likelihood	Consequence	Risk	Controls	Likelihood	Consequence	Risk	Comments
...						...				...
Culturally significant artefact uncovered on-site	Preliminary site investigations indicates potential for cultural significance; but go-ahead likely	Further reapproval required by EPA/EIA if added significance uncovered	5	2	4	Nominal at this time, as relocation to other site requires location redesign	5	2	4	EPA mitigating measures required as a result of EIA recommendations; relocation requires policy not technical solution
Geosurvey indicates site not suitable	In situ ground conditions unsuitable for construction	Site use compromised	5	1	3	Project moved to more appropriate site if required	5	2	4	Very unlikely; site adaptability via tech. design improvements/ amendments very possible
EPA recommends higher level assessment than anticipated, progress slows	More thorough examination of environmental impact carried out	Approval takes longer, process may become more public	5	4	5	Proper consultation as above	5	4	5	Design adaptations to address EPA mitigating measures dictated explicitly as a result of EIA
...	...	...			.	...			.	(continued)

Table 7.6 Risk examples (Continued)

Event	Explanation	Impact	Construction (sample)							Comments
			Likelihood	Consequence	Risk	Controls	Likelihood	Consequence	Risk	
Cyclone—Blue alert maximum	'Tie-down' of site required	Loss of time	3	4	4	Practices put into place; as induction training	3	5	5	Likely 2-day loss for tie-down and storm dissipation
Cyclone—Yellow alert maximum	Tie-down of site required to address likely severe storm	Loss of time, possible minor impact on plant and work completed	4	3	4	Practices put into place; as induction training	4	4	4	Likely 3-day loss for tie-down and addressing strong winds rework
Cyclone—Red alert maximum	Tie-down of site required to address severe winds on-site	Loss of time, work damage, extensive rework and temporary structures/sheet (re)piling	5	2	4	Practices put into place; as induction training	5	2	4	Likely 2-week (minimum) loss due to violent storm/cyclone. Tie-down insufficient to protect work thus extensive rework

A job hazard analysis (JHA) is carried out by using the above method in the design and operation phases to identify the greatest threat to worker safety on-site. Two risks identified as category 2 (requiring specific JHA and management/response plans): sheet piling failure and temporary structure failure (assisting the sheet piling). Both risks acknowledge potential seawater rush/ingress to voids at backfilling to sheet piling; thus, worker exit becomes a high-category priority; this likelihood is limited with proper survey and design, although the consequences would be in a major category. Table 7.7 is indicative of identification and acknowledgment of safety issues.

Safety compliance measures (such as the brief examples tabulated in Table 7.7) might be expected to address the following generic issues:

- Health, Safety and Environmental (HSE) training mandatory for all prior to commencing on-site; classes/expenses by company; assessment pass required.
- Constructors' Blue Card, mandatory for workers, checked prior to deployment towards assistance in appropriate updates, training and assessment.
- Take 5s: small 5-min personal assessments (daily/weekly) of on-site activities; forwarded to line managers/supervisors to clarify details towards
    - S(ee what could threaten safety)
    - A(sses how big of a threat it is)
    - F(unction or type of threat)
    - E(valuate what do be done to improve the risk)
    - T(ake action and put into place measure above)
    - Y(ell and tell others about the risk)
- JHA are carried out prior to the commencement of each new task or activity to compile major threats to workers.
- Occupational Safety and Health Management Plan as recommended by WorkSafe Western Australia requires analysis of all activities involving five or more personnel to include quantification of the risk, consequences and methods for assessing effectiveness of controls.
- Maximum working temperatures are set at more than 35°C at 10 a.m. or anytime at 40°C and as such result in a call to stop work. Hydration advice applies at all times with short breaks as necessary in demountable kitchen/lunch areas with air-conditioning.
- Personal protective equipment (PPE) is supplied and to be worn by all on-site, notwithstanding avoidance of any potentially harmful occurrence.
    - Failure to use PPE will be reported as a near-miss incident, an accumulation of such reports for any one worker may result in disciplinary action

Table 7.7 Safety and job hazard sample

Construction of barge landing: Job hazard analysis (sample)

Event	Explanation	Impact/equipment	Likelihood	Consequence	Risk	Controls	Likelihood	Consequence	Risk	Comments
Sheet pile failure	Sheet piling fails due to poor construction or excessive pressure	Workers near sheet piling/ barge/temp structure: increased when back-filled as sheet light	4	1	1	Proper installation of piles and isolation until back-filling completed	4	2	2	All workers near piling need to be inducted in proper use: site arrival induction/ training
...						...				...
Crane failure	Crane boom or jib fails	Crane (gravity hazard for workers)	4	1	1	Workers should avoid elevated object(s)	4	3	4	Inducted in proper procedure: site arrival induction/ training
Crane overturning	Crane over-turns due to localised soil failure (adverse weather), or overloading	Crane (large structure) possibly falling if incorrectly placed/used	4	1	1	Considered site location/placing crane adheres to lifting limits applied	4	4	4	Induction/ training for mobilisation and operators
Lacerations	Worker severe skin cuts by sharp object on site	Loss of blood, infections	2	3	3	Take 5, PPE for hands especially (gloves/ handling)	4	3	4	Inducted in proper procedure: site arrival induction/ training? ID of support First-Aid
...										

- PPE's include long-sleeved shirts and trousers, steel-capped safety boots, hard hat to be worn outside of vehicles and buildings, safety glasses to be worn outside, sunscreen on uncovered parts of body if outside and gloves for manual handling (at all times)
- Mechanical and electrical equipment to have daily safety compliance tests before commencing work and also inspection upon site arrival
  - Any nonconformance use will be recorded as an incident and will be considered as a breach of the operator's contract, resulting in disciplinary action.
- Local emergency response authorities (St. John Ambulance, State Emergency Services) to be notified of the nature and scale of work being completed
- Safety officers on-site/all workers on-site to be valid holders of first aid qualifications with emergency wardens trained to a higher level in emergency response; additional safety meetings held
- Evacuation plan and drill for each team on periodic (but random) basis
- All incidents, no matter how small/near-miss, to be reported to supervisor and logged on company intranet; site safety officer to process/report/take action
- Safety meetings are to be held for all teams each week, incidents reviewed, new threats discussed and old threats refreshed to eliminate complacency becoming the catalyst for injury. This ensures that safety remains in the forefront of all workers' thinking and the first priority for any days' work

Step 7(iii): Environmental impact
This barge landing facility is a small component of a much larger mining and associated infrastructure project. The Environmental Protection Authority (EPA) requires consultation on large-scale projects requiring a high level of assessment (public enquiry, unless a specific impact is seen as publicly unacceptable). Hence, this part of the project needs more detailed assessment than might be considered appropriate for a similar project completed in isolation.

An environmental impact matrix (similar to [i] risk and [ii] safety assessment above) is used for an impact assessment, advised by AS-14042. Marine environment(s) adjacent to the site require acknowledgement of land, water, air, and noise pollution assessment. Detailed consideration of adverse impacts identifies as important: dust, emission, coastal erosion, and topsoil removal/re-instatement, all in line with an overarching company Environmental Management Policy as part of Health Safety and Environmental Considerations. Where applicable, a series of community meetings will be held throughout project progression, with open forums providing an opportunity for concerns to be raised and addressed. Table 7.8 is indicative of the identification and acknowledgment of environmental impact concerns.

Table 7.8 Environmental Impact Assessment (EIA) sample

Event	Explanation	Impact	Likelihood	Consequence	Risk	Controls	Comments
Oil/chemical spill: general (parts per million to be cross-referenced)	Drum/container of noxious liquid... leaks whilst being moved through the facility	Toxic liquid enters ocean affecting marine life	4	2	3	Due diligence; any noted spills need cleanup, management, and isolation plan for major spills	Landing deck must drain and thus is cambered, so threat of spill will flow toward ocean, thus must be identified and dealt with by following procedure
Dust and particulate matter emissions (parts per million to be cross-referenced)	Excessive, as identified, particulate matter emission from construction works	Escape into atmosphere/ immediate effect plus settlement in port and out to sea	2	2	2	Wetting of site during works, appropriate treatment to landing deck road base for operation phase	Road base cement dust... results in high consequence if road wetting procedures not followed
Loss of specific breeding or hatching grounds for marine life (as detailed within EIA/EPA documents)	Possible/expected partial use of site for marine life breeding and/or hatching; unable to be used at initial period	Fauna require (specialist) relocation elsewhere during spawning periods (as identified in EIA documents)	5	3	4	Mitigation measures as noted in environmental survey (EIA documents) completed prior to design phase submission	May need to relocate site if found to be an ongoing issue, but requires preconsideration prior to design, overall technical project solutions minimal, unless travel distance vastly increased to jetty site
Concrete footing remains after project decommissioning	Footing deemed unobtrusive buried but remaining	May be exposed as beach erodes further back naturally	2	3	3	Will take many years to become exposed, removal is an option if local authority or EPA deems it unacceptable; current EIA expected to leave as is for potential reef location purposes	
Toxic oil spills or other noxious liquids from spills remain in sediment after decommissioning		Coastal effect regeneration of vegetation near impossible	4	2	3	Careful cleaning of spills at outset; disaster mitigation of wider state authority requires the coordination of current quality management system operation/maintenance documentation	

Note: Environmental impact summary table (sample). To be read in conjunction with EIA/EPA report/mitigation documentation.

Step 8: Prepare BQ from drawings/detailed take-offs (as Section 5.3)
An illustrative example of a BQ take-off using CESMM3/4 is deemed similar to Step 8 of Section 7.1.1.

Step 9: Generate a detailed estimate; input unit rates into the final BQ (as Section 2.2.3)
Parts 1 through 7 of Step 9 are illustrated in Figure 7.10.

Reference	Description	Units	Add	Ddt	Qnts	Unit Cost	Cost
A	General items						
A100	Contractual Requirements						
A110	Performance Bond	unit	1		1	180000	180000
A120	Insurance of Works	unit	1		1	400000	400000
A200	Specific Requirements						
A211	Accomodation for eng staff: offices	unit	1		1	941.06	941.06
A212	Accomodation for eng staff: laboratorites	unit	1		1	5500	5500
A213	Accomodation for eng staff: cabins (Accomodation)	unit	2		2	5000	10000
A221	Services for eng staff: vehicles	unit	2		2	3340	6680
A222	Services for eng staff: phones (mobiles)	unit	2		2	174	348
A231	Services for eng staff: office (site)	unit	1		1	400	400
A232	Services for eng staff: laboratories	unit	1		1	200	200
A233	Services for eng staff: Survey Equipment	unit	1		1	2000	2000
A242	Services for eng staff: chainmen	unit	1		1	21600	21600
A250	Testing Materials	hr	100		100	64.2	6420
A260	Testing Works	hr	150		150	1000	150000
A300	Method Related Charges						
A311	Accomodation & Buildings: Offices	unit	1		1	4500	4500
A313	Accomodation & Buildings: Cabins	unit	8		8	4500	36000
A314	Accomodation & Buildings: Stores	unit	2		2	4500	9000
A315	Accomodation & Buildings: Messroom	unit	1		1	9000	9000
A321	Services: Electrical	hr	4368		4368	0.17	742.56
A322	Services: Water	L	10000		10000	0.01	100
A325	Services: Site Transport	unit	1		1	100000	100000
A327	Services: Welfare	unit	1		1	20000	20000
A331	Plant: Cranes	unit/hr	20		20	150	3000
A332	Plant: Transport	unit/hr	1		1	500	500
					Total This Page		966931.62
					Running Total		$ 966,932

Figure 7.10 Class A3** method–related charges and unit rates given above are illustrative. Class D–Z unit rates usually deemed to encompass labour + plant + materials supply/installation + overheads + profit, unless otherwise stated.

Reference	Description	Units	Add	Ddt	Qnts	Unit Cost	Cost
B400	Samples						
B410	From the surface or from trial pits and trenches: undisturbed soft material	N	10		10	50	500
B423	From Boreholes: groundwater	N	10		10	100	1000
B500	Site tests and observations						
B512	Groundwater	N	4		4	400	1600
B513	Standard penetration	N	6		6	250	1500
B516	Pressure meter	N	6		6	400	2400
B521	Plate bearing	N	6		6	650	3900
B525	In-situ density	N	6		6	450	2700
B600	Instumental observations						
B631	Settlement guages: installations	N	4		4	1200	4800
B632	Settlement guages: readings	N	4		4	120	480
B700	Laboratory tests						
B711	Classification: moisture content	N	6		6	120	720
B712	Classification: atterberg limits	N	4		4	180	720
B714	Classification: particle size analysis by sieve	N	4		4	100	400
B721	Chemical content: organic matter	N	6		6	360	2160
B722	Chemical content: sulphate	N	8		8	200	1600
B723	Chemical content: pH value	N	6		6	200	1200
B724	Chemical content: contaminants	N	8		8	300	2400
B731	Compaction: standard	N	6		6	150	900
B732	Compaction: heavy	N	4		4	130	520
B742	Consolidation: triaxial test	N	4		4	200	800
B752	Permaebility: falling head	N	4		4	200	800
B762	Soil strength: consolidation undrained triaxial, with pore water pressure measurement	N	4		4	200	800
B765	Soil strength: shearbox - peak and residual	N	6		6	250	1500
B766	Soil strength: california bearing ratio	N	6		6	150	900
					Total This Page		34300
					Running Total		$1,077,682

Figure 7.10 (Continued) Class A3** method-related charges and unit rates given above are illustrative. Class D–Z unit rates usually deemed to encompass labour + plant + materials supply/installation + overheads + profit, unless otherwise stated.

	Description	Units	Add	Ddt	Qnts	Unit Cost	Cost
B771	Rock Strength: unconfined compressive strength of samples	N	6		6	260	1560
B800	Professional services						
B810	Technician	hr	200		200	60	12000
B832	Engineer/geologist: chartered	hr	100		100	120	12000
B850	Overnight stays in connection with visits to site	N	5		5	100	500
C	Geotechnical and other specialists						
C600	Diaphram walls						
C611	Excavation in materials other than rock: depth not exceeding 5m	m³	576		576	5.2	2995.2
C621	Excavation in rock: depth not exceeding 5m	m³	576		576	15.56	8962.56
						Total This Page	38017.76
						Running Total	$1,115,699

Figure 7.10 (Continued) Class A3** method-related charges and unit rates given above are illustrative. Class D–Z unit rates usually deemed to encompass labour + plant + materials supply/installation + overheads + profit, unless otherwise stated.

	Description	Units	Add	Ddt	Qnts	Unit Cost	Cost
D	Demolition and Site clearance						
D100	General Clearance	ha	2		2	1200	2400
E	Earthworks						
E110	Excavation by Dredging:	m³	1980		1980	8.5	16830
	Topsoil (sand)						
E216	Excavation by Cutting	m³	5940		5940	3.66	21740.4
	Topsoil (sand)						
	5.5m Maximum depth						
E345	Excavation for Footings	m³	396		396	2.33	922.68
	Artificial hard material exposed at surface						
	2m to 5m depth						
E648	Filling	m³	12312		12312	2.72	33488.64
	To a depth of 9.5m						
	Imported Ballast						
E627	Filling	m³	18150		18150	2.72	49368
	Embankments						
E721	Filling Ancillaries	m²	1740		1740	0.02	34.8
	Preparation of Topsoil surface						
E724	Filling Ancillaries	m²	1296		1296	0.02	25.92
	Preparation of Ballast surface						
F	Insitu Concrete						
F263	Provision of Concrete	m³	72.45		72.45	231	16735.95
	Grade C30						
F624	Placement of Reinforced Concrete	m³	72.45		72.45	20	1449
	Footings						
	Exheeding 0.5m width						
G	Concrete Ancillaries						
G115	Formwork	m²	16.8		16.8	21.5	361.2
	Horizontal						
G145	Formwork	m²	120.3		120.3	21.5	2586.45
	Vertical						
G540	Reinforcement	t	33.9		33.9	1971	66816.9
	Assorted Reobar sizes						
	Approximately 6% of concrete weight						
					Total This Page		212759.94
					Running Total		$1,328,459

Figure 7.10 (Continued) Class A3** method-related charges and unit rates given above are illustrative. Class D–Z unit rates usually deemed to encompass labour + plant + materials supply/installation + overheads + profit, unless otherwise stated.

Reference	Description	Units	Add	Ddt	Qnts	Unit Cost	Cost
G811	Concrete Accessories	m²	4.5		4.5	1.5	6.75
	Wood float finish to top surface						
N	Miscellaneous Metalwork						
N160	Miscellaneous Plate	m	360		360	23.5	8460
	150x150x16mm Flat Plates						
	136 plates						
N162	Miscellaneous Framing	m	360		360	44.7	16092
	180 PFCs						
	20x 18m Lengths						
							0
N240	Tie Rods	Units	440		440	7.7	3388
	25mm Diameter						
	36m						
N240	Tie Rods	Units	88		88	8.5	748
	32mm Diameter						
	36m						
N240	Tie Rods	Units	176		176	10.2	1795.2
	38mm Diameter						
	36m						
P	Piles						
P800	Interlocking Sheet Piles	Units	176		176	2400	422400
	CMI Sheet Piles						
	15m long GG-95's						
						Total This Page	452889.95
						Running Total	$1,781,349

Figure 7.10 (Continued) Class A3** method-related charges and unit rates given above are illustrative. Class D–Z unit rates usually deemed to encompass labour + plant + materials supply/installation + overheads + profit, unless otherwise stated.

Reference	Description	Units	Add	Ddt	Qnts	Unit Cost	Cost
R	Roads and Pavings						
R117	Limestone 300 deep	m²	1680		1680	8.78	14750.4
R146	Roadbase 200 deep	m²	1680		1680	12.15	20412
R351	Bituminous Spray	m²	1680		1680	1.67	2805.6
R381	Wearing course 7mm aggregate	m²	1680		1680	1.16	1948.8
R728	Hardcore base - roadbase for landing	m²	1296		1680	12.15	20412
R812	Road and entry signs	nr	11		11	300	3300
R824	Continuous luminous lines	m	500		500	5.8	2900
V	Painting						
V316	Painting of beams on sheet piling, less than 300 mm wide	m	540		540	18	9720
W	Waterproof						
W121	Sheet piles - gaps in longitudinal joins	m²	1183		1183	10	11830
X	Miscellaneous Work						
X146	3m high security fence	m	300		300	8.5	2550
X247	Boom gate @ entrance	nr	2		2	3000	6000
						Total This Page	96628.8
						Running Total	$1,877,978

Figure 7.10 (Continued) Class A3** method-related charges and unit rates given above are illustrative. Class D–Z unit rates usually deemed to encompass labour + plant + materials supply/installation + overheads + profit, unless otherwise stated.

Step 10: Project value reflection (as Sections 2.2.2 and 2.2.3; see also illustrative Step 10 of Section 7.1.1)

The total cost for this project is Aus$1,878,000, of which some Aus$800,000 is applicable to materials costs, representing 44% of the total. Considerable transportation costs are likely due to the remote location, as are labour costs for skilled and semiskilled workers accommodated near the site for the duration of the works.

Material supply is expected to encompass both local and imported: fibre-reinforced composites are most economically sourced from overseas (via CMI's Australian distributor); steel supply, on the other hand, should ideally be sourced from Australia for both structural purposes (walers and rods) and reinforcing (bollard concrete footings); local crushed rock products should be utilised for ballast backfill of the sheet piling, for rock armour around the facility and for the steep slopes. Similarly, road construction aggregates in the form of natural gravels and crushed limestone and rock bases are plentiful locally. Illustrative materials' unit rates/costs are shown in Table 7.9.

Table 7.9 Material costing

Material	Quantity	Unit	Supplier	Unit cost ($)
Sheet piling structure				
GG-95 Sheets FRP sheet piling (12 m length)	60	Each	CMI	2,400.00
180 PFC36 m long each	720	$/m	Onesteel	44.70
25 to 38 mm rod diameter ranging, average price indicated	15,480	m	Onesteel	9.50
GG-240 Capping, capping for GG-95 sheets	108	m	CMI	15.00
GG-75/GG-95 corner panel	4	m	CMI	3,000.00
Steel work				
Plates for placing in concrete 150 mm square plates	880		Onesteel	2.00
Fill				
Road base backfilling of sheets, 0.5 m	648	cu m	WA Limestone	77.00
Gravel backfilling of sheets, 1 m depth	1296	cu m	WA Limestone	20.00
30 mm ballast backfilling of sheets, 4.5 m	5832	cu m	WA Limestone	20.00
50 mm ballast backfilling of sheets, 5 m	6480	cu m	WA Limestone	15.00
Road works	1350	sq m	Pioneer	1.50
Binder, 10 mm seal				
Aggregate, 10 mm cover	1350	sq m	WA Limestone	1.50

(continued)

Table 7.9 Material costing (Continued)

Material	Quantity	Unit	Supplier	Unit cost ($)
Road base, 200 mm layer CBR >100%	1350	sq m	WA Limestone	15.40
Limestone/gravel 300 mm layer CBR > 60%	1350	sq m	WA Limestone	6.60
Road furniture				
Signs various large signs	11	Each	Safety Sign Service	140.00
Line Marking access roads and crane/truck exclusion zones	500	M		10.00
Bollards				
DBB1-80 double bollard 80- to 100-tonne bollard	2	each	PM&I	950.00
SBC1-20 Single 20–30 tonne bollard	2	each	PM&I	500.00
Cement	10	tonnes		1,100.00
Aggregate 7 and 18 mm aggregate, average cost indicated	40	cu m	WA Limestone	15.00
Sand for concrete mix	30	cu m	WA Limestone	8.00
N40 bars tensile reinforcing	480	m		6.00
N24 bars ligs at 250 mm centres	461	m		5.00
Reinforcing chairs	100	each		
Holding down bolts, various sizes	32	each		
GuardRail maxi rail, resistant to 36 t at 80 km/h, 32 m side 28 m	92	m	Ingal Civil	80.00
Handrail miniguard steel barrier—36 m on all sides	108	m	Ingal Civil	28.00
Boom Gate				
6 m Boom Gate 240V	2	each	Boom-Gate Australia	3,000.00
Offices—Demountables 12 m*3 m*2 room	3	unit	Cavalier	9,152.00
Offices—Demountables 12 m*3 m*1 room	1	unit	Cavalier	4,576.00
Foundations Gravel	3200	cu m	WA Limestone	20.00
Cement (3%)	0.25	t		1,100.00
Total				801,186.00

Step 11: Construction method-statement resources review for future cost-monitoring and future feedback (comparing expected with actual) towards method-statement continuity (Sections 2.2.3, 2.2.4, 3.1, 3.1.1 and 3.2.2; illustrative Step 11 of Section 7.1.1)

Construction method examples might pertain to site handover and preparation, construction of barge landing, bollard installation, construction of laydown area and access road and rockwall armour installation as below:

Rockwall armour installation

Start/end date:	Monday, 4/04/*, Friday, 15/07/*
Resources:	Engineers, periodic HSE staff, surveyor and assistant, site supervisor(s), plant operators, labourers
Plant:	CAT 315c excavator, Hino tandem axle tipper truck, site and transportation vehicles
Method:	The construction of the rockwall armour occurs throughout the project's duration but has no direct support to the barge landing, and thus represents opportunities for float resource redistribution and resource-aggregation/smoothing by maintaining the 'spare' workers for site-tasks completed/idle; and conversely, allowing labour relocation when critical path tasks require help (see charting opportunities for resources [labour]; aggregation/smoothing peaks and troughs detailed in step 5 above). Tandem trucks will be used to position rocks from the coast to build the rockwall, where an excavator levels discrepancies and site labour assists in rock placement and construction.

Step 12: Value engineering review: Alternative specification options (as Sections 2.2.2 and 2.2.2.1)

The feasibility of the structural system chosen is dependent on the method used for the retaining wall on the barge mooring. Three options were considered for this design element; namely, caissons, piles with concrete slabs and sheet piling.

- Caissons were considered too large/expensive for use in a short-term structure and too difficult to remove at the end of the structure's life and so were deemed unsuitable.
- Piling was an attractive option due to the large depth of cohesionless sand that exists over the site. Displacement piles, although sound,

require additional complex system designs resulting in potentially more costly pile installation.

- Sheet piling was chosen due to its light weight, relatively simple installation (purchased as proprietary equipment) and being an environmentally attractive solution because its design allows the reuse of nearly all members of the sheet pile wall structure.

Among the options considered for crane provision to unload the barge were mobile or tower cranes, but the most flexible and least time-consuming option identified was a crawler crane, with the ability to unload the barge in halves, allowing heavier loads to be placed further from the near side of the barge.

Steps 13 and 14: Cash flow/revenue 'S-curve' analysis and tender-pack dissemination
For Step 13, see Section 2.3.1, as well as the illustrative example given in Step 13 of Section 7.1.1; for Step 14, see Chapter 5, as well as the illustrative listing given in Step 14 of Section 7.1.1.

7.1.3 Practice example: Design solution for a landing facility proposal

Step 1: Review the client's brief; clarify problems/needs/scope (as Section 5.4.1)
To transport construction material, a barge landing facility* was proposed. Approximately 76-m-long vessels are anticipated; a crawler crane will pick up construction resources from the barge for loading onto trucks and transportation to the site. This landing facility shall facilitate a mining company development, which, in totality, involves the construction of crushing plants, a conveyor system, rail transportation from the mine to a port and port facilities in which iron ore materials are mined, transported and processed; crushed materials are stockpiled, loaded and transported/loaded onto ships that are berthed and moored along a wharf against dolphin structures.

Step 2: Generate a working concept for the design solution (as Sections 5.4.1 and 5.4.2)
Road design: The road component shall satisfy the following: access for delivery of a crawler crane (Liebherr LR1200 or equivalent); truck access from the existing road to the working section of the wharf;

---

* This hypothetical/academic/educational example acknowledges with thanks contributions by Cong Bui, Tim Bird, M.B. DeGersigny and Glenn Hood; all designs, specs, timings and costings are provided for illustrative purposes only.

gradients not greater than 15° to allow trucks to traverse when loaded; temporary structure with an expected working life of 3 years; 500 mm of compacted crushed road base subgrade (topping of the compacted road base not required); crane transport width of 7 m and clearance on each side of 1 m, giving a minimum road width of 9 m plus further 0.5 m for the installation of a steel guard; road design gradients minimised to allow easier movement of heavily laden trucks; and crane operation up to 1% gradient.

Levels and design gradients: Caisson area, 0% flat; wharf working/storage area, 1% towards wharf; and roads, 2% as a compromise between drainage and stability of the vehicles using the road. Existing rock armour to be removed (and stockpiled/reused) to allow for placement of rock fill for base of road and working areas of the wharf; earth-moving equipment to a distance of 31 m requires long-reach excavators.

Caisson platform design: Loading conditions include 200-tonne crawler crane; live load of 30 kPa on deck, independent of vehicular loads, cyclonic winds, wave loads, current loads, ship berthing loads and mooring loads. It has been determined that the height of the platform should be 1.0 m above the maximum sea level of RL 6.0 m. The height of the platform will be at RL 7.0 m, making the height of the caissons 10.0 m tall taken from the seabed level. The design size of the caisson itself requires the weight of the (crawling) crane and the live load over the deck caused by moving vehicles to be spread over the caisson at a uniform pressure of approximately 30 kPa. Pressure calculations under the track require: maximum distance of centre of gravity to load, 30 m; maximum load to be lifted by crane at any one time, 15 t; worst pressures under the track caused by boom hanging, 90° and 45° over the tracks; soil pressure profile for a 10-m height and distribution of 30 kPa calculates required wall thickness/stability of the caissons and the associated steel reinforcement.

Bollard design: Fifty-tonne bollard foundations were designed using deadman anchors because the low friction coefficient of the soil necessitates a large mass of concrete to hold the bollard in place with coastal lateral earth pressure to restrain the foundation deemed unavailable. Two deadman anchors, radiating out at 40° behind the foundation and declined at 30°, provide lateral restraint with an angle of pull on the bollards limited to 30° from the horizontal. The 15- and 50-tonne bollards are proprietarily bolted on/cast-in caissons with a point of pull on the bollard at less than or equal to 350 mm above the foundation's surface.

Steps 3 and 4: Generate working/reference-specifications and design drawings (as Sections 2.2.2, 2.2.3, 4.1, 5.2 and 5.4)
All material testing and work is to be carried out in accordance with ISO 9001:2008. Steel shall be from a certified Australian supplier or if imported shall meet the requirements of Australian Standards and test certificates.

Components of the facility shall as a minimum comply with the following Australian Standards and Guidelines:

- AS4997 Guidelines for the Design of Maritime Structures
- AS1726 Geotechnical Site Investigations
- AS3798 Guidelines on Earthworks
- AS4678 Earth Retaining Structures
- AS1348 Roads and Traffic Engineering
- AS3600 Concrete Structures
- AS4100 Steel Structures
- AS1163, AS1710 Hollow Sections
- AS3678 Hot-Rolled Steel Plates
- AS1110, AS1111, AS1252 Bolts
- AS1112 Nuts
- AS1237 Flat Washers
- AS1553 Structural Steel Welding
- AS/NZS 3845 Road Safety Barrier Systems
- AS2550.1 Cranes, Hoists and Winches—Safe Use; Part 1 General Requirements
- Main Roads WA, Specification 603, Safety and Traffic Barrier System
- Main Roads WA, Specification 501, Pavements
- Main Roads WA, Specification 100 General Requirements
  - Section 101 Description of Works
  - Section 102 Survey Information
  - Section 103 Site Facilities
  - Section 104 Entry to Land
  - Section 105 Water Supplies
  - Section 106 Utilities and Services
  - Section 107 Contract-Specific Requirements

Design drawings as follows in example 7.1.3 (excluding typical details for services/proprietary systems via easy-access standard drawings downloads from Local Authority/suppliers, etc.).

**Example 7.1.3:   Drawing 1: Plan of barge landing facility and
Section A-A**

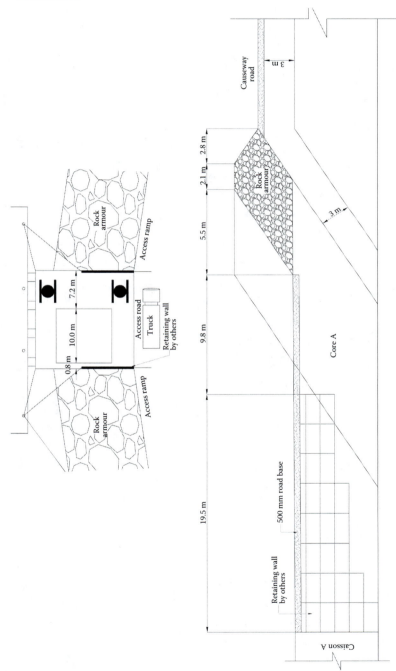

Example 7.1.3:   Drawing 2: Plan of landing area and section of access road

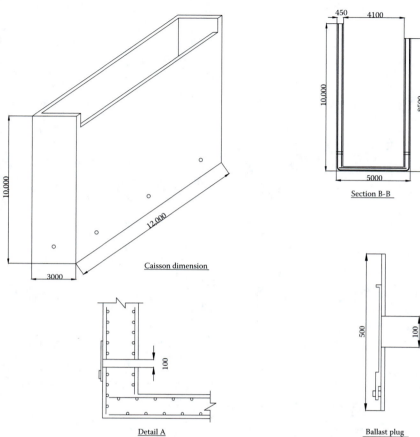

Caisson dimension

Section B-B

Detail A

Ballast plug

Example 7.1.3:   Drawing 3: Bollard foundation arrangement plan, Section A-A and details

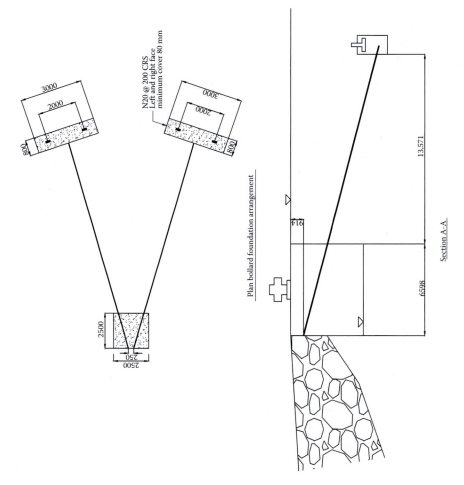

Step 5: Establish construction methodology towards method-statement(s) compilation (as Sections 2.2.3, 2.2.4, 3.1, 3.1.1 and 3.2.2)
A suitable design solution is argued to be the application of closed-base concrete caissons, allowing the caisson to float rather than being craned into position by ship. To allow the caissons to float with the aid of a crane lifting them into the water, the caissons will be constructed in a float-out area. The float-out area will be similar in concept to that used by ship-building companies to float-out their large ships once constructed.

Once the constructed caissons are floating, a tug will tow the caisson into position during high tide. After correct alignment with the concrete foundations, ballast plugs located at the base of the caisson will be released to allow a controlled flow of water into the void of the caisson. This will allow the caisson to sink slowly until it reaches its concrete foundations. The ballast plugs will be reconnected and extra water will then be pumped into the caisson to add extra weight to avoid any movement caused by wave and current loads.

The excavation for the bollard foundations and the placement of the earth anchors takes place after the removal of the required amount of armour, then an excavator sets about digging the foundations in one day (scheduled on the 14th of July), with formwork and reinforcement for the first bollard being in place the following day. The pour takes place the day after, with the formwork then stripped and prepared for the second bollard foundation during the next 2 days. For the construction of the 50-tonne bollards, each respective component of the construction is allotted a day in the construction programme. The construction of the bollards is not part of the critical path for the project and thus has programme float. The installation of the deadman anchors occurs 1 week after the construction of the bollard foundations; however, it can be placed earlier albeit the post stressing cable must not be strained for at least a week to allow for sufficient concrete hardening. The 15-tonne bollards are installed into the caisson by cast-in bolts placed during manufacture and in accordance with the suppliers' specifications.

Step 6: Prepare the project timeline schedule (as Sections 3.1, 3.1.1 and 3.2.2)

The individual breakdown of the number of days planned per task is displayed in the pie chart in Figure 7.11. Construction is scheduled to begin no later than the 6th of June and be finished by the 19th of October. A detailed schedule of all works is displayed (somewhat illustratively) in

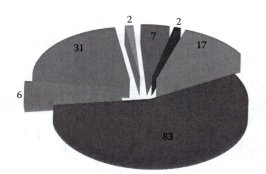

- Preliminaries (7)
- Procurement (2)
- Construction of concrete works (17)
- Earth works and road (83)
- 50-tonne bollards (6)
- Construct landing (31)
- Site clean up (2)

Figure 7.11 Activities.

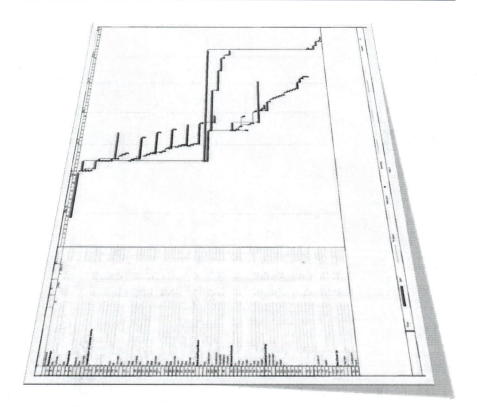

Figure 7.12 Schedule example.

the Gantt chart in Figure 7.12 where individual tasks, group tasks and critical paths are typically shown.

Step 7: Assess risk, confirm quality management systems and occupational health and safety compliance and address environmental impact (all as Sections 3.3, 4.1 and 5.1.7)
Maritime construction and engineering requires risk identification and ranking, alongside associated mitigation measures. A risk matrix coding below adapted from AS/NZS ISO 31000 seeks to assess risk for labour, plant, environment and overall. Examples are illustrated in Figure 7.13.

Steps 8 and 9: The BQ using CESMM3 for the project with input unit rates (as Sections 5.3 and 2.2.3)
Parts 1 through 5 of step 9 are illustrated in Figure 7.14.

Variable/probability	Risk impact/Consequence				
	1 Insignificant	2 Minor	3 Moderate	4 Major	5 Catastrophic
Labour	First aid injury	Medical treatment injury	LTI (lost time injury)	Injury, disability	Fatality
Plant/material/structure	Minor damage to plant or system	Damages impact on budget and program	Significant damage to plant or system	Extensive damage to plant or system	Virtual complete loss of plant or system
Environment	Limited impact to minimal area	Minor, short-medium term impact to local area of limited	Moderate but reversible impact	Medium to long term widespread impact	Long term, widespread, significant impact
Project overall	Minor impact to project	Short-medium term impact to project	Moderate impact to project	Medium to long term impact	Long term, significant impact
A: Almost certain common/ frequency occurrence (anytime)	1A (High)	2A (High)	3A (Extreme)	4A (Extreme)	5A (Extreme)
B: Likely will probably occur in most circumstances (10 per year)	1B (Moderate)	2B (High)	3B (High)	4B (Extreme)	5B (Extreme)
C: Possible might occur at some time (1 per year)	1C (Low)	2C (Moderate)	3C (High)	4C (Extreme)	5C (Extreme)
D: Unlikely to occur (1 per 10 years)	1D (Low)	2D (Low)	3D (Moderate)	4D (High)	5D (Extreme)
E: Rare occur in exceptional circumstances (1 per 100 years)	1E (Low)	2E (Low)	3E (Moderate)	4E (High)	5E (High)

| 2.0 Design and construction risk | | | | |
Project risks				
2.1 Design	Design of bridge is incapable of delivering the services at the anticipated cost	2C	Increased cost and time to fix problems	The contractor will be responsible
2.2 Construction	The risk that events occur during construction which prevent the barge landing facility being delivered on time and cost	2C	Delay and cost	TMG Engineering will access each delay and using AS4000 conditions of contract decide on the cause of action
Labour risks				
2.3	Incorrect use of equipment due to insufficient training	3B	LTI/property damage	Training
2.4	Faulty equipment due to insufficient maintenance	3C	LTI/property damage	Regular maintenance check/tagging
2.5	Moving vehicles due to reduced visibility	4C	Permanent injury/death	Spotters/training
2.6	Incorrect manual handling of materials	4C	Permanent injury or disability	Safety procedures/training
2.7	Dust due to traffic and movement of fill material	2C	Medical treatment injury	PPE/dust suppression
2.8	Dry heat around 30 to 37 degrees due to incorrect personal water management	4D	Injury	Correct training/PPE
2.9	Workers falling into water	1C	First aid injury	Correct use of barricades

Figure 7.13 Risk example illustrations.

3.0 Financial Project risks		
3.1 Interests rates pre-completion	3C	The risk that prior to completion, interest rates may move adversely	Increased project costs	Interest rate hedging may occur under the Project Development Agreement

4.0 Operating Project risks		
4.1 Operator failure	2D	Risk that a subcontractor may fail financially or fail to provide contracted services to specification	Failure may result in service unavailability. Alternative arrangements may be needed with delay and increased costs	The contractor chosen must be diligent about which subcontractors are used in the project. Subcontractors must be checked for both reputation and financial position. Any additional costs will be passed on to the contractor

Figure 7.13 (Continued) Risk example illustrations.

Ref	Item Description	Unit	Quantity	Rate	Totals
A	**CLASS A:** **GENERAL ITEMS**				
A100	Contractual Requirements				
A120	Insurance of the works (includes 10% stamp duty)	LS	1	0.28%	$20,235
A130	Third Party Insurance (includes 10% stamp duty)	LS	1	0.33%	$23,848
A200	Specified Requirements				
A221	Services for the Engineer's staff Transport vechicles (x 2)	Wk	22	$387.50	$17,050
A233	Equipment for use by the Engineer's staff. Surveying Equipment	Wk	22	$1,550.00	$34,100
A300	**Method-Related Charges**				
A310	Accommodation and Buildings				
A311	Offices				
	12x3m including Transport (x2)	Wk	23	$194.00	$8,924
	Chemical Toilet (20 man unit)	Wk	23	$116.00	$2,668
A313	Accommodation for the Supintendant 2 bedroom building	Wk	26	$1,550.00	$40,300
A315	Canteen and Mess rooms				
	12x3m (cold water A/C etc)	Wk	23	$420.00	$19,320
A320	Services				
A321	Electricity				
	Generator Including Transport	LS	1	$3,875.00	$2,500
A322	Water				
	5000L Water Tank with pump	LS	1	$2,480.00	$2,480
A327	Traffic Control	Wk	22	$4,000.00	$88,000
A330	Plant				
A331	Cranes				
	25T Franner	Hr	150	$225.00	$33,750
A332	Work Boat	Day	5	$20,000.00	$100,000

Figure 7.14 Bills of quantities. Class A3** method-related charges and unit rates given above are illustrative. Class D–Z unit rates usually deemed to encompass labour + plant + materials supply/installation + overheads + profit, unless otherwise stated.

A333	Earthmoving				
	Water Truck for dust suppression	Hr	376	$155.00	$58,280
A334	Compaction				
	Hand Compactor (x3)	Day	7	$165.00	$3,465
A335	Concrete Mixing				
	Mobile Batching Plant	Hr	120	$465.00	$55,800
A370	Supervision and Labour				
A371	Supervision				
	Project manager x 1	Month	5.5	$7,750.00	$42,625
	Site Manager x 1	Month	5.5	$9,700.00	$53,350
	Engineers x 2	Month	5.5	$6,200.00	$68,200
B	Class B: Ground Investigation				
B513	Standard Penetration	nr	12	$35.00	$420
B523	California Bearing Ratio	nr	12	$38.00	$456
B515	In-situ density	nr	12	$39.00	$468
C	Class C: Geotechnical services				
C700	Ground Anchorages				
C715	50 tonne bollard anchor in soil material with permanent single corrosion protection	nr	4	$7,000.00	$28,000
D	Class D: Demolition and Site Clearance	N/A			
E	Class E: Earthworks				
E400	General Excavation				
E437	Rock (Includes disposal)	m^3	9200	$22.60	$207,920
E500	Excavation Ancillaries				
E563	Excavation of material below the Final Surface and replacement with stated material	m^3	57700	$30.70	$1,771,390
E600	Filling				
E617	Filling in caisson with imported natural material	m3	3483	$48.80	$169,970
E647	Fill to stated depth with imported rock	m^3	35,500	$48.80	$1,732,400

Figure 7.14 (Continued) Bills of quantities. Class A3** method-related charges and unit rates given above are illustrative. Class D–Z unit rates usually deemed to encompass labour + plant + materials supply/installation + overheads + profit, unless otherwise stated.

F	**Class F:** **In Situ Concrete**				
F200	Placing and Provision of Concrete				
	Grade 50MPa, 20mm agreggate				
F283	Caissons				
	12m x 4m x 10m	m^3	500	$325.50	$163,350
F624	Bollard foundations				
	(2.5m x 2.5m x 2.5m) x 2	m^3	75	$325.50	$24,773
G	**Class G:** **Concrete Ancillaries**				
G100	Formwork: rough finish				
G145	Plane Vertical				
	Width exceeding 1.22m	m^2	1650	$173.60	$286,440
G500	Reinforcement				
G515	Plain round steel bars				
	Nominal size 16mm	t	1.75	$3,332.50	$5,832
G516	Plain round steel bars				
	Nominal size 20mm	t	19.05	$3,332.50	$63,484
G700	Post tensioning prestressing				
G713	Prestressed cable between deadman and foundation (7.2m long)	nr	4	$4,650.00	$18,600
H	**Class H:** **Precast Concrete**				
H500	Slabs				
H526	Concrete slab, 4.5 m2, 9 tonne	m3	14.4	$325.00	$4,680
H600	Segmental Units				
H650	Retaining Wall	nr	10	$16,300.00	$163,000
I	**Class I:** **Pipework - Pipes**				
		N/A			
J	**Class J:** **Pipework -Fittings & Valves**				
		N/A			

Figure 7.14 (Continued) Bills of quantities. Class A3** method-related charges and unit rates given above are illustrative. Class D–Z unit rates usually deemed to encompass labour + plant + materials supply/installation + overheads + profit, unless otherwise stated.

K	**Class K:** **Pipework- manholes and pipework** **ancillaries**	N/A			
L	**Class L:** **Pipework Supports & Protection,** **ancillaries to laying and excavation**	N/A			
M	**Class M:** **Structural Metalwork**	N/A			
N	**Class N:** **Miscellaneous Metalwork**				
N130	Ladder	m	10	$113.30	$1,133
N140	15 Tonne Bollard	nr	6	$810.00	$4,860
N140	50 tonne bollard anchor in soil	nr	2	$2,325.00	$4,650
O	**Class O:** **Timber**				
O100	**Hardwood componants**				
O137	Timber edging to platform	m	116	$94.25	$10,933
P	**Class P:** **Piles**	N/A			
Q	**Class Q:** **Piling Ancillaries**	N/A			
R	**Class R:** **Roads and Pavings**				
R100	Sub-bases, flexible road bases and surfacing				
R128	500mm Compacted Rock Base	m²	3,100	$58.90	$182,590
R800	Ancillaries				
R811	Non Iluminated Road Signs	nr	12	$650.00	$7,800

Figure 7.14 (Continued) Bills of quantities. Class A3** method-related charges and unit rates given above are illustrative. Class D–Z unit rates usually deemed to encompass labour + plant + materials supply/installation + overheads + profit, unless otherwise stated.

S	Class S: Rail Track		N/A		
T	Class T: Tunnels		N/A		
U	Class U: Brickwork, Blockwork & Masonary		N/A		
V	Class V: Painting		N/A		
W	Class W: Waterproofing		N/A		
X	Class X: Miscellaneous work				
X100	Fences				
X173	metal Guard rails inc posts and ends	m	104	$307.00	$31,928
Y	Class Y: Sewer and water Main renovation and ancillary works		N/A		
Z	Class Z: Simple Building works incidental to civil engineering works		N/A		
	Subtotal				$5,559,972
	Contingency Plan		10%		$555,997
	Profit Margin		10%		$555,997
	Gst		10%		$555,997
	Total				$7,227,963

Figure 7.14 (Continued) Bills of quantities. Class A3** method-related charges and unit rates given above are illustrative. Class D–Z unit rates usually deemed to encompass labour + plant + materials supply/installation + overheads + profit, unless otherwise stated.

Step 10: Project value reflection and elemental/class cost comparison (as Sections 2.2.2 and 2.2.3)
Using CESMM3, the proposed barge landing facility (utilising a float-out technique) will cost approximately $7.2 million (including government service tax, contingency and profit).

The major costs associated with the construction of the barge facility are the earthworks, which accounts for 55% of the total costs because there is a considerable amount of 'armour' to remove. The concrete works also make up a significant 10% portion of the total. Figure 7.15 displays smaller components as nonzero% values, albeit prediction to this degree of accuracy is unlikely.

Steps 11 and 12: Method-statement resources review for future cost-monitoring and future feedback towards method-statement continuity, as well as value engineering exercises might be, to an extent, deemed applicable via review of Section 7.1.2 (in conjunction with review of Sections 2.2.3, 2.2.4, 3.1, 3.1.1, 3.2.2 and Sections 2.2.2 and 2.2.2.1)

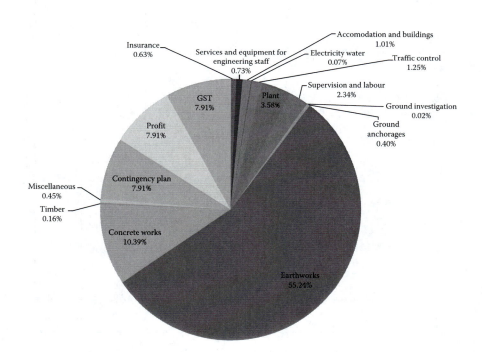

Figure 7.15 Project breakdown.

Step 13: Cash flow/revenue 'S-curve' analysis (as Section 2.3.1)

S-Curves allow progress tracking of the project at the precontract stage as well as at the all-important on-site construction stage. The S-curve for the barge landing facility is shown in Figure 7.16. The initial start-up of the project only takes 10% of the amount of time and costs 7% of the project budget. After week 2, the major part of the project commences with costs climbing linearly through to week 12. This represents 69% of the projected costs. It is noted that any variations that occur at this time frame will significantly add to the project cost and may result in extension of time claims. From week 12 to week 19, the costs increase at a reduced rate with this time frame of the project representing 25% of the estimated financials. A contingency amount and the accumulation of retention fees somewhat protects client interests.

Step 14: Tender-pack, towards contractors' tender-offer, and client selection (as Chapter 5 and practice Step 14 of Section 7.1.1)

Having prepared the information discussed in the 13 steps above as part of a traditional procurement approach, the design team then distribute tender documents to facilitate a selective/competitive/negotiated tendering process. Builders prepare a tender-bid/offer (alongside a detailed account of respective expertise, experience, quality systems, management structure(s), work method, time schedule, and the like); following review of the available submitted tender-bids, a builder is recommended for selection, and a contract is subsequently signed and work commences.

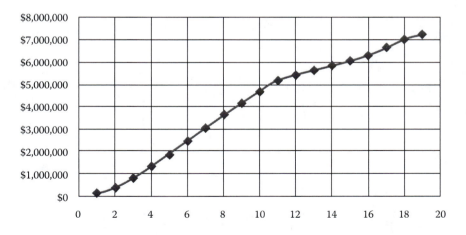

Figure 7.16 S-curve.

7.2 REPRESENTATIVE CIVIL ENGINEERING COST AND OUTPUT-EFFICIENCY INFORMATION

The data shown in Tables 7.10 through 7.25 have been adapted from a number of sources.*

- Australian currency (Aus$) applies.
- All costs and output efficiency data provided purely for academic purposes.
- Recycling rates/efficiencies towards sustainable specification options.
- Pollution prevention costs (Delft eco-costs) provided for background.
- Projected landfill, tipping and aggregate levy charges are shown assumed.
- Rate buildups are presented for land remediation and soil processing, with rubble processing charges described for both mobile plant applications as well as fixed sites that include annual set-up and processing costs.
- Incinerator (energy recovery plant) costs are similarly described/assumed in terms of a facility's yearly costs for capital, operation and maintenance, alongside a suggested incinerator gate fee per tonne.
- Materials' prices are presented in a unit-rate format (Aus$ rate per unit).
- Approximate densities are listed for principal building materials.
- Plant efficiencies are described in litres per hour.
- It is restated that prices/rates are provided for academic reference only.

---

* Costs/prices/plant efficiencies and recycling rates are provided purely for academic purposes only. HM Customs and Excise; DERFA (ne-DoE); Davis, Langdon/Spon's Price Book(s); London Metal Exchange; Deutsche Bank; Salvo-Web; Dundee City Council; CTU University of Dundee; DERL Baldovie Incinerator; Tayside Contracts; UoD CTU LCA Cullet studies; Letsrecycle.com; Delft UoT, NL; a wide range of plant and Web-based supplier catalogues; Rawlinson's Price Books; Cordell's Price Books.

Table 7.10 Unit rates

Resource/material	Unit rates	
	$/tonne	$/unit
Brick and block	—	—
Clay commons	$117/tonne	$221/1000
Brick; clay facings	$148/tonne	$280/1000
Brick concrete common	$68/tonne	$128/1000
Brick concrete facings (BS 6073: 225 × 112.5 × 75 mm)	$135/tonne	$255/1000
Block; concrete; aerated; 2.8 N/m²; 100 mm	$180/t	$14/m²
Block; concrete; dense; 7.01 N/m²; 100 mm	$94/t	$9/m²
Brick: recycled cleaned salvaged bricks	$89/t	$170/1000
Concrete	—	—
Concrete grade C7.5; 20 mm aggregate; approximate standard mix ST1	$45/tonne	$106/m³
Cement ordinary portland cement to BS12	$174/tonne	$5/25 kg bag
Processing (crushing/screening) demolition arisings/materials on-site	$13/t	
Housekeeping transport; origin to stockpile	$1.20/tonne	
Housekeeping transport; stockpile to processing plant	$1.00/tonne	
Establishing plant and equipment on-site; removing	$0.80/tonne	
Maintaining and operating plant	$3.50/tonne	
Crushing hard material on-site	$5.20/tonne	
Screening material on-site	$0.80/tonne	
Quarry product; material + delivery >5,000 tonnes; excluding tax	—	—
Graded granular material; 1 A, B, C; general fill	$13.5/tonne	
Reclaimed material; 2 E; blast furnace slag; general fill	$9/tonne	
Well-graded granular; 6 A	$16/tonne	
Selected granular/graded material; NG505, 6 F, I, J, N, P fill	$14/tonne	
Wet/dry cohesive material; 2 A, B general	$14/tonne	

(continued)

Table 7.10 Unit rates (Continued)

Resource/material	Unit rates	
	$/tonne	$/unit
Filter material to drain trenches, type A, B, C	$14/tonne	
Graded granular crushed rock, nat-l sand, gravels; NG803/4 type 1/2	$17/tonne	
Graded granular crushed concrete (reclaimed)	$9/tonne	
Aggregates; all-in for concrete production; 40, 20, 10 mm	$16/tonne	
Fine aggregates, screen sand for concrete production	$15/tonne	
Transport quarry prod; add extra 9 p/mile >20 mile catchment	$3/t	
Transport quarry; 20 tonne tipper lorry; + journey time $1/hour		$3/t/h; $34/h
Landscaping/soil	—	—
Landscaping: subsoil	$5/t	$8/m³
Landscaping: topsoil	$12/tonne	$18/m³
Landscaping: woodland timber mulch	$14/tonne	$22/m³
Landscaping: compost seeding	$12/tonne	$27/m³
Gate fee and cost to deposit green waste at a recycling facility	$38/t	
Landscaping: grass-seed, BSH ref. A3; sowing rate 25–50 g/m²	$15/tonne	$118/25 kg bag
Bitumen macadam	—	—
Bitumen macadam of 10, 20 and 40 mm aggregate	$58/tonne	
Coated roadstone DBM roadbase; class 903; HRM base wearing	$58/tonne	
Timber	—	—
Softwood sections for carpentry	$840/t	$420/m³
Softwood for joinery	$802/t	$555/m³
Hardwood for joinery (Iroko)	$3150/t	$1600/m³
Hardwood for piers jetties, Groynes; 150 × 75 mm	$2510/t	$15/m³
Plywood; exterior quality (18 mm)	$1200/t	$21/m²
Plywood; interior joinery (6 mm)	$1402/t	$8/m²
Chipboard sheet flooring (18 mm)	$308/t	$6/m²
Used timber beams, baulks, pit props, kerbing, barriers	$442/t	$222/m³
mixed wood delivered to a wood recycler (disposal equiv. cost)	$18/t	
Chipping plant + labour	$17–170/t	$8–70/m³
Door; softwood internal		$160/nr.
Particleboard/plasterboard	—	—
Timberboard to roofs	$/tonne	$11/m²

(continued)

Table 7.10 Unit rates (Continued)

Resource/material		Unit rates	
		$/tonne	$/unit
Gypsum plasterboard; 2700 × 2400 × 1000 × 9.5 mm thick at $10/m		$1500/t	$3/m²
Process plasterboard		$13/t	
Transport material to stockpile and processing	2		
Establishing plant and equipment	1		
Maintaining and operating plant	4		
Processing material	5		
Screening material	1		
Metal/steel		—	—
Steel pipes to BS 3601; 125 mm excluding delivery		$1200/t	$38/m
Steel structure universal beam 1016 × 305 mm (222, 249) excluding delivery		$830/t	
Steel plain ended; diameter, 110 mm		$2900/t	$11/m
Steel cost average; US$250–350 per imperial tonne		$310/t	
Steel production cost; US$240/tonne with energy making up 15%–20%		$240/t	
Steel stainless solids, 18/8		$1300/t	
Steel bar; steel reinforced		—	—
Rebar; grade 250 plain round R6; mild steel; BS 4449; straight		$510/ tonne	
Transport steel, 45–50 mile radius		$20/t	
Copper			
Copper price; US$1.40/lb London Metals Exchange		$2900/t	
Copper production; mining, milling and processing US$0.50–0.80/lb		$1400/t	
Copper recycling; 60% of production		$800/t	
Aluminium			
Aluminium price; US$1750/tonne, London Metals Exchange		$1600/t	
Aluminium production cost breakdown of US$1100/ tonne:		$110/t	
Alumina (34%)			
Energy (28%)			
Other raw materials (14%)			
Other costs (13%)			
Labour (11%)			

(continued)

Table 7.10 Unit rates (Continued)

	Unit rates	
Resource/material	$/tonne	$/unit
Ferrous scrap metal	—	—
Ferrous: 0A plate and girder scrap	$160/t	
Ferrous: No. 1 old steel scrap	≈	
Ferrous: No. 2 new steel scrap		
Ferrous: 4A low residual bale scrap		
Ferrous: 4C new production bale scrap		
Ferrous: light iron scrap		
Ferrous: 7B mixed steel turning scrap		
Ferrous: 8B mixed steel cutting scrap		
Ferrous: foundry heavy cast iron scrap		
Ferrous: No. 9 cast scrap		
Ferrous: No. 10 light cast scrap		
Ferrous: 11 cast iron boring scrap		
Ferrous: 12A new production heavy steel scrap		
Nonferrous scrap metal	—	—
Dry bright wire scrap	$2600/t	
Copper heavy scrap	≈	
Copper No. 2 wire scrap		
Copper braziery scrap		
Copper cable insulated 45% scrap		
Brass cuttings scrap	$1800/t	
Brass swarf scrap	≈	
Brass and copper radiators scrap		
Brass mixed scrap		
Brass heavy scrap		
Brass gunmetal scrap	$1800/t	
Aluminium pure/alloy cuttings scrap	≈	
Aluminium old rolled scrap		
Aluminium clean cast scrap		
Aluminium turnings scrap		
Aluminium foil scrap		
Lead scrap	$650/t	
Lead batteries scrap	$43/t	
Zinc mixed/cuttings scrap	$380/t	

(continued)

Table 7.10 Unit rates (Continued)

Resource/material	Unit rates	
	$/tonne	$/unit
Glass/ceramics	—	—
Float glass (4 mm)	$4300/t	$43/m²
Sealed double glazing units	$9900/t	$102/m²
Ceramic wall tiles (198 × 65 × 6 mm)	$2220/t	$39/m²
Glass cullet		
Glass: mixed colour cullet for recycling into aggregates	$8/t	
Glass: mixed colour cullet for recycling by container glass industry	$18/t	
Glass: green glass cullet	$18/t	
Glass: clear glass cullet	$35/t	
Glass: brown/amber glass cullet	$40/t	
Glass collection/handling (average cost/t from bottlebanks)	$17/t	
Glass recycled as sand replacement (processing costs)	$60/t	
Plaster and paint	—	—
Plaster (carlite bond)	$222/ tonne	
Paint; internal, white masonry; $8/L		$17/m²
Paint; external		$3/m²
Tiles and paviors	—	—
Clay floor tiles (150 × 150 × 12 mm)	$650/t	$18/m²
Vinyl floor tiles (300 × 300 × 2 mm)		$5/m²
Precast concrete paving slabs (200 × 100 × 60 mm)		$4/m²
Clay roof tiles (plain 265 × 165 mm)		$350/1000
Precast concrete roof tiles (419 × 330 mm)		$800/1000
Insulation	—	—
Glass fibre quilt; 80 mm thick	$1200/t	$4/m²
Polystyrene board/slab; 40 mm thick	$2100/t	$4/m²
Polystyrene grade EHD(N); 100 mm thick		$2/m²
Processing	$7/t	
Transport material to stockpile		
Transportation material stockpile to processing		
Establishing plant and equipment		
Maintaining and operating plant		
Screening material on-site		
Processing insulation to compost		$6–14/test

(continued)

Table 7.10 Unit rates (Continued)

Resource/material		Unit rates	
		$/tonne	*$/unit*
	Chemical composition for agricultural enhancement: K, phosphate, nitrate, organic		
Rainwater pipework/gutters		—	—
Aluminium pipes; dia 63.5 mm		$2000/t	$17/m
Cast iron pipes; dia 65 mm		$3300/t	$21/m
Clay pipes; vitrified; 100 mm		$620/t	$9/m
Piped supply systems		—	—
Copper pipes; 15 mm		$4300/t	$2/m
Copper pipes; 35 mm		$4800/t	$12/m
PVC drain pipes fittings		—	—
uPVC; BS 5255; 50 mm pipes		$1250/t	
uPVC; BS 4576; 50 mm pipes		≈	
uPVC BS 4576; 112 mm half-round			
Plastic; bulk value			
Plastic waste		—	—
Clear and light blue PET		$190/t	
Coloured PET		≈	
HDPE single col			
HDPE mixed col			
PVC			
Mixed			
Plastic road furniture: cleaned road cones, bollards, polyethylene			
Plastic recycle process		$250+/t	
Transport plastic waste; $100/70 mile radius; $20/15 mile radius		$35/t	$180/t/140 m
Electrical systems		—	—
Electrical cable, 2.5 mm; 250 V blue cable; 415 kg/km		$3400/t	$2/m
Wired-in PVC-insulated cables in welded conduit			$180/nr.
See also nonferrous metal above		—	—
Cable scrap, salvage value		$85/t	
Cable recycling	30% of $471 rate for copper processing	$410/t	
	70% of $140 rate for plastic processing		
Transportation cable waste (pro rate 70% plastic and 30% nonferrous metal)		$30/t	
General			
Additional services labour handling rate (general)		7/t	

Table 7.11 Material densities

Product		Approximate density (tonne/unit)
Landscaping/subsoil	1.6 tonne/m^3	
Bricks	1.0 tonnes/m^3	59 bricks/m^2; 526 bricks/m^3 (215 × 102.5 × 65 mm)
Concrete blocks	2.4 tonnes/m^3	solids, 99/m^3, 450 × 225 × 100 mm; hollows, 69/m^3, ht 140 mm
Tiles	0.75 tonnes/m^3	
Concrete	2.4 tonnes/m^3	
Timber	0.5 tonnes/m^3	150 × 75 × 1000 mm; 88 nr timbers/m^3; 175 nr timbers/t
Plasterboard	0.75 tonnes/m^3	9.5 mm thick; 105 nr
Metal steel	2–4 tonnes/m^3	
Metal steel pipes	2 tonnes/m^3	@ dia 125 × 1000 mm; 64 nr pipe/m^3; 32 nr pipes/t
Roof sheeting	0.4 tonnes/m^3	
PVC u	0.3 tonnes/m^3	@ dia 110 × 1000 mm; 81 nr pipe/m^3; 270 nr pipes/tonne
Cardboard packaging	0.05 tonnes/m^3	
Glass	2.4 tonnes/m^3	
Road plainings	2.4 tonnes/m^3	Remix proportions 350 mm repair depth; 0.84 tonnes/m^2
Bitumen material	0.75 tonnes/m^3	
Electrical cable	0.415 tonne/km	

Table 7.12 Plant efficiency: Fuel consumption

Plant	Consumption
16 tonne 360° tracked excavator	1.8 L/h
3 m cub loading shovel	0.7 L/h
1 m jaw crusher	22.5 L/h
Mobile cone crusher	21.5 L/h
Three-deck mobile screen	9.0 L/h
50 tonne low loader	1.0 L/mile
Tractor (Ford 660)	9.0 L/h
Paver (Blaw Knox PF-191)	16.0 L/h
Backacter 11.5 t, 19 t	12.0 L/h
Backhoe JCB 3CX, 56 kW	7.5 L/h
Roller BW 90 AD	2.0 L/h
Dozer CAT D6E	21.0 L/h
Loader CAT 953B	18.0 L/h
Scraper CAT 621F	41.5 L/h
Skidsteer Bobcat 553	3.4 L/h
Dumper 2 tonne	3.0 L/h
Dumper 36 tonne	25.0 L/h
Pump	1.1 L/h

Table 7.13 Landfill tax and tipping charges

Classification	Rate ($/tonne)
Landfill tax inactive inerts (soil, rubble, etc.)	$4/t
Landfill tax other waste (timber, painted items, etc.)	$26/t ($130 per load)
Nonhazardous wastes	$16/t
Landfill tax charge (contaminated)	±$55/t
Contaminated liquid	$50/t
Contaminated sludge	$80/t
Special waste	$32/t
Hazardous wastes	$20/t
Special waste regulation charge	$2/t
Gate fee	$9/t
Transportation cost	$9/t

Table 7.14 Projected/assumed aggregate levy: If/when fully in effect

Item	Actual	Overall	Knock-on costs
Aggregate levy	$3/m³	$5/m³	Project cost increases 9%
Precast concrete block	$3/m³	$5/m³	Dense aggregate levy rate increase of $0.5/m²
Concrete roof tiles	$3/m³	$5/m³	$1 levy rate increase per square metre of roofing
Precast concrete	$3/m³	$5/m³	2% to 5% levy cost increase for pavings, flags, blocks and kerbs
Bitumen macadam paving	$3/m³	$5/m³	2.5% levy rate increase of on-cost materials

Table 7.15 Land remediation techniques

Remediation technique	Material	Rate ($/tonne)	Rate ($/unit)
Removal	Disposed material	$120/tonne	$190/m³
Clean cover	Surface area of site	$41/tonne	$66/m³
On-site encapsulation	Encapsulated material	$60/tonne	$100/m³
Bioremediation/soil washing	Treated material	$110/tonne	$180/m³

Table 7.16 Land remediation: Dig-and-dump costs (≤10,000 m³)

Operation	Rate ($/t)	Rate ($/unit)
Site investigation and testing	$5/t	$7/m³
Excavation and backfill	$18/t	$30/m³
Disposal costs	$35/t	$60/m³
Haulage	$14/t	$20/m³
Rotavation of material in stockpiles	$1/t	$2/m³
Blending contaminated sludge/liquid with clean for transport off-site 1:1	$5/t	$7/m³
Crushing and screening of excavated material	$6/t	$9/m³
Transportation off-site to landfill (75 round-trip)		$5/km

Table 7.17 Soil processing

Soil processing	Project cost	Unit rate
Soil; ≈30%–80% of site remediation project cost of $9–40/m³	5%	$3/m³
Particle/pore size distribution (PSD) test to determine reuse suitability of clay, silt, gravel and cobble boulder	$35/test	
Chemical composition to determine agricultural enhancement: potassium, phosphate, nitrate and organic matter	≈$13	
Moisture content to determine transport, handling and storage	$3/test	

Table 7.18 Crushing (mobile plant): Costs specimen unit rates

Item		$/t	$/Unit
Establishing plant and equipment on-site; removing upon completion			
Crushing plant set-up/removal			$850/nr
Screening plant set-up/removal			$430/nr
Maintain and operate plant			
Crushing plant			$5200/wk
Screening plant			$1300/wk
Transportation from demolition site to stockpiles		$1/t	$1/m³
Transportation from stockpile to crushing plant		$2/t	$2/m³
Breaking up material on-site using impact breakers			
Mass concrete		$5/t	$12/m³
Reinforced concrete			
Brickwork			
Crushing material on-site			
Mass concrete	ne 1000 m³	$5/t	$10/m³
	1000–5000 m³	≈	≈
	>5000 m³		
Reinforced concrete	ne 1000 m³		
	1000–5000 m³		
	>5000 m³		
Brickwork	ne 1000 m³	$9/t	
	1000–5000 m³	≈	
	>5000 m³		
Screening material on-site		$1/t	$1/m³

Table 7.19 Crushing (mobile plant): Cost of importing and crushing

Item	$/tonne	$/m³
Importing new fill materials		
Cost of new aggregate delivered to site		
Addition for new aggregate tax		
cost of importing fill materials	$20/t	$34/m³
Disposing of site material		
Related cost of removing materials from site		
Related cost of disposal of site material	$12/t	$18/m³
Crushing old/existing site materials and reuse/recycling		
Transport material from excavation/demolition site to stockpile		
Transportation of material from stockpile to crushing plant		
Establishing plant and equipment on-site, removing upon completion		
Maintaining and operating plant		
Crushing hard material on-site		
Screening material on-site		
Cost of crushing site old/existing materials for reuse/recycling	$22/t	$13/m³

Table 7.20 Crushing/processing (fixed site) demolition recycling costs

Fixed site recycling plant: Annual cost	$
Capital payback on plant costs of $2 million over a 5-year period	
Recovery of site engineering costs over 5 years	
Oversite sealant: concrete 37,200 m²	
Service electrical cost	
Planning EIA	
Total financing costs for plant and siteworks payable at 12%/year	
Ground rent/acre for 14 acres (5.5 ha) annually	
Maintenance of facilities at 1% of capital payback	
Fuel and power costs	
Quality regime at wages	
Seven operatives × 2000 man-hours	
One manager × 2000 man-hours	
Additional labour handling rate	
Overheads at 10% of wages bill	
Purchase of feedstock and/or landfill of reject material 40,000 t	
Cumulative yearly cost	$1.5M

Table 7.21 Incinerator energy generation/waste recycling: Capital, operating and maintenance costs

Capital, operating, maintenance costs	$ p.a.
Maintenance costs, including equipment, emission testing, lubricants…	
Consumables, sand, lime, carbon…	
Basic administration costs, stationery…	
Staffing (26 staff + social costs)	
Professional fees	
Property taxes	
Insurance	
Ash disposal	
External disposal of surplus clinical waste	
Technical support	
Combined contingency	
Total yearly operating costs	$4 million
Gate fee charged on delivered mixed waste contracted bulk quantities	$55/tonne
(Noncontract small amounts	$85/t)

Table 7.22 Eco-costs pollutant prevention (Delft UoT): Procurement of river gravel

Pollutant	Cost per kilogram
Prevention of acidification	$10/kg ($SO_x$ equivalent)
Prevention of eutrophication	$5/kg (phosphate equivalent)
Prevention of heavy metals	$1020/kg (calculation based on Zn)
Prevention of carcinogenic	$20/kg (PAH equivalent)
Prevention of summer smog	$75/kg (calculation based on VOC equivalent)
Prevention of winter smog	$120/kg (calculation based on fine dust)
Prevention of global warming	$0.50/kg ($CO_2$ equivalent)

Source: Adapted from Delft University of Technology (Delft UoT), body of work at: http://www
.io.tudelft.nl/onderzoek/onderzoeksprogrammas/technology-transformation/design-for-sustainability
-emerging-markets/sub-theme-4-knowledge-exploration-and-decision-support/.

Table 7.23 Eco-costs (Delft UoT): River gravel and RCA/t

	Eco-costs variables	New river gravel ($)	Crushed concrete ($)
Materials	Virtual pollution prevention costs	0.2	1.8
	Depletion of fossil fuels	0.3	0.0
	Depreciation	3.6	3.6
	Labour	0.0	0.0
Land use	Species richness	0.8	0.0
	Total	5.0	5.5

Source: Adapted from Delft University of Technology (Delft UoT), body of work at: http://www
.io.tudelft.nl/onderzoek/onderzoeksprogrammas/technology-transformation/design-for-sustainability
-emerging-markets/sub-theme-4-knowledge-exploration-and-decision-support/.

Table 7.24 Eco-costs (Delft UoT) concrete excluding steel, river gravel and RCA/t

	Eco-costs variable (concrete t)	New river gravel ($)	20% Crushed concrete ($)
Materials	Virtual pollution prevention costs	44	44
	Depletion of fossil fuels	0	0
	Depreciation	22	22
	Labour	0	0
30 km transport	Truck (including diesel and road) Truck driver	3	3
	Total	70	70

Source: Adapted from Delft University of Technology (Delft UoT), body of work at: http://www
.io.tudelft.nl/onderzoek/onderzoeksprogrammas/technology-transformation/design-for-sustainability
-emerging-markets/sub-theme-4-knowledge-exploration-and-decision-support/.

Table 7.25 Eco-costs (Delft UoT) extra burden of fixed-site plant compared with mobile plant

	Per tonne mixed aggregate		
Item	Quality	Transport	Total
Eco-costs	$17	$2	$20
CO_2 emission	2.1 kg	4.5 kg	6.6 kg

Source: Adapted from Delft University of Technology (Delft UoT), body of work at: http://www
.io.tudelft.nl/onderzoek/onderzoeksprogrammas/technology-transformation/design-for-sustainability
-emerging-markets/sub-theme-4-knowledge-exploration-and-decision-support/.

References and Further Reading

CHAPTER I

1. Available at http://www.washingtonaccord.org/.
2. Available at http://www.architecture.com/NewsAndPress/News/RIBANews/News/2012/RIBAcallsformajorprocurementreform.aspx.

CHAPTER 2

1. Available at http://www.thecommonwealth.org/Internal/191086/142227/Members and http://www.chogm2011.org/CHOGM2011/The_Commonwealth.html.
2. Scott, J. (1991) *The Penguin Dictionary of Civil Engineering*. UK: Penguin.
3. AS/NZS 4536:1999. (1999) Australia/New Zealand Standards. Life-cycle costing—An application guide.
4. Royal Institution of Chartered Surveyors, Building Cost Information Service. (2012) RICS elemental standard form of cost analysis principles, instructions, elements and definitions. Available at http://www.bcis.co.uk/downloads/BCIS_Elemental_Standard_Form_of_Cost_Analysis_4th__NRM__Edition_2012.pdf.
5. Seeley, I. (1996) *Building Economics*. UK: Plagrave Macmillan.
6. David Langdon & Seah International (ed). (2010) *Spon's Asia-Pacific Construction Cost Handbook*.
7. Rawlinsons. (2014) *Rawlinson's Construction Cost Guide*. Australia: Rawlinsons.
8. *Cordell's Building Cost Guide*. Australia: Cordell Building Publications, 2014.
9. Davis Langdon. (2013) *Spon's Civil Engineering and Highway Works Price Book*. UK: Spon's, Taylor & Francis.
10. Franklin + Andrews (2003) *Spon's Railway Construction Price Book*. UK: CRC Press.
11. Royal Institution of Chartered Surveyors, Building Cost Information Service, BCIS Wessex. (2013) UK; Johnson V & Judd P, Laxton's Publishing Limited, 2012, UK.
12. United National Economic and Social Commission for Asia and the Pacific (UNESCAP). (2014) Available at http://www.unescap.org/stat/data/.
13. Australian Bureau of Statistics/International Statistics. (2014) Available at http://www.abs.gov.au/websitedbs/D3310114.nsf/home/Links+to+Other+Statistical+Agencies+&+Related+Sites?opendocument#from-banner=LN.

14. ICE. (2012) *Civil Engineering Standard Method of Measurement: CESMM4*. London: Thomas Telford.

15. Smith, R. C. (2013) *Estimating and Tendering for Building Work*. UK: Routledge.

16. Available at http://www.commerce.wa.gov.au/worksafe/.

17. Available at http://www.ato.gov.au/businesses/pathway.aspx?pc=001/003/103&alias=gst.

18. Australian Institute of Quantity Surveyors (AIQS) (2003) *Australian Cost Management Manual*, Australian Institute of Quantity Surveyors, Canberra.

19. Brook, M. (2004) *Estimating and Tendering for Construction*. UK: Elsevier Butterworth-Heinemann.

20. Walker, I. and Wilkie, R. (2002) *Commercial Management in Construction*. UK: Blackwell Science Ltd.

21. Allen, S., Grant, F., Fortune, C., Marks, P. and Fleming, F. et al. (2008) *Herriot-Watt University Cost Management Module Packs*.

22. Park, C. S. (2011) *Contemporary Engineering Economics*. USA: Prentice-Hall.

23. Vajpayee, S. K. (2001) *Fundamentals of Economics for Engineering Technologists and Engineers*. USA: Prentice-Hall.

24. Sullivan, W. G., Bontadelli, J. A., and Wicks, E. A. (2001). *Engineering Economy*. USA: Prentice-Hall.

25. Available at http://www.pta.wa.gov.au/.

26. Kirk, J. and Dell'Isola, A. J. (1995) *Life Cycle Costing for Design Professionals*, 2nd ed. New York: McGraw-Hill.

27. Whyte, A. (2011) *Life-Cycle Cost Analysis of Built Assets: An LCCA Framework*. Germany: VDM Publishing.

28. Flanagan, R., Norman, G. and Furbur, D. assisted by a Steering Committee appointed by the RICS. (1983) *Life Cycle Costing for Construction*. UK: Royal Institution of Chartered surveyors publications.

29. Available at http://www.bentley.com/en-AU/Products/.

30. Whyte, A. and Pham, A. (2012) Life-cycle cost analysis for infrastructure project pavement design. *Proceedings ASEC-2012*, July 11–13, Perth, Western Australia.

31. Whyte, A. and Pham, A. (2012) Life-cycle analysis to evaluate infrastructure pavement design in Western Australia. *Proceedings of ICCOEE-2012*, June 12–14, UTP, Malaysia, ISBN 978-983-2271-77-2.

32. American Concrete Pavement Association ACPA. (2002) *Life Cycle Cost Analysis: A guide for Comparing Alternative Designs*, Volume 1. Illinois: American Concrete Pavement Association.

33. Barringer, H. P. (2003) A life cycle cost summary. *International Conference of Maintenance Societies*, Perth.

34. Bloomerberg, L. D. and Volpe, M. (2008) *Completing Your Qualitative Dissertation: A Roadmap from Beginning to End*. California: Sage Publications.

35. BTCE. (1990) *Pavement Management: Development of a Life Cycle Costing Technique*, B. o. T. a. C. Economics (ed.). Canberra: Australian Government Publishing Service.

36. Creswell, J. W. (2005) *Educational Research: Planning, Conducting, and Evaluating Quantitative and Qualitative Research*, 2nd ed. New Jersey: Pearson Education.

37. Demos, G. P. (2006) *Life Cycle Cost Analysis and Discount Rate on Pavements for the Colorado Department of Transportation*. Colorado: Colorado Department

of Transportation. Available at http://www.dot.state.co.us/publications/PDFFiles/discountrate.pdf (accessed May 15, 2009).

38. Alliance. (1996) Investment Management and Strategic Financial Advice. Available at http://allianceinvestments.com.au/files/16KH315JV7/Investment%20Wise%20Vol%208%Issue%201.pdf (accessed August 22, 2009).

39. Life Cycle Costing. (1996) National AustStab Guidelines. Available at http://www.auststab.com.au/pdf/nat02.pdf (accessed August 22, 2009).

40. New South Wales Treasury. (2004) *Total Asset Management: Life Cycle Costing Guideline*, 1st ed. Australia: New South Wales Treasury.

41. Sharp, D. R. (1970) *Concrete in Highway Engineering*. Sydney: A. Wheaton and Co.

42. Smith, M. R. and Walls, J. / Federal Highway Administration. (1998) Life cycle cost analysis in pavement design. In *Search of Better Investment Decisions*, Volume 1. Washington: US Department of Transportation.

43. Vanier, D. J. (2001) *Why Industry Need Asset Management Tools*. USA: Institute for Research in Construction. Available at http://irc.nrc-cnrc.gc.ca/pubs/fulltext/nrcc44702/nrcc44702.pdf (accessed July 14, 2009).

44. Wallace, H. A. and Martin, J. R. (1967) *Asphalt Pavement Engineering*. New York: McGraw-Hill Book Company.

45. Words, F. (2007) Consumer Price Index (CPI)—Australia. Available at http://www.fxwords.com/c/consumer-price-index-cpi-australia.html (accessed August 22, 2009).

46. Whyte, A. and Gayner, L. (2013) Life-cycle analysis of infrastructure pavement applications in Western Australia. In *New Developments in Struc. Engineering & Construction*, Yazdani, S. and Singh, A. (eds.), USA. pp. 1331–1336, ISBN-13:978-981-07-5354-2.

47. Ashworth, A. (1999) *Cost Studies of Buildings*. Harlow: Longman.

48. CoC City of Canning WA. (2011) *Road Resurfacing and Rehabilitation Future Directions*. MRRG. WA, Australia.

49. DoD Department of Defence Aus. (1998) *Life-cycle Costing in the Department of Defence*. Australia: Department of Defence publications. ISSN 1036-7632.

50. DoS&R Dept for Sport and Recreation Western Australia. (2005) *Life Cycle Cost Guidelines*. WA Publishing Service.

51. Flanagan, R. and Jewell, C. (2005) *Whole Life Appraisal for Construction*. Oxford, UK: Blackwell Publishing, 182 pp.

52. Olubodun, F., Kangwa, J., Oladapo, A. A. and Thompson, J. (2010) Appraisal of level of application of LCC construction industry of the UK. *Structural Survey*, 28, 4, pp. 254–265, Emerald.

53. Konig, H., Kohler, N., Kreißig, J. and Lutzendorf, T. (2010) *A Life Cycle Approach to Buildings*. Munich: Institut fur Internationale Arch-Dok.

54. USA DoC Dept of Commerce. (1995) *Life-Cycle Costing Manual for Energy Manag Program*. USA: Dept. of Commerce Publications.

55. Villacres, J. (2005) *Pavement Life-cycle Cost Studies Using Actual Cost Data*. UK: Asphalt Pavement Alliance.

56. WAERA Western Australian Economic Regulation Authority. (2006) *A Guide for Preparing the Financial Information Component of an Asset Management Plan*. Australia: WAERA Publications.

57. Whyte, A. (2011) *Life-Cycle Cost Analysis of Built Assets: LCCA Framework*. Germany: VDM, 116 pp, ISBN-10: 3639336364.

58. Whyte, A. and Pham, A. (2012) Life-cycle cost analysis to evaluate pavement design in Western Australia. *Proceeding of ICCOEE2012*, Malaysia, 1569537709.

59. Whyte, A. and Gajda, P. (2011) Life-cycle costing analyses of offshore substructure options: Pile-on-water V floating substructure traditional land option. *Proceedings of Advances in Steel & Aluminium Conference*, Malaysia, pp. 348–359.

60. Dept. of Planning, Perth, Government of Western Australia. (2010) *Perth Waterfront Plan*. WA, Australia: City of Perth Publications.

61. Calafell, J. G. and Bosque, R. P. D. (2001) *The Ratio of International Reserves to Short-term External Debt as an Indicator of External Vulnerability: Some Lessons from the Experience of Mexico and Other Emerging Economies*. Mexico: Banco de México, p. 25.

62. Donald, B., Wright, J. et al. (2001) *Unified Facilities Criteria: Maintenance and Operation of Waterfront Facilities*. USA, U.S. Army Corps of Engineers, p. 219.

63. Ker, P. (2010) Home where the harbour is? An idea worth floating. The Age. National, Fairfax Digital. Australia: The Age Newspaper Publications.

64. Poprzeczny, J. (2010) $300m Waterfront project moving ahead. *West Australian Business News*. Perth: WA Business News.

65. Watanabe, E., Utsunomiya, T. et al. (2004) Hydroelastic analysis of pontoon-type VLFS: A literature survey. *Engineering Structures*, 26, 2, pp. 245–256.

66. Rawlinson's Construction (Costs) Handbook 2010; Cordell's Building Cost Guide 2010. Local materials suppliers including: Rocla Concrete, Next Generation Concreting Services, BGC Cement, Formstruct, In-Situ Construction & Maintenance, Central Systems, Australian Marine Complex Co., & Fremantle Ports Online.

67. Whyte, A. and Scott, D. (2010) Life-cycle costing analysis to assist design decisions: Beyond 3D building information modeling. *ICCCBE2010 & EG-ICE10*, July 1–3, Nottingham, UK.

68. Al-Hajj, A., Whyte, A. and Aouad, G. (2000) The development of a framework for life-cycle costing in object-orientated & VR technologies. *ALES 200 Annual Convention*, July 20–22, Beirut, Lebanon.

69. Kishk, M. and Al-Hajj, A. (1999) An integrated framework for life-cycle costing in building. *Proceedings of RICS COBRA 1999*, September 1–2, UK: Univ of Salford.

70. Whyte, A. and Al-Hajj, A. (2000) Participation as the synthesis for computer-integrated-construction and innovation in design. *International Journal of Construction Information Technology*, 8, 2, Winter, pp. 51–62.

CHAPTER 3

1. Snook, K. (1995) *CPI—Co-ordinated Project Information*. UK: Chartered Institution of Building.

2. Whyte, A. (2011) *Life-Cycle Cost Analysis of Built Assets: LCCA Framework*. VDM, 116 pp, Chapter 7, ISBN-10: 3639336364, ISBN-13: 978-3639336368, Germany.

3. Institution of Civil Engineers. (2012) *Civil Engineering Standard Method of Measurement*. UK: CE Publications.

4. Available at http://www.primavera.com/; http://www.oracle.com/us/products /applications/primavera/overview/index.html; http://www.artemispm.com/lang _en/default.asp; http://www.microsoft.com/office/project/default.htm; http://www .bentley.com/en-AU/Products/.

5. Wilson, J. (2003) Gantt charts: A centenary appreciation. James M. Wilson. *European Journal of Operational Research*, 149, 2, pp. 430–437.

6. Calvert, R. (1995) *Introduction to Building Management*. UK: Butterworth-Heinemann.

7. Heizer, J. and Render, B. (2010) *Principles of Operations Management*. USA: Prentice-Hall.

8. Sullivan, W., Bontadelli, J. and Wicks, E. (2002) *Engineering Economy*. USA: Prentice-Hall.

9. Larson, E. and Gray, C. (2010) *Project Management the Managerial Process*. USA: McGraw-Hill.

10. Hartley, S. (2008) *Project Management Principles Processes and Practice*. Australia: Pearson Education Australia.

11. Nicholas, J. and Steyn, H. (2011) *Project Management for Engineering, Business, and Technology*. USA: Butterworth-Heinenmann.

12. Available at http://www.commerce.wa.gov.au/worksafe/.

13. Available at http://www.bom.gov.au/.

14. Available at http://www.bbc.co.uk/news/business-21852427; http://www.bbc .co.uk/news/business-22549710.

15. Pierce, D. (2004) *Project Scheduling and Management for Construction*. USA: RSMeans.

16. Oberlender, G. (1993) *Linear Scheduling of Highway Construction Projects*. Project management for engineering and construction, USA: McGraw-Hill.

17. Smythe, C. (2012) *The Continuity of Method Statements Throughout the Tendering and Construction Stages of A Project*. Unpublished Research Report, A. Whyte (ed.), WA: Curtin University.

18. Ashworth, A. (2002) *Pre-Contract Studies*, 2nd ed. Oxford: Blackwell Science.

19. Australian and New Zealand Standard. (2009) *Risk Management - Principles and Guidelines*.

20. AS/NZS ISO31000:2009. (2009) Standards Australia Online. Available at http://www.saiglobal.com.

21. Bennett, F. L. (2003) *The Management of Construction: A Project Life Cycle Approach*. Oxford: Butterworth-Heinemann.

22. Borys, D. (2012) The role of safe work method statements in the Australian construction industry. *Safety Science*, 50, pp. 210–220.

23. Calvert, R. E., Bailey, G. and Coles, D. (1995) *Introduction to Building Management*, 6th ed. Oxford: Laxton's.

24. Cooke, B. and Williams, P. (1998) *Construction Planning, Programming and Control*. Hampshire: Palgrave.

25. Dolan, S. P., Williams, G. J., Barrows, W. J., Dickson, J. W., Torry, D. and Drury, R. F. (2003) Performance improvement techniques used on a goodwyn a platform, North-west Shelf, Australia. *SPE Drilling and Completions Journal*, 18, 2, pp. 138–145.

26. Flanagan, R. and Norman, G. (1993) *Risk Management and Construction.* Oxford: Blackwell Scientific Publications.
27. Illingworth, J. R. (1993) *Construction Methods and Planning.* London: E and FN Spon.
28. Mak, S. and Picken, D. (2000) Using risk analysis to determine construction project contingencies. *Journal of Construction Engineering Management*, 126, 2, pp. 130–136.
29. Mak, S., Wong, J. and Picken, D. (1998) The effect on contingency allowances of using risk analysis in capital cost estimating: A Hong Kong case study. *Construction Management and Economics*, 16, pp. 615–619.
30. Mawdesley, M., Askew, W. and O'Reilly, M. (1997) *Planning and Controlling Construction Projects: The Best Laid Plans.* Essex: Adison Wesley Longman.
31. Oxley, R. and Poskitt, J. (1996) *Management Techniques Applied to the Construction Industry*, 5th ed. Oxford: Blackwell Science.
32. Potts, K. F. (1995) *Major Construction Works: Contractual and Financial Management.* Essex: Longman Group.
33. Smith, A. J. (1995) *Estimating, Tendering and Bidding for Construction.* London: Macmillan Press.
34. Smith, N. J. (1999) *Managing Risk in Construction Projects.* Oxford: Blackwell Science.
35. Smith, R. C. (1986) *Estimating and Tendering for Building Work.* Essex: Pearson Education.
36. St John, Holt, A. (2001) *Principles of Construction Safety.* Oxford: Blackwell Science.
37. Thompson, P. and Perry, J. (1992) *Engineering Construction Risks: A Guide to Project Risk Analysis and Risk Management.* London: Thomas Telford.
38. Tweeds. (1996) *Laxton's Guide to Risk Analysis and Management.* Oxford: Laxton's.
39. Uher, T. E. and Zantis, A. S. (2011) *Programming and Scheduling Techniques*, 2nd ed. Sydney: University of New South Wales Press.
40. Whyte, A. and Cammarano, C. (2012) Value management in infrastructure projects in Western Australia: Techniques and staging. In *Procs 28th Annual ARCOM Conference*, Smith S. D. (ed.), September 3–5, Edinburgh, UK, pp. 797–806, ISBN: 978-0-9552390-6-9 (2-vols).
41. Armstrong, M. (1992) *Management Processes and Functions.* London: Institute of Personnel and Development.
42. Arditi, D. (2002) Constructability analysis in the design firm. *Journal of Construction Engineering and Management*, 128, 2, pp. 117–126.
43. Fong, S. W. and Ashworth, A. (1997) Cost engineering research in the Pacific Basin–A value based research framework. *The Australian Institute of Quantity Surveyors*, Reffered Journal, 1, 1, pp. 6–10.
44. Bowen, P. (2010) The awareness and practice of value management by South African consulting engineers: Preliminary research survey findings. *International Journal of Project Management*, 28, pp. 285–295.
45. Cheah, C. Y. J. (2005) Appraisal of value engineering in construction in Southeast Asia. *International Journal of Project Management*, 23, pp. 151–158.
46. Flyvbjerg, B. (2008) Curbing optimism bias and strategic misrepresentation in planning. *European Planning Studies*, 16, 1, pp. 3–21.

47. Harvey, D. I. (2008) *Innovation in Design/Build Bridges*. Canada: Associated Engineering.

48. IVM: Institute of Value Management. (2011) *Value Management Techniques*. Available at http://www.ivm.org.uk/techniques.php.

49. Kelly, J., Male, S. and Graham, D. (2004) *Value Management of Construction Projects*. UK: Blackwell Science.

50. Leeuw, C. P. (2001) Value management: An optimum solution. *International Conference on Spatial Information for Sustainable Development*, Kenya.

51. Norton, B. and McElligott, W. C. (1995) *Value Management in Construction: Practical Guide*. UK: Macmillan.

52. Palmer, D. M. A. (2002) *Construction Management: New Directions*, 2nd ed. Oxford: Blackwell Publishing.

53. Qiping, S. (2004) Applications of value management in the construction industry in China. *Engineering Construction and Architectural Management*, 11, pp. 9–19.

54. Selg, R. A. (2006) Value engineering for hazardous waste projects. *AACE International Transactions,* 1, pp. 4060–4070.

55. Smith, J. (1998) *Building Cost Planning for the Design Team*. Australia: UNSW Press.

56. Spaulding, W. (2005) The use of function analysis as the basis of value management in the Australian construction industry. *Construction Management and Economics*, 23, 7, pp. 723–731.

57. Stewart, R. B. (2010) *Value Optimization for Project and Performance Management*. Hoboken, New Jersey: Wiley.

58. Wixson, J. and Heydt, H. J. (1991) The human side of value engineering. *Society of American Value Engineers*, 1, pp. 30–38.

59. Whyte, A. and Crew, A. (2011) Critical-chain project management in Western Australia: Towards construction project duration reduction. *RICS Construction & Property Conference*, UK, September 12–13.

60. Australian Constructors Association Report. (2008) *Scope for Improvement*. Australia: Blake Dawson Publications.

61. Goldratt, E. (1997) *Critical Chain*. NY, USA: North River Press.

62. Goldratt, E. (1984) *Theory of Constraints: The Process of Ongoing Improvement*. NY, USA: North River Press.

63. Maylor, H. (2002) *Project Management*, 3rd. ed. London: Prentice Hall.

64. Srinivasan, M., Best, W. and Sridhar, C. (2007) Warner Robins air logistics centre streamlines aircraft repair and overhaul. *Interfaces*, 37, 1, pp. 7–21. Available at http://interfaces.journal.informs.org/cgi/reprint/37/1/7 (accessed August 12, 2010).

65. Uher, T. (2003) *Programming and Scheduling Techniques–Construction Management*. New South Wales, Australia: UNSW Press.

66. Creswell, J. (2007) *Qualitative Inquiry and Research Design: Choosing Among Five Approaches*, 2nd. ed. Thousand Oaks: Sage Publications.

67. Abdulaziz, A. and Whyte, A. (2012) Synthesis of trad. project eng. man. in construction projects with agile approaches towards efficiency gains. In *Research, Dev. & Practice in Structural Engineering & Con.*, Vanissorn V. et al. (eds.), Singapore: Research Publishing Services, pp. 921–926.

68. Cadle, J. and Yeates, D. (2004) *Project Management for Information Systems*. UK: Prentice Hall.

69. Chin, G. (2004) *Agile Project Management: How to Succeed in the Face of Changing Project Requirements*. USA: Amacom Books.

70. Cleden, D. (2009) *Managing Project Uncertainty*. Farnham, England; VT: Gower.

71. Eastman, C., Teicholz, P., Sacks, R. and Liston, K. (2008) Frontmatter. In *BIM Handbook: A Guide to Building Information Modeling*. Hoboken, New Jersey: John Wiley & Sons, Inc.

72. Flyvbjerg, B., Bruzelius, N. et al. (2003) *Megaprojects and Risk: An Anatomy of Ambition*. UK: Cambridge Univ Press.

73. Goodpasture, J. C. (2009) *Project Management the Agile Way: Making it Work in the Enterprise*. USA: J. Ross Publishing.

74. Highsmith, J. (2003) Agile project management: Principles and tools. Agile Project Management Advisory Service Executive Report. USA: Cutter Consortium Publications.

75. Liker, J. K. and Online, S. B. (2004) *The Toyota Way: 14 Management Principles from the World's Greatest Manufacturer*. New York: McGraw-Hill.

76. Miller, R., Lessard, D. and IMEC Research Group. (2000) *The Strategic Management of Large Engineering Projects: Shaping Institutions, Risks, and Governance*. USA: MIT Press, 237 pp.

77. Naim, M. and Barlow, J. (2003) An innovative supply chain strategy for customized housing. *Construction Management and Economics*, 21, 6, pp. 593–602.

78. Owen, R. and Koskela, L. (2006b) An Agile Step Forward in Project Management. *2nd Specialty Conference on Leadership and Management in Construction and Engineering*, May 4–6, 2006 Grand Bahama. CIB/ASCE.

79. Owen, R. L. and Koskela, L. (2006a) Agile Construction Project Management. *6th International Postgraduate Research Conference in the Built and Human Environment*, 6/7 April 2006 Delft, Netherlands. Research Institute for the Built and Human Environment, University of Salford.

80. Ribeiro, F. L. and Fernandes, M. T. (2010) Exploring agile methods in construction small and medium enterprises: A case study. *Journal of Enterprise Info Management*, 23, 2, pp. 161–180.

81. U.S. Department of Transportation, F. H. A. (2005) Public Roads Megaprojects - Are you ready? FHWA-HRT-2006-002, http://www.fhwa.dot.gov/publications /publicroads/06jan/02.cfm, accessed 2013.

82. Whyte A. and Scott D. (2010) Life-cycle costing analysis to assist design decisions: Beyond 3D building information modelling. In *Computing in Civil and Building Engineering*, Tizani W. (Ed), p.173, UK: Nottingham University press.

83. Toussia, M. and Whyte, A. (2012) Factors influencing delay in construction civil engineering projects in the Persian Gulf Countries. In *Research, Dev & Practice in Structural Engineering & Constrn.*, Vanissorn V. et al. (eds.), Singapore: Research Publishing Services, pp. 1011–1014.

84. Assaf, S. A. and Al-Hejji, S. (2006) Causes of delay in large construction projects. *International Journal of Project Management*, 24, 4, pp. 349–357.

85. Crosthwaite, D. (2000) The global construction market: A cross-sectional analysis. *Construction Management and Economics*, 18, 5, pp. 619–627.

86. Baloia, D. and Price, A. (2001) Modelling global risk factors affecting construction cost performance. *International Journal of Project Management*, 21, 4, pp. 261–269.

87. El-Sayeghb, A. S. F. S. M. (2006) Significant factors causing delay in the UAE construction industry. *Construction Management and Economics*, 24, 11, pp. 1167–1176.

88. Lopes, J. (2003) The relationship between construction outputs and GDP: Long-run trends from Portugal. *Association of Researchers in Construction Management*, 1, pp. 309–317.

89. Raz. (2011) Afazaayesh-e-Budge-ye-Umrani; natije-ye-sabok-shodan-e-baar-e -yaaraane-haa. Available at http://raznews.ir/fa/pages/?cid=18029.

90. Shada. (2011) Afzaayesh-e-20 darsadi-ye-budje-ye-omrani-ye-keshvar dar saal e 89. Available at http://shada.ir/2559-fa.html (accessed April 11, 2011).

91. Shazly, S. E. H. M. R. E. (2012) Oil dependency, export diversification and economic growth in the Arab Gulf States. *European Journal of Social Sciences*, 29, 3, pp. 397–404.

92. Shiri. (2010) Vakaavi-ye-dalaayel-e-ta'khir dar ejra-ye-proje-haaye-omraani-yeostaan. Khorasan. Mashha: Persian Publications.

CHAPTER 4

1. Available at http://www.iso.org/iso/home.html.

2. Available at http://www.jas-anz.com.au/.

3. Available at http://www.abcb.gov.au/~/media/Files/Download%20Documents /Product%20Certification/codemark_scheme_rules_v2009-1_27-03-20091 .pdf.

4. Available at http://www.abcb.gov.au/~/media/Files/Download%20Documents /Product%20Certification/codemark_scheme_rules_v2009-1_27-03-20091 .pdf & Building Code of Australia.

5. Deming, W. E. (2000) *Out of the Crisis*. USA: MIT Press.

6. AS/NZS ISO 90001:2008; ISO 9000-1 - Quality Management & Quality Assurance Standards – Guidelines for Selection & Use.

7. AS/NZS ISO 90001:2008.

8. Available at http://www.abcb.gov.au/; http://www.abcb.gov.au/; https://services .abcb.gov.au/abcbshop/; http://www.abcb.gov.au/~/media/Files/Download%20 Documents/Product%20Certification/codemark_scheme_rules_v2009-1_27 -03-20091.pdf.

9. WFTAO (World Federation of Technical Assessment Organisations).

10. Available at http://www.dfat.gov.au/facts/legal_system.html.

11. AS4000-1997. General Conditions of Contract; AS2124-1992. General Conditions of Contract.

12. Available at http://www.epa.gov/lean/environment/methods/kanban.htm.

13. Available at http://www.bechtel.com/six_sigma.html.

14. Work-Safe-Australia. Government of Australia. (2011) Available at http://www .safeworkaustralia.gov.au/sites/SWA/about/Publications/Documents/686 /AnnualNotifiedFatalitiesReport2010-11.pdf.

15. Work-Safe-Australia. Government of Australia. (2011). Available at http:// www.safeworkaustralia.gov.au/sites/swa/about/publications/pages/notified fatalitiesmonthlyreport.

16. The National Standard for Construction Work was declared by the NOHSC, in accordance with section 38 of the National Occupational Health and Safety Commission Act 1984 (Commonwealth), on April 27, 2005.

17. Available at http://www.safeworkaustralia.gov.au.

18. Nikraz, O., Dejahang, M. and Vimonsatit, V. (2009) Construction Industry Safety: Fall and Fall Protection issues in the Building and Construction Industry: A Worldwide Issue. *Proceedings for 2nd International Conference of Health Safety Environment, 1–2nd November, 2009, Tehran, Iran.*

19. Available at http://www.structural-safety.org/.

20. Available at http://www.structural-safety.org/how-to-report/.

21. Available at http://www.ice.org.uk/topics/structuresandbuildings/Related-Groups.

22. Available at http://www.structural-safety.org/view-report/cross410/.

23. Whyte, A. and Brandis, P. (2011) Concrete construction safety: Investigating (Western) Australian formwork practice. *Proceedings of RICS COBRA*, September 12–13, School of the Built Environment, University of Salford, pp. 1024–1032.

24. Whyte, A. and Brandis, P. (2012) Concrete formwork: An investigation of safe local practice. *Proceedings of ASEC-2012*, July 11–13, Perth, Australia.

25. Engineers Australia EA. (2010) *Code of Ethics*. Australia: Institute of Engineers Australia.

26. Standards Australia SA. (2010a) *Part One: Documentation and Surface Finish*. AS3610.1-Formwork for Concrete. Australia: Standards Australia International Ltd.

27. Ferguson, S. and Crawford, D. (2010) *Formwork Procedures: Formwork Design Handbook*. Unpublished Development Committee Report. Australia: Standards Australia International Ltd.

28. Pallet, P. (2001) *Investigations into Aspects of Falsework*. UK: University of Birmingham.

29. Hurd, M. K. (2005) *Formwork for Concrete*, 7th ed. USA: American Concrete Institute.

30. Hinze, J. (2000) The need for academia to address construction site safety through design. *Construction Congress VI*, February 20–22, American Society of Civil Engineers.

31. Travers, I. (2009) Caged ladders – A case of mistaken identity. Available at http://www.safetysolutions.net.au (accessed September 16, 2010).

32. Main Roads Western Australia MRWA. (1966) A report on the collapse of falsework at Welshpool road overpass. Curtin University. Available at http://www.library.curtin.edu.au (accessed April 2, 2010).

33. Snowy Mountains Engineering Corporation SMEC. (2010) Structural engineer's report on the collapse of falsework. Department of Territory and Municipal Services. Available at http://www.tams.act.gov.au (accessed September 8, 2010).

34. State Coroner Victoria. (2000) Case No: 3315/100. Coroner's Court of Victoria, AUS. Available at http://www.coronerscourt.vic.gov.au (accessed September 18, 2010).

35. Construction, Forestry, Mining and Energy Union of Western Australia CFMEU. (2009) *Report–Christchurch Grammar School*. Australia: North West Corner Development, CFMEU–Construction and General. Available at http://www.cfmeu.asn.au (accessed March 16, 2010).

36. Office of the Federal Safety Commissioner OFSC. (nd) Formwork incident. *Safety and Health Lessons Learnt*. Australia: Department of Education, Department of Education, Employment and Workplace Relations. Available at http://www.fsc.gov.au (accessed August 21, 2010).

37. Caswell, A. (2010) *Personal Communication with Standards Australia Senior Project Management*. SA/Curtin U., unpublished, August 15.

38. Standards Australia SA. (2010b) *Pathways for Standards Development*. Australia: Standards Australia International Ltd. Available at http://www.standards.org.au (accessed September 16, 2010).

39. Ferguson, S. (2010b) *Personal Communication with the SA Co-Authors of Formwork Design Handbook*. SA/Curtin U., unpublished, October 1.

40. Sparvell, A. (2010) *Personal Communication with the Engineers Australia National Industry Relations Management*. EA/Curtin U., unpublished, October 6.

41. AS 3610-1995, Formwork for concrete; Under Revision: DR 02319 CP; DR 05029 CP; DR 08167; Superseded By: AS 3610.1-2010 (in part); Supersedes: AS 3610-1990, DR 93275; Amendments: AS 3610-1995/Amdt 1-2003; Supplements: AS 3610 Supp 1-1995; AS 3610 Supp 2-1996 http://infostore.saiglobal.com/store/Details.aspx?ProductID=298503&gclid=CIanm6KDw7k CFed_QgodnywAnQ.

42. Faisal, A. and Whyte, A. (2012) Towards assessing productivity in off-site building methods for engineering and construction projects. In *Research, Dev & Practice in Structural Engineering & Constrn*, Vanissorn V. et al. (eds.), Singapore: Research Publishing Services. pp. 915–920.

43. Abdel-Wahab, M. et al. (2011) Trends of productivity growth in the construction industry across Europe, US and Japan. *Construction Management and Economics*, 29, pp. 635–644.

44. Alvanchi, A. et al. (2011) Off-site construction planning using discrete event simulation. *Journal of Architectural Engineering*, 1, p. 26.

45. Bernstein, H. M. et al. (2011) *Prefabrication and Modularization Increase Productivity*. Details Outlined in New McGraw-Hill Construction Report. PR Newswire.

46. Blismas, N. and Wakefield, R. (2009) Drivers, constraints and the future of offsite manufacture in Australia. *Construction Innovation*, 9, pp. 72–83.

47. Blismas, N. et al. (2005) Constraints to the use of off-site production on construction projects. *Architectural Engineering and Design Management*, 1, pp. 153–162.

48. Construction Industry Institute. (2002) *Develop a Decision Support Tool for Prefabrication, Preassembly, Modularization, and Off-Site Fabrication*. USA: Construction Institute publications CII Publication No. RR171-12.

49. Doloi, H. (2007) Twinning motivation, productivity and management strategy in construction projects. *Engineering Management Journal; EMJ*, 19, p. 30.

50. Durdyev, S. and Mbachu, J. (2011) On-site labour productivity of New Zealand construction industry: Key constraints and improvement measures. *The Australasian Journal of Construction Economics and Building*, 11, p. 18.

51. Eastman, C. M. and Sacks, R. (2008) Relative productivity in the AEC industries in the US for on-site and off-site activities. *Journal of Construction Engineering & Management*, 134, pp. 517–526.

52. Egan, J. (1998) Rethinking construction. Construction Task Force Report. London: HMSO.

53. Enshassi, A. et al. (2007) Factors affecting labour productivity in building projects in the Gaza strip. *Journal of Civil Engineering and Management*, 13, pp. 245–254.

54. Finnimore, B. (1989) *Houses from the Factory: System Building and the Welfare State 1942–1974*. London: Rivers Oram Press.

55. Gibb, A. and Isack, F. (2003) Re-engineering through pre-assembly: Client expectations and drivers. *Building Research & Information*, 31, pp. 146–160.

56. Gibb, A. G. F. (1999) Off-site fabrication: Prefabrication, pre-assembly and modularisation. UK: Whittles Publishing.

57. Gibb, A. F. (2001) Standardization and pre-assembly- distinguishing myth from reality using case study research. *Construction Management and Economics*, 19, pp. 307–315.

58. Goodier, C. and Gibb, A. (2007) Future opportunities for offsite in the UK. *Construction Management and Economics*, 25, pp. 585–595.

59. Goodier, C. I. and Gibb, A. G. F. (2005) The value of the UK market for offsite. Buildoffsite. Available at http://www.buildoffsite.com (accessed April 9, 2012).

60. Hampson, K. D. and Brandon, P. (2004) Construction 2020-A vision for Australia's Property and Construction Industry. Brisbane, Australia: CRC Construction Innovation.

61. Housing Forum. (2002) *Homing in on Excellence – A Commentary on the Use of Offsite Fabrication Methods for the UK Housebuilding Industry*. London: Housing Forum.

62. Lu, N. and Liska, R. W. (2008) Designers' and general contractors' perceptions of offsite construction techniques in the United State construction industry. *International Journal of Construction Education and Research*, 4, pp. 177–188.

63. Mohamed, S. et al. (2008) Trends of skills and productivity in the UK construction industry. *Engineering, Construction and Architectural Management*, 15, p. 372.

64. Nadim, W. et al. (2010) Offsite production in the UK: The way forward? A UK construction industry perspective. *Construction Innovation*, 10, pp. 181–202.

65. Pan, W. et al. (2007) Perspectives of UK housebuilders on the use of offsite modern methods of construction. *Construction Management and Economics*, 25, pp. 183–194.

66. Pasquire, C. and Gibb, A. (2002) Considerations for assessing the benefits of standardisation and pre-assembly in construction. *Journal of Financial Management of Property & Construction*, 7, pp. 151–161.

67. Shen, Z. et al. (2011) Comparative study of activity-based construction labor productivity in the United States and China. *Journal of Management in Engineering*, 27, pp. 116–124.

68. Song, L. and Abourizk, S. M. (2008) Measuring and modeling labor productivity using historical data. *Journal of Construction Engineering & Management*, 134, pp. 786–794.

69. Tatum, C. B. and Vanegas, J. A. (1987) *Constructability Improvement Using Prefabrication, Preassembly, and Modularization*. Modularization (CII Publication No. SD-25). Austin: Construction Industry Institute, The University of Texas at Austin.

70. Taylor, M. D. (2010) A definition and valuation of the UK offsite construction sector. *Construction Management and Economics*, 28, pp. 885–896.

71. Venables, T. et al. (2004) *Manufacturing Excellence: UK Capacity in Offsite Manufacturing*. London: The Housing Forum.
72. Whyte, A. and Bikaun, G. (2012) Design specification decision-making in rural Australia: Precast V in-situ concrete. *Proceedings of ICCOEE-2012*, June 12–14, UTP, KL, Malaysia.
73. Welfare, S. (2009) Interview on Construction Resource Management. Perth, June 30.
74. Richardson, J. G. (1991) *Quality in Precast Concrete - Design, Production and Supervision*. London: Longman Group.
75. Bosak, A. (2009) Interview on Construction Resource Management. Perth, July 22.
76. Tashakkori, A. and Teddlie, C. (1998) *Mixed Methodology, Applied Social Research Methods Series*. London: SAGE Publications.
77. Olomolaiye, P. O., Harris, F. C. and Jayawardane, A. K. W. (1998) *Construction Productivity Management*. England: Addison Wesley Longman, Chartered Institute of Building.
78. Australian Constructors Association Report. (2008) *Scope for Improvement*. Australia: Blake Dawson Publications.
79. National Bureau of Economic Research. (2009) Available at http://www.nber .org/ (accessed May 13, 2009).
80. Richardson, J. G. (1983) *Precast Concrete Production*. UK: Viewpoint publications.
81. Illingworth, J. R. (2000) *Construction Methods and Planning*, 2nd ed. Canada: EFN SPON.
82. Harris, F. and McCaffer, R. (2001) *Modern Construction Management*, 5th ed. UK: Blackwell Publishing.
83. Faniran, O. (2005) *Engineering Project Management*. Australia: Pearson Education.
84. Project Management Institute (PMI) (2004) *A Guide to the Project Management Body of Knowledge*, 3rd ed., Pennsylvania, USA: Project Management Institute, Inc.
85. Kazaz, A. and Ulubeyli, S. (2007) Drivers of productivity among construction workers: A study in a developing country. *Building and Environment*, 42, 5, pp. 2132–2140. Available at http://www.sciencedirect.com/science/article/B6V23-4K5HWB1-1/2/6d4947c7ca6738a68c6fc349e4c725f6 (accessed May 12, 2009).
86. Navon, R. (2005) Automated project performance control of construction projects. *Automation in Construction*, 14, 4, pp. 467–476. Available at http://www.sciencedirect.com/science/article/B6V20-4DVW5YY-1/2/847f9a2c81a7 5fa40361a873a4f3fd18 (accessed April 21, 2009).
87. Creswell, J. W. (1995) *Research Design: Qualitative and Quantitative Approaches*. Thousand Oaks, CA: SAGE.
88. Almusharrafa, A. and Whyte, A. (2012) Defects prediction towards efficiency gains in construction projects. In *Research, Dev & Practice in Structural Engineering & Constrn*, Vanissorn V. et al. (eds.), Singapore: Research Publishing Services. pp. 985–990.
89. Mills, A., Love, P. and Williams, P. (2009) Defect costs in residential construction, *Journal of Construction Engineering and Management*, ASCE, 135: 1, 12.
90. Bogus, S. et al. (2006) Strategies for overlapping dependent activities. *Construction Management & Economics*, 24, 8, pp. 829–837.

91. Bogus, S. et al. (2011) Simulation of overlapping design activities. *Journal of Construct Engineering and Management*, 137, 11, pp. 950–957.
92. Bogus, S. et al. (2005) Concurrent engineering approach to reducing delivery time. *Journal of Construct Engineering and Management*, 131, 11, pp. 1179–1185.
93. Burati, B. et al. (1992) Causes of quality deviations in design and construction. *Journal of Construct Engineering and Management*, 118, 1, p. 16.
94. Cheetham, D. W. (1973) Defects in modern buildings. *Building*, 225, pp. 91–94.
95. Dehghan, R. et al. (2011) Mechanism of design activity overlapping in construction projects and the time-cost tradeoff function. *Procedia Engineering*, 14, pp. 1959–1965.
96. Fayek, A. et al. (2003) *Measuring and Classifying Construction Field Rework* (C. a. E. Engineering, Trans.). Edmonton, Alberta, Canada: University of Alberta.
97. Forbes, L. H. and Ahmed, S. M. (2011) *Modern Construction: Lean Project Delivery and Integrated Practives*. Boca Raton, Florida: Taylor & Francis Group.
98. Krishnan, V. et al. (1997) A model-based framework to overlap product development activities. *Management Science*, 43, 4, pp. 437–451.
99. Krishnan, V. et al. (1995) Accelerating product development by the exchange of preliminary information. *Journal of Mechanical Design*, 117, 4, pp. 491–498.
100. Knocke, J. (1992) *Post Construction Liability and Insurance*. London: E & FN Spon.
101. Josephson, P.-E. et al. (1999) The causes and costs of defects in construction 7 building projects. *Automation in Construction*, 8, p. 7.
102. Love, P. E. et al. (1997) Understanding rework in construction. *Construction Process Re-Engineering*. Australia: Griffith University, pp. 269.
103. Love, P. E. & Li, H. (2000) Quantifying the causes and costs of rework in construction. *Construction Management and Economics*, 18, 4, pp. 479–490.
104. Love, P. E. et al. (2000) Modelling dynamics of design error induced rework in construction. *Construction Management and Economics*, 18, 5, p. 9.
105. Love, P. E. and Smith, J. (2003) Bench-marking, bench-action: Rework. *ASCE Journal of Management in Engineering*, 19, 4, pp. 147–159.
106. Mills, A. et al. (2009) Defect costs in residential construction. *Journal of Construction Engineering & Manage*, 135, 1, pp. 12–16.
107. Said, M. (2009) Improved schedule analysis considering rework impact and optimum delay mitigation. Canada: Univ of Waterloo. MSc.
108. Peña-Mora, F. et al. (2003) *Introduction to Construction Dispute Resolution*. USA: Prentice Hall.
109. Roemer, T. A., Ahmadi, R. et al. (2000) Time-cost tradeoffs in overlapped product development. *Operations Research*, 48, 6, pp. 858–865.
110. Terwiesch, C. et al. (1999) Measuring the effectiveness of overlapping development activities. *Management Science*, 45, 4, pp. 455–465.

CHAPTER 5

1. Bailey, I. (1998) *Construction Law in Australia*. Australia: Law Book Co of Australia.
2. Available at http://www.thecommonwealth.org/.

3. Royal Institute of British Architects (2012) Building Ladders of Opportunity, UK: RIBA Publications, (http://www.architecture.com/Files/RIBAHoldings /PolicyAndInternationalRelations/Policy/PublicAffairs/2012/BuildingLadders ofOpportunity.pdf)

4. AS 2125. (1992) Tender form (OFFER).

5. Available at http://www.jctltd.co.uk/.

6. Hackett, M., Robinson, I. and Statham, G. (2007) *The Aqua Group Guide to Procurement, Tendering & Contract Administration.* UK: Blackwell Publishing.

7. The Joint Contracts Tribunal Limited. (2011) Deciding of the Appropriate JCT Contract by the Joint Contracts Tribunal of the UK; Relationship Contracting: Optimising Project Outcomes by the Australian Constructors Association. UK: Sweet & Maxwell/ Thomson Reuters (Professional) UK Limited.

8. AS 4120. (1994) Code of Tendering from Australian Standards.

9. Available at http://www.apcc.gov.au/SitePages/Home.aspx; http://australia.gov .au/topics/business-and-industry/industry/building-and-construction-industry.

10. Available at http://www.constructors.com.au/publications/tendering_guide /ACA%20Guidelines%20for%20Tendering%2011%20August%202006%20 Final%20.pdf.

11. Available at http://www.constructors.com.au/publications/pp_partnerships /22175%20PP%20Partnerships%2024pp%20d5P0.pdf.

12. Archived documents from the UK Governmental website, http://webarchive .nationalarchives.gov.uk/20100503135839/http:/www.ogc.gov.uk/ppm_docu ments_construction.asp.

13. Archived documents from the UK Governmental website, http://webarchive .nationalarchives.gov.uk/20100503135839/http:/www.ogc.gov.uk/documents /CP0066AEGuide6.pdf.

14. Bailey, I. and Bell, M. (2008) *Understanding Australian Construction Contracts*, Australia: Lawbook Co. Thomson Reuters.

15. Cooke, J. (2001) *Architects, Engineers and the Law.* Australia: Federation Press.

16. Uher T., (2010) *Fundamentals in Building Contract Management*, Australia: University of New South Wales Press.

17. Available at http://www.neccontract.com/international/New_Zealand.asp.

18. Available at http://fidic.org/; http://fidic.org/node/149.

19. Available at http://web.worldbank.org.

20. Available at http://web.worldbank.org/WBSITE/EXTERNAL/PROJECTS /PROCUREMENT/0,pagePK:84271~theSitePK:84266,00.html; http://web .worldbank.org/WBSITE/EXTERNAL/PROJECTS/PROCUREMENT/0,content MDK:20103444~isCURL:Y~menuPK:84284~pagePK:84269~piPK:6000155 8~theSitePK:84266,00.html.

21. AS 2124. (1992) General Conditions of Contract.

22. AS 4000. (1997) General Conditions of Contract AS4000, Australian Standards.

23. Knowles, R. (2005) *150 Contractual Problems and their Solutions*, UK: Wiley.

24. Available at http://www.bom.gov.au/.

25. Available at http://www.slp.wa.gov.au/legislation/statutes.nsf/main_mrtitle_2320 _homepage.html; http://www.epa.wa.gov.au/; http://www.austlii.edu.au/au /legis/wa/consol_act/epa1986295/; http://www.epa.wa.gov.au/docs/1139_EIA _Admin.pdf.

26. Whyte, A. and Macpherson, E. (2011) Standard forms of contract selection criteria: A qualitative analysis of the Western Australian construction industry. *Proceedings of RICS COBRA*, September 12–13, School of the Built Environment, University of Salford, UK: RICS Publication. pp. 651–661.
27. Australian Constructors Association Report. (2008) *Scope for Improvement*. Australia: Blake Dawson Publications.
28. Turner, A. (1997) *Building Procurement*, 2nd ed. London: Macmillan.
29. Cox, S. and Clamp, H. (1999) *Which Contract?: Choosing an Appropriate Building Contract*, 2nd ed. London: RIBA Publications.
30. Chua, D. and Loh, P. (2006) CB-contract: Case-based reasoning approach to construction contract strategy formulation. *Journal of Computing in Civil Engineering*, 20, 5, pp. 339–350.
31. Wang, W., Hawwash, K. and Perry, J. (1996) Contract type selector (CTS): A KBS for training young engineers. *International Journal of Project Management*, 14, 2, pp. 95–102.
32. Luu, D., Ng, S. and Chen, S. (2003) Parameters governing the selection of procurement system-An empirical survey. *Engineering, Construction and Architectural Management*, 10, 3, pp. 209–218.
33. Oyetunji, A. and Anderson, S. (2006) Relative effectiveness of project delivery and contract strategies. *Journal of Construction Engineering and Management*, 132, 1, pp. 3–13.
34. Love, P., Skitmore, M. and Earl, G. (1998) Selecting a suitable procurement method for a building project. *Construction Management and Economics*, 16, pp. 221–233.
35. Wong, K. and So, A. (1995) A fuzzy expert system for contract decision making. *Construction Management and Economics*, 13, pp. 95–103.
36. Skitmore, R. and Marsden, D. (1988) Which procurement system? Towards a universal procurement selection technique. *Construction Management and Economics*, 6, pp. 71–89.
37. Creswell, J. (2007) *Qualitative Inquiry and Research Design: Choosing Among Five Approaches*, 2nd. ed. Thousand Oaks: Sage Publications.
38. Macpherson, E. (2011) *Standard Forms of Contract Selection Criteria: Qualitative Analysis of the WA Construction Industry*. Unpublished Research Report, A. Whyte (ed.), Perth WA: Curtin University.
39. Atmodjo, T. S. (2011) *Examination of Forms of Contract for Civil Engineering: Standard Forms & Bespoke Comparison*. Unpublished Research Report, A. Whyte (ed.), Australia: Curtin University (unpublished).
40. Calver, R. (2011) The Australian Building Industry Contract (ABIC): Principles and Practice. Australia: Master Builders Australia Publications, http://www.masterbuilders.com.au/Content/ViewAttachment.aspx?id=1101&attachmentNo=105.
41. IACCM. (2009) *Report on the Standard Forms of Contract Questionnaire Sent to IACCM Members*. UK: IACCM.
42. Main Roads. (2010) *Book 1 Tender Submission Document*. Perth: Main Roads, 2011—. *Book 2 Conditions of Contract*. Perth: Australian Standards & Main Roads.
43. Water Corporation. (2010) *Water Corporation General Conditions of Contract based on Australian Standard AS4000-1997*. Perth: Australian Standards & Water Corporation.

44. The Thomson Corporation. (2012) WestLaw. Available at http://www.westlaw
.com.au (accessed August 13, 2012).

45. Shnookal, T. and Charrett, D. (2010) Standard form contracting; The role for
FIDIC contracts domestically and internationally. *Society of Construction Law
Conferences*, p. 30.

46. Available at http://www.watercorporation.com.au/about-us.

47. Available at https://www.mainroads.wa.gov.au/AboutMainRoads/AboutUs
/Pages/AboutUsHome.aspx.

48. Ting, S. N. and Whyte, A. (2010) An analysis of payment effected by standard
forms of construction contract in Sarawak, Malaysia: A comparative study
of PWD-1975, CIDB-2000 and PAM-1998. *Malaysian Construction Research
Journal MCRJ*, 6, 1, pp. 59–65, ISSN 1985–3807.

49. Ashworth, A. (1998) *Civil Engineering Contractual Procedures*. United Kingdom:
Addison Wesley Longman Limited.

50. Bockrath, J. T. (2000) *Contracts and the Legal Environment for Engineers and
Architects*, 6th ed. CA, USA: McGraw-Hill Higher Education.

51. Collier, K. (2001) *Construction Contracts*, 3rd ed. New Jersey: Prentice
-Hall.

52. Ong, S. L., Tan, S. I. and Oon, C. K. (2000) *Guide on the CIDB Standard Form
of Contract for Building Work*. Malaysia: Construction Industry Development
Board (CIDB).

53. Danuri, M. S., Hussain, S. M. N. and Mustaffa, N. E. (2008) Viability of dispute
avoidance procedure in the Malaysian construction industry. *RICS Construction
and Building Research Conference*, September 4–5, Dublin Institute of
Technology. Available at http://www.rics.org/NR/rdonlyres/80684B62-78A1
-46E5-8E64-3B2B47281DE8/0/Danuri.pdf.

54. Harbans, K. S. (2002) *Engineering and Construction Contracts Management:
Post-Commencement Practice*. Malaysia: LexisNexis Business Solutions.

55. Rajoo, S. (1999) *The Malaysian Standard Form of Building Contract (The
PAM 1998 Form)*, 2nd ed. Malaysia: Malayan Law Journal.

56. Sia, P. S. (2008) Evaluation on PWD 75, Standard Form of General Conditions
of Contract in Sarawak. Master Theses, Faculty of Engineering, UNIMAS Kota
Saramahan, Sarawak, Malaysia.

57. Williams, K. (1992) *Civil Engineering Contracts–Volume 1: The Law,
Administration, Safety*. England: Ellis Horwood Limited.

58. Vogt, W. P. (2007) *Quantitative Research Methods for Professionals*. USA:
Pearson Education/Allyn & Bacon.

59. CIDB. (2000) *CIDB Standard Form of Contact for Building Works*. UK:
Construction Industry Development Board.

60. JKR. (1961) *Public Works Department (PWD 75) General Conditions of
Contract*. Jabatan Kerja Raya Sarawak, Government of Malaysia.

61. PAM. (1998) *Agreement and Conditions of Building Contract*. Malaysia:
Pertubuhan Arkitek.

62. MLJ 444, Lee Kam Chun v Syarikat Kukuh Maju Sdn Bhd (Syarikat Perumahan
Pegawai Kerajaan Sdn Bhd, Garnishee [1988] 1 MLJ 444).

63. MLJ 16, Shen Yuan Pai v Dato Wee Hood Teck [1976] 1 MLJ 16.

64. Available at http://edit.epa.wa.gov.au/EPADocLib/Environmental%20Impact
%20Assessment%20Administrative%20Procedures%202012.pdf.

65. Available at http://edit.epa.wa.gov.au/EPADocLib/Environmental%20Impact%20Assessment%20Administrative%20Procedures%202012.pdf; superseding Administrative Procedures 2002, http://www.epa.wa.gov.au/docs/2192_gg026.pdf; http://www.iso.org/iso/home/standards/management-standards/iso14000.htm.

66. Bates, G. (2013) *Environmental Law in Australia*. Australia: Lexis Nexis Butterworths.

67. Bates, G. (2002) *Environmental Assessment: Australia's New Outlook under the Environmental Protection & Biodiversity Conservation Act 1999 (Cth), Environmental Law Review,* UK: Vathek Publishing, Maidstone, 4, 4, pp. 203–224.

68. Available at http://www.planning.wa.gov.au/; http://www.planning.wa.gov.au/publications/6561.asp; http://www.planning.wa.gov.au/dop_pub_pdf/State__PlanningStrategy2.pdf.

69. Available at http://www.environment.gov.au/epbc/protect/wetlands.html.

70. Available at http://edit.epa.wa.gov.au/EPADocLib/EAG6-Timelines%20March%202013.pdf.

71. Available at http://www.epa.wa.gov.au/EIA/EPAReports/Pages/default.aspx?cat=EPA%20Reports&url=EIA/EPAReports; http://www.epa.wa.gov.au/EIA/EPAReports/Pages/default.aspx?cat=EPA%20Reports&url=EIA/EPAReports; http://www.epa.wa.gov.au/docs/2192_gg026.pdf.

72. Available at http://edit.epa.wa.gov.au/EPADocLib/EAG6-Timelines%20March%202013.pdf; http://edit.epa.wa.gov.au/EPADocLib/Environmental%20Impact%20Assessment%20Administrative%20Procedures%202012.pdf.

73. Available at http://www.epa.wa.gov.au/docs/2192_gg026.pdf.

74. Kirby, R. S. (1913) *The Elements of Specification Writing: A Text-Book for Students in Civil Engineering*. USA: Archive Publication.

75. Scott, J. (1991) *The Penguin Dictionary of Civil Engineering*, UK: Penguin.

76. Available at http://www.NATSPEC.com.au; http://www.natspec.com.au/Products_Services/Specifications.asp; http://www.natspec.com.au/Documents/Specifications/Sample/0331%20Brick%20and%20block%20construction%20without%20guidance.pdf.

77. Available at http://www.abcb.gov.au/about-the-national-construction-code/the-building-code-of-australia.

78. Available at http://www.abcb.gov.au/~/media/Files/Download%20Documents/Product%20Certification/codemark_scheme_rules_v2009-1_27-03-20091.pdf; http://www.abcb.gov.au/about-the-national-construction-code/the-building-code-of-australia.

79. Available at https://www.mainroads.wa.gov.au/BuildingRoads/TenderPrep/Pages/TenderPrep.aspx; https://www.mainroads.wa.gov.au/BuildingRoads/TenderPrep/Specifications/Pages/specifications.aspx.

80. Available at http://www.thenbs.com/.

81. Available at https://www.mainroads.wa.gov.au/BuildingRoads/TenderPrep/Specifications/Pages/specifications.aspx.

82. Available at http://www.natspec.com.au/Products_Services/Specifications.asp; http://www.natspec.com.au/Documents/Specifications/Sample/0331%20Brick%20and%20block%20construction%20without%20guidance.pdf.

83. Available at http://www.thenbs.com/products/index.asp.

84. Available at http://www.icis.org/siteadmin/rtdocs/images/10.pdf.

85. McGregor, C. (2001) *A Description & Comparison of National Specification Systems*. UK: NBS Services Representative to International Construction Information Society.

86. Available at http://news.bbc.co.uk/2/hi/europe/3469823.stm; http://news.bbc .co.uk/2/hi/in_depth/photo_gallery/3454863.stm; http://news.bbc.co.uk/2/hi /europe/3764046.stm.

87. Available at http://www.concretecentre.com/codes__standards/eurocodes/euro code_2/european_concrete_standards/european_standards_concrete.aspx.

88. Sharif, M. and Whyte, A. (2012) Towards improved infrastructure design & construction in rough terrain & inclement environments. In *Research, Dev. & Practice in Structural Engineering & Constrn*, Vanissorn V. et al. (eds.), Singapore: Research Publishing Services, pp. 991–996.

89. Singh, A. and Shoura, M. M. (2006) A life cycle evaluation of change in an engineering organization: A case study. *International Journal of Project Management*, 24, 4, pp. 337–348, Elsevier.

90. Newnan, D. G. (1983) *Engineering Economic Analysis*, 2nd ed. San Jose, CA: Engin'g Press, Inc.

91. Bullen, F. (2003) Design and construction of low-cost, low-volume roads in Australia. *Transportation Research Record: Journal of the Transportation Research Board*, 1819, 1, pp. 173–179.

92. Hearn, G. J. and Massey, C. I. (2009) Engineering geology in the management of roadside slope failures: Contributions to best practice from bhutan and ethiopia. *Quarterly Journal of Engineering Geology and Hydrogeology*, 42, 4, p. 511.

93. Khawlie, M. R. and A'war, R. (1992) Terrain evaluation for assessment of highways in the mountainous Eastern Mediterranean of Lebanon. *Bulletin of Engineering Geology and the Environment*, 46, 1, pp. 71–78.

94. Lav, A. H. (2003) Method for balancing cut-fill and minimizing the amount of earthwork in the geometric design of highways. *Journal of Transportation Engineering*, 129, p. 564.

95. SD-LTAP. (2000) *Gravel Roads, Maintenance and Design Manual, South Dakota Local Transportation Assistance Program (SD LTAP)*, USA: U.S. Department of Transportation, Federal Highway Administration, November.

96. US Agency for International Development (USAID). (2007–2011) Afghanistan Infrastructure & Rehabilitation Program by The Louis Berger Group Inc. /Black & Veatch Special Projects Corp. Joint Venture (LBG/B&V), http://pdf .usaid.gov/pdf_docs/pdacw228.pdf.

97. Transportation Research Board National Research Council, 2009, 'Transportation Research Program for Mitigating and Adapting to Climate Change and Conserving Energy, A', National Academies Press USA http://www.nae.edu/Publications.aspx.

98. Available at http://www.gbca.org.au/green-star/.

99. Available at http://www.breeam.org/.

100. Available at http://www.dgbc.nl/wat_is_dgbc/dgbc_english; http://www.breeam .org/.

101. Available at http://www.usgbc.org/leed.

102. Available at http://www.ibec.or.jp/CASBEE/english/.

103. Available at http://www.bsria.co.uk/news/article/bream-or-leed//.

104. Whyte, A. (2011) *Life-Cycle Assessment of Built-Asset Waste Materials: Sustainable Waste Options*. Germany: Lambert Academic Publishing.

105. Available at http://www.aggregain.org.uk/; http://www.wrap.org.uk/.
106. Available at http://www.greenspec.co.uk/.
107. Available at http://www.aggregain.org.uk/; http://www.wrap.org.uk/; http://www.greenspec.co.uk/.
108. Available at https://www.mainroads.wa.gov.au/BuildingRoads/Standards Technical/MaterialsEngineering/Pages/Pavements_Technology.aspx; http://www.wasteauthority.wa.gov.au/news/pavement-specifications-501-launched; http://www.dft.gov.uk/ha/standards/mchw/vol1/; http://www.dft.gov.uk/ha/standards/mchw/.
109. Available at https://www.mainroads.wa.gov.au/BuildingRoads/Standards Technical/MaterialsEngineering/Pages/Pavements_Technology.aspx; http://www.wasteauthority.wa.gov.au/news/pavement-specifications-501-launched.
110. Available at http://www.dft.gov.uk/ha/standards/mchw/vol1/; http://www.dft.gov.uk/ha/standards/mchw/.
111. Collins, R. (2001) *Fit for Purpose Specifications: Using Recycled/Secondary Aggregates*. Watford: BRE.
112. Available at http://www.netregs.gov.uk; http://www.commerce.wa.gov.au/worksafe/content/services/Certification_registration_and/Important_information_for_Demolition_licence.html; http://www.hse.gov.uk/pubns/indg220.pdf.
113. Whyte, A. and Marshall, N. (2011) Demolition protocol & best local practice for the Western Australian construction and demolition industry. *Proceedings of RICS COBRA*, September 12–13, School of the Built Environment, University of Salford, pp. 1398–1406, ISBN: 978-1-907842-19-1.
114. Cardno, B. (2008a) Assessment of waste disposal and material recovery infrastructure for Perth. Available at http://www.zerowastewa.com.au/documents/waste_disposal_recovery_infrastructure_for_perth2020.pdf (accessed May 16, 2010).
115. Cardno, B. (2008b) Detailed investigation into existing and potential markets for construction and demolition material. Available at http://www.zerowastewa.com.au/documents/investigation_markets_for_recycled_cnd_materials.pdf (accessed March 31, 2010).
116. Hyder Consultants. (2009) *Recycling Activity in Western Australia 2007/08 & 2008/09*. Available at http://www.zerowastewa.com.au/documents/external_docs/WA_Recycling_Activity_0708_0809.pdf (accessed July 22, 2010).
117. ICE. (2008) Demolition Protocol 2008. Available at http://www.ice.org.uk/getattachment/eb09d18a-cb12-4a27-a54a-651ec31705f1/Demolition-Protocol-2008.aspx (accessed February 23, 2010).
118. Life Cycle Strategies Pty. Ltd. (2010) Executive summary for LCA program investigating the recycling of construction waste in Perth, WA. Available at http://www.zerowastewa.com.au/documents/external_docs/SWIS_2008_Evaluation_Report_Earthcarers_recycling.pdf (accessed July 22, 2010).
119. Whyte, A., Dyer, T. D. and Dhir, R. K. (2005) Best practicable environmental options (BPEOs) for recycling demolition waste. In *Achieving Sustainability in Construction*, R. Dhir et al. (eds.). London: Thomas Telford, pp. 245–251.
120. Whyte, A., Dyer, T. and Lau, H. (2006) Best practicable environmental options to manage residual construction materials. *International Symposium: Sustainable Energy, ISESEE 2006*, Malaysia, pp. 199–204.

121. Dhir, R. et al. (2004) *Promoting Best Practicable Environmental Options for the Commercially Sustainable Use of Glass Cullet in Construction*. Oxon: WRAP, 118 pp.

122. Scottish Environment Protection Agency / Scottish Executive. (2003) *The National Waste Plan*. UK: Charlesworth, 135 pp.

123. Scottish Environment Protection Agency. (2000) *Best Practicable Environmental Option—Decision Making Guidance*. Stirling: SEPA, 68 pp.

124. 2000/532/EC. (2000) *European Waste Catalogue*. Consleg Office for Official Publications of the European Communities, Belgium, 31 pp.

125. Dickie I. and Howard, N. (2000) *Assessing Environmental Impacts of Construction: Industry Consensus, Breeam and UK Ecopoints*. UK: British Research Establishment Publication BRE Digest 446, Watford, 12 pp.

126. Department of the Environment, Transport and the Regions Energy Consumption Guide 70. (1998) *Energy Use in the Mineral Industries of Great Britain*. UK: HMSO, pp. 2–19.

127. United Nations Environmental Programme Industry and Environment (1996) *Life Cycle Assessment: What is it and How to Do it,* USA: UNEP Publications.

128. AEA. (2001) National atmospheric emissions inventory UK, emissions of air pollutants 1970–99. Available at http://www.aeat.co.uk/netcen/airqual/naei/annreport/annrep99/.

129. Brown, S. (2003) Sustainable road construction: A local authority's experience, in 'Recycling and reuse of waste materials. Dhir R. (ed.) (2003), UK: Thomas Telford, pp. 761–768.

130. Davis Langdon. (2004) *Spon's Architects' and Builders' Price Book*, 123rd ed. UK: Spon Press.

131. BREWEB/ Ecological Development/ Sassi. (2000) Outline specification and cost comparison for recycled materials. Report to RMC Environmental Fund and Circle 33 Housing Trust, 33 pp.

132. Whyte, A. and Laing, R. (2012) Deconstruction and reuse of building material, with specific reference to historic structures. In *Research, Dev. & Practice in Structural Engineering & Constrn*, Vanissorn V. et al. (eds.), Singapore: Research Publishing Services. pp. 171–176.

133. BREWEB/ Ecological Development/ SASSI. (2000) Outline specification and cost comparison. RMC Environl Fund and Housing Trust, 33 pp.

134. Dhir, R., Newlands, M. and Dyer, T. (eds) (2003) *Sustainable Waste Management*, UK: Thomas Telford.

135. Building Research Establishment. (1993) Efficient use of aggregates and bulk construction materials. BRE Report 243.

136. Chen, Z., Li, H. and Wong, C. T. C. (2002) Bar-code system for reducing construction wastes. *Automation in Construction*, 11, pp. 521–533.

137. Chong, W. K. and Hermreck, C. (2010) Understanding transportation energy and technical metabolism of construction waste recycling. *Resources, Conservation and Recycling*, 54, pp. 579–590.

138. Construction (Health, Safety and Welfare) Reg. (1996) Environl Protection Act 1990 (c. 43).

139. Available at http://www.hse.gov.uk/pubns/indg220.pdf.

140. Dann, N., Hills, S. and Worthing, D. (2006) Assessing how organizations approach the maintenance management of listed buildings. *Construction Management and Economics*, 24, pp. 97–104.

141. Dhir, R. K., Dyer, T. D., Duff, M. J. and Csetenyi, L. J. (2004) *Promoting BPEO Glass Cullet in Construction*. Oxon: WRAP, 118pp.

142. Gluch, P. and Baumann, H. (2004) The life cycle costing (LCC) approach. *Building and Environment*, 39, 5, pp. 571–580.

143. DETR. (1998) *Energy Consumption Guide 70: Energy Use in the Minerals Industries of Great Britain*. URL: http://www.carbontrust.co.uk/publications /pages/home.aspx, UK: DETR Energy Efficiency Best Practive Programme.

144. Kishk, M. et al. (2003) Effective feedback of whole-life. *Journal of Financial Management of Property and Construction*, 8, 2, pp. 89–98.

145. Kneifel, J. (2010) Life-cycle carbon and cost analysis. *Energy and Buildings*, 42, 3, pp. 333–340.

146. Li, H., Chen, Z., Yong, L. and Kong, S. C. W. (2005) Application of integrated GPS and GIS tech. *Automation in Construction*, 14, pp. 323–331.

147. NHTG. (2007) *Traditional Building Craft Skills: Scotland*. UK: National Heritage Training Group.

148. NHTG. (2008) *Traditional Building Craft Skills: Skills Needs Analysis of the Built Heritage Sector in England*. UK: National Heritage Training Group.

149. Reser, J. P. and Bentrupperbäumer, J. M. (2005) What and where are environmental values? *Journal of Environmental Psychology*, 25, pp. 125–146.

150. Soronis, G. (1992) An approach to the selection of roofing materials for durability. *Construction and Building Materials*, 6, 1, pp. 9–14.

151. Tam, V. W. Y. and Tam, C. M. (2006) A review on the viable technology for construction waste recycling. *Resources, Conservation and Recycling*, 47, pp. 209–221.

152. Whyte, A. (2012) *Life-Cycle Assessment of Built-Asset Waste Materials: Sustainable Disposal Options*. Germany: Lambert KG.

153. Young, M. E. et al. (2000) An investigation of the consequences of past stone-cleaning intervention on future policy and resources. Report to Historic Scotland, UK: Historical Scotland Publications.

154. Young, M. and Budge, C. (2001) Cost effective salvage and re-use of granite (CESAR). Report to Aberdeenshire Council, Masonry Conservation Research Group, UK: Robert Gordon University.

155. Available at http://www.bentley.com/en-AU/Products/.

156. Available at http://bim.natspec.org/; http://bim.natspec.org/index.php /natspec-bim-documents/national-bim-guide.

157. Available at http://www.thenbs.com/topics/BIM/index.asp; http://www .nationalbimlibrary.com/; http://www.nce.co.uk/news/nce-it/national-bim -library-launches-this-week/8627938.article#.

158. Available at http://www.thenbs.com/topics/BIM/articles/costSavingBenefits OfBIM.asp.

159. Whyte, A. and Luca, M. (2013) Building Information Modelling (BIM) & Integrated Project Delivery (IPD): Workplace utilization. In *New Developments in Struc. Engineering & Construction*, Yazdani, S. and Singh, A. (eds.), Singapore: Research Publishing Services. pp. 1167–1172.

160. AGDIISR. (2010a) *Issues Paper: Digital Modelling and the Built Environment*. Canberra: A.C.T.

161. AGDIISR. (2010b) *Building a Culture of Innovation in the Built Environment.* Canberra: A.C.T.

162. AIA. (2010) Report on BIM/IPD forums. Available http://www.architecture .com.au/i-cms?page=10832.

163. Aranda, G. (2009) BIM demystified. *International Journal of Managing Projects in Business*, 2, 3, pp. 419–434.

164. Arayici, Y. (2008) Towards BIM for existing structures. *Structural Survey*, 26, 3, pp. 210–222.

165. Howard, R. (2008) BIM experts views. *Advanced Engineering Informatics*, 22, 2, p. 271.

166. Kaye, A. (2009) Integrated project design: Technical & practice. *Architect's Journal*, 230, 4, p. 32.

167. Kent, D. (2010) Understanding construction industry experience. *Journal of Construction Engineering and Management*, 136, 8, p. 815.

168. Kiviniemi A., Tarandi V., Karlshoj J., Bell H., Karud O. (2008) Review of the development and implementation of IFC compatible BIM, Final report of the Erabuild project, Erabuild 2008, http://www.senternovem.nl/mmfiles/Erabuild%20 BIM%20Final%20Report%20January%202008_tcm24-253611.pdf

169. Lesniewski, L., Krygiel E. and Berkebile B. n.d. Roadmap for integration: American Institute of Architects Report on Integrated Practice, United States of America.

170. Malkin, R. (2010) BIM for efficient sustainable design. *Architecture Australia*, 99, 5, p. 105.

171. Mesh Conference Series. (2011) Building-SMART, http://buildingsmart.org .au/national-strategy-for-bim-adoption.

172. MHC McGraw Hill Construction. (2007) Interoperability in the construction industry. Available http://www.fiatech.org/images/stories/research /mcgrawhill2007report-interoperability.pdf.

173. MHC McGraw Hill Construction. (2008) Building Information Modelling (BIM); SmartMarket report series; http://www.bim.construction.com/research/.

174. MHC McGraw Hill Construction. (2010) Green BIM: How building information modelling is contributing. Available at http://www.wbdg.org/pdfs /mhc_smartmarket_rep2010.pdf.

175. Novitski, B. (2010) IPD contracts. *Architectural Record*, 198, 10, p. 49.

176. Sacks, R. and Barak, R. (2010) Teaching building information modeling as an integral part of freshman year civil engineering education, *Journal of Professional Issues in Engineering Education and Practice*, 136, 1, pp. 30–38.

177. Barnes, M. (1992) *Examples of Civil Engineering Standard Method of Measurement (CESMM).* UK: ICE Publications.

178. Barnes, M. (1992) *CESMM Handbook: A Guide to the Financial Control of Contracts Using the civil Engineering Standard Method of Measurement.* UK: ICE Publications.

179. Sierra, J. (1998) Civil engineering measurement: A guide to the Australian standard 1181–1982: Method of measurement of civil engineering works and associated building works, Australia: Australian Institute of Quantity Surveyors Publications.

180. Sierra, J. *Method of Measurement of Civil Engineering Works and Associated Building Works Australian standard.* AS 1181–1982, Australia: Australian Institute of Quantity Surveyors Publications.

181. Seeley, I. H. and Murray, G. (2001) *Civil Engineering Quantities*. 6th edn. UK: Palgrave MacMillan.
182. Tweeds. (1995) *Taking off Quantities: Civil Engineering*. UK: E&FN Spon.
183. ICE (2012) *CESMM4: Civil Engineering Standard Method of Measurement, 4th edn. by the Institution of Civil Engineers UK*. ICE Publications/Thomas Telford.
184. AS 1181-1982. Method of measurement of civil engineering works and associated building works; Australian Standard Method of Measurement of Building Works Fifth Edition by the Australian Institute of Quantity Surveyors and the Master Builders.
185. Available at https://www.mainroads.wa.gov.au/BuildingRoads/TenderPrep/Pages/smm.aspx.
186. Available at http://www.bentley.com/en-AU/Products; http://www.buildsoft.com.au/.
187. ICE (2012) *CESMM4: Civil Engineering Standard Method of Measurement, 4th edn by the Institution of Civil Engineers UK*. UK: ICE publications/Thomas Telford.
188. Seeley, I. H. and Murray, G. (2001), *Civil Engineering Quantities*, 6th edition, UK: Palgrave Macmillan.
189. Smith, S. et al. (1998) *Builder's Detail Sheets*. UK: Routledge.
190. Bertoline, G. (2008) *Introduction to Graphics Communications for Engineers*, USA: MacGraw-Hill.
191. Seeley, I. (1995) *Building Technology*, UK: Palgrave Macmillan.
192. Seeley, I. (1995) *Building Economics*; Mitchell's building series, which covers many components and essential detail/section drawings. UK: Palgrave Macmillan.
193. Rich, P. and Dean, Y. (1999) *Principals of Element Design*. UK: Oxford Architectural Press.
194. Clough, R., Sears, G. and Sears, S. (2000) *Construction Project Management*, USA: Wiley.
195. Austroads Guide to Road Design; https://www.onlinepublications.austroads.com.au/collections/agrd/guides; MRWA Supplement to Austroads Guide to Road Design- Part 3; https://www.mainroads.wa.gov.au/BuildingRoads/StandardsTechnical/RoadandTrafficEngineering/GuidetoRoadDesign/Pages/MRWA_Supplement_to_Austroads_Guide_to_Road_Design___Part_3.aspx.
196. Austroads Guide to Road Design; https://www.onlinepublications.austroads.com.au/collections/agrd/guides; MRWA Supplement to Austroads Guide to Road Design - Part 3; https://www.mainroads.wa.gov.au/BuildingRoads/StandardsTechnical/RoadandTrafficEngineering/GuidetoRoadDesign/Pages/MRWA_Supplement_to_Austroads_Guide_to_Road_Design___Part_3.aspx.
197. Heath (treatment of sewage and disposal of effluent and liquid waste) regulations 1974 part 4a regulation 42 (and associated updates); Department of Health, Government of Western Australia. Code of practice for the design, manufacture, installation and operation of Aerobic treatment units (ATUs).
198. Cocks, G. and Teague, M. (1987) *Disposal of Stormwater Runoff Soakage*. 4th nat Local Government Engineering Conference, Australia: IEA Perth, http://www.housing.wa.gov.au/aboutus/projects/brownlie/Documents/Land_use_planning/Bentley_Regen_Part4.pdf.

199. ArcelorMittal Commercial RPS (2008), *Piling Handbook*, UK/Luxembourg: ArcelorMittal Commercial RPS Publications, http://sheetpiling.arcelormittal .com/uploads/files/ArcelorMittal%20Piling%20Handbook_rev08.pdf.

200. ThyssenKrupp GfT Bautechnik (2010), *Sheet Piling Handbook Design*, Germany: ThyssenKrupp GfT Bautechnik Publishing, http://www.thyssenkrupp -bautechnik.com/fileadmin/Leistungen/01_Spundwandprofile/_media/english /sheet_piling_handbook_3rd.pdf.

201. McAnally, P. and Boyce, B. (1980) *Geomechanics Design*, Australia: Queensland Institute of Technology publications.

202. Mahtab, M. and Grasso, P. (1992) *Geomechanics Principles in the Design of Tunnels and Caverns in Rocks (Developments in Geotechnical Engineering)*, USA: Elsevier Science.

CHAPTER 6

1. Available at http://www.engineersaustralia.org.au/professional-development /chartered-status/chartered-status_home.cfm; http://www.engineersaustralia .org.au/sites/default/files/shado/About%20Us/Overview/Governance /codeofethics2010.pdf; http://www.engineersaustralia.org.au/sites/default /files/professional_engineer_stage2competency_approved.pdf.

2. Available at http://www.ipenz.org.nz/ipenz/.

3. Available at http://www.ice.org.uk/; http://www.ice.org.uk/getattachment/1ebe1f7 e-7b36-43a2-a4eb-520901cc01cc/Code-of-professional-conduct-for-members.aspx.

4. Available at http://www.washingtonaccord.org/; http://www.washingtonaccord .org/GradProfiles.cfm; http://www.washingtonaccord.org/IEA-Grad-Attr-Prof -Competencies-v2.pdf.

5. Philosophy of Engineering Proceedings from the Royal Academy of Engineering Volumes 1 & 2. (2010) Available at http://www.raeng.org.uk/news/publications /list/reports/philosophy_of_engineering_volume1.pdf; http://www.raeng.org.uk /news/publications/list/reports/Philosophy_of_engineering-Vol_2.pdf.

6. Available at http://www.raeng.org.uk/events/pdf/Public_Attitude_Perceptions _Engineering_Engineers_2007.pdf.

7. Available at http://www.universitystory.gla.ac.uk/chair-and-lectureship/?id=790.

8. Available at http://www.engineersaustralia.org.au/sites/default/files/shado /About%20Us/Overview/Governance/codeofethics2010.pdf.

9. Whyte, A. (2011) *Life-Cycle Assessment of Built-Asset Waste Materials: Sustainable Waste Options*, Germany: Lambert Academic Publishing.

10. Kirk, J. and Dell'Isola, A. J. (1995) *Life Cycle Costing for Design Professionals*, 2nd ed. New York: McGraw-Hill.

11. Whyte, A. (2011) *Life-Cycle Cost Analysis of Built Assets: An LCCA Framework*. Germany: VDM, 116 pp, ISBN-10: 3639336364.

12. Faniran, O. *Engineering Project Management: An Introductory Text*. Australia: Pearson Education Australia.

13. Available at http://www.myersbriggs.org/.

14. BBC News. (2012) Personality tests: Can they identify the real you? Available at http://www.bbc.co.uk/news/magazine-18723950.

15. BBC News. (2012) What's your personality type? Find out with the 'What Am I Like?' Personality test. Available at http://www.bbc.co.uk/science/humanbody /mind/surveys/whatamilike/static_quiz.shtml.

16. BBC News. (2012) Personality Test. Available at http://www.bbc.co.uk/news /magazine-18723950.

17. Blake, R. and Mouton, J. (1964) *The Managerial Grid: The Key to Leadership Excellence*. USA: Gulf Publishing.

18. Available at http://www.bumc.bu.edu/facdev-medicine/files/2010/10/Leadership -Matrix-Self-Assessment-Questionnaire.pdf.

19. Buchanan, D. and Huczynski, A. (2010) *Organizational Behaviour*. UK: Financila Times/Prentice Hall.

20. Hersey, P. and Blanchard, K. (2012) *Management of Organizational Behaviour*. USA: Prentice Hall.

21. Fiedler, F., Chemers, M. and Mahar, L. (1976) Improving leadership effectiveness: The LEADER MATCH Concept. New York: Wiley.

22. Vroom, V. and Yetton, P. (1973) *Leadership and Decision Making*. USA: Pittsburgh University Press.

23. Freud, S. (1900) *The Interpretation of Dreams/ The Interpretation of Dreams*, 3rd edition, by Sigmund F.; translated by A. A. Brill. 1913. New York: The Macmillan Company.

24. Adair, J. (1973) *Action Centred Leadership*. New York: McGraw-Hill.

25. Calvert, R. E. (1976/2011) *Introduction to Building Management*. UK: ButterworthHeinemann (1976) and Routledge (2011).

26. Shannon, C. and Weaver, W. (1971) *The Mathematical Theory of Communication*. USA: University of Illinois Press.

27. Oakland, J. (2003) *Total Quality Management*. USA: Routledge.

28. Morris, D. (1999) *The Naked Ape*. UK: Delta.

29. AIA, Report on BIM/IPD. Available at http://www.architecture.com.au/.

30. AECOM's 2013 Blue Book. Available at http://www.aecom.com/deployedfiles /Internet/Geographies/Australia-New%20Zealand/AECOM%20Blue%20 Book%202013.pdf.

31. Whyte, A. and Scott, D. (2010) Life-cycle costing analysis to assist design decisions: Beyond 3D building information modelling. In *Computing in Civil and Building Engineering, Proceedings of the International Conference*, W. Tizani (ed.), June 30–July 2, Nottingham, UK: Nottingham University Press, Paper 87, p. 173, ISBN 978-1-907284-60-1.

32. Whyte, A. (2010) The role of tertiary education in integrating professional contribution in the multi-disciplinary building design team. *Proceedings of ACEN Aus Collaborative Education Network National Conference*, Publication Number PUB-SE-DCE-SL-57594, pp. 507–521.

33. Whyte, A. (2012) Construction educational interdisciplinary project-work as an antecedent for improved integration in BIM. In *Research, Dev. & Practice in Structural Engineering & Construction*, Vannisorn, V., Singh, A. and Yazdani, S. (eds), Singapore: Research Publishing Services. pp. 1125–1130.

34. Whyte, A. and Al-Hajj, A. (2000) Participation as the synthesis for computer-integrated-construction and innovation in design. *International Journal of Construction Information Technology*, 8, 2, Winter 2000, pp. 51–62.

35. Al-Hajj, A., Whyte, A. and Aouad, G. (2000) The development of a framework for life-cycle costing in object-orientated & VR technologies. *ALES 200 Annual Convention*, July 20–22, Beirut, Lebanon.

36. Arayici, Y. and Aouad, G. (2005) Computer integrated construction: An approach to requirements engineering. *Engineering, Construction and Architectural Management*, 12, 2, pp. 194–215, Emerald.

37. CRC. (2009) Cooperative Research Centre for Construction Innovation. National Building Information Modelling (BIM) Guidelines & Case Studies, Brochure, April 2009.

38. Dennis, A. et al. (1988) Information technology to support electronic meetings. *MIS Quarterly*, 11, pp. 591–624.

39. Gallupe, R. et al. (1988) Computer based support for group problem finding: An empirical investigation. *MIS Quarterly*, 12, pp. 277–296.

40. Green, M. (2009) Issues paper: Digital modelling & the built environment. Built Environ. Digital Modelling Working Group Discussion Paper Nov, Dept. of Innovation, Industry, Science & Research, Australian Gov.

41. Hu, W. (2008) Information life-cycle modeling framework for construction project life-cycle management. *International Seminar on Future Info. Tech. and Management Engineering*, November 20, pp. 372–375.

42. Jarvenpaa, S. et al. (1988) Computer support for meetings of groups working on unstructured problems: A field experiment. *MIS Quarterly*, 12, pp. 646–666.

43. Kull, D. (1982) Group decisions: Can a computer help. *Computer Decisions*, 15, 5, pp. 64–70.

44. Powell, M. and Newland, P. (1995) An integrating interface to data. Cited in Brandon, P. and Betts, M. (1995) *Integrating Construction Information*. UK: E&FN Spon, pp. 70–96.

45. Watson, R. et al. (1988) Using GDSS to facilitate group consensus: Some intended and some unintended consequences. *MIS Quarterly*, 12, pp. 463–478.

46. Shannon, E. and Weaver W. (1949) *The Mathematical Theory of Communication*. USA: University of Illinois Press.

47. Australian Institute of Architects, Cooperative Research Centre. (2009) *National Building Information Modelling (Conceptual Framework)*. Australia: AIA Publications.

CHAPTER 7

1. Available at http://www.constructors.com.au/publications/tendering_guide /ACA%20Guidelines%20for%20Tendering%2011%20August%20 2006%20Final%20.pdf.

2. Available at http://www.mainroads.wa.gov.au; http://www.mainroads.wa.gov .au/BuildingRoads/TenderPreparation/Pages/TenderPrep.aspx; http://www .mainroads.wa.gov.au/BuildingRoads/TenderPreparation/MajorWorks/Pages /MajorWorks.aspx.

3. Delft University of Technology (Delft UoT). Body of work. Available at http:// www.io.tudelft.nl/onderzoek/onderzoeksprogrammas/technol ogy-transformation/design-for-sustainability-emerging-markets /sub-theme-4-knowledge-exploration-and-decision-support/.

Index

Page numbers followed by f and t indicate figures and tables, respectively.